Mission to Mars:
2025

Mission to Mars:
2025

VOLUME I: WHAT WE CAN DO

DANIEL F. GOERKE

iUniverse, Inc.
New York Lincoln Shanghai

Mission to Mars: 2025
Volume I: What We Can Do

Copyright © 2007 by Daniel F. Goerke

iUniverse books may be ordered through booksellers or by contacting:

iUniverse
2021 Pine Lake Road, Suite 100
Lincoln, NE 68512
www.iuniverse.com
1-800-Authors (1-800-288-4677)

ISBN-13: 978-0-595-41587-8 (pbk)
ISBN-13: 978-0-595-85935-1 (ebk)
ISBN-10: 0-595-41587-3 (pbk)
ISBN-10: 0-595-85935-6 (ebk)

Printed in the United States of America

To Danielle,

With love

Three-Volume Table of Contents

Volume I: What We Can Do

Volume II: Why We Should Go
(And Why You Should Not Wait For NASA)

Volume I—Detailed Table of Contents

Part III—Rounding Up the Usual Suspects

Part IV—On Mars

Part V—Of Humans and Histories

Table of Figures

Part I

Introductory Material For Volume I

[What's to say about] someone [who] claims to be a Martian—or at least professes a deep affinity for the Red Planet and persists in a belief that he or she might have once lived there? ...

Ray Bradbury had written that he had always looked on himself as a Martian; indeed, that he amiably chose to believe that he had once lived on Mars, if one considered reincarnation a viable possibility. He had gone on to explain that a certain Dr. Electrico had informed him they had met on a battlefield in the Argonne in France during World War I, where he had died to be re-treaded as Ray Bradbury. And if he could have died in the Argonne and come back, then he could just as easily have lived on Mars and come back ...

[We sat down for] lunch [at] ... a table in the sun near the front door, and after a lengthy interval the menus arrived ... I no longer recall what I ate, nor what Ray ordered, but I do remember the conversation. It was about Mars, naturally ... I said we would live there in the times of my daughter, or perhaps my daughter's daughter. He said, "Oh, no, much sooner than that ..."

The sun was headed toward the beaches when we left the table ... "I love that planet," he said. And in the flyleaf of my book he had written, "For Bob, who loves Mars almost more than I do." On the way to my hotel, I wondered about this business of loving a planet. Clearly Ray saw nothing odd there. Nor did I.

What is it we love about a painting? A sculpture? An automobile as art, like a Bugatti or Ferrari? Talk to a Russian about Mother Russia, and he's not speaking of politics. He's talking about the land itself. These are all inanimate, and yet people profess to love them. Indeed, books are inanimate, too, and we love them, cherish them, reread them; we buy them at auctions, preserve them in expensive libraries ...

How can you love a planet, truly? ...

—Robert Morris in *Mars: Our Future on the Red Planet*

Preface to Volume I

The three volumes of this work have taken on lives of their own. Initially envisioned as a letter, then a 200-page book with a few general ideas, its 1000 pages finally became the three volumes that make up this project. Beyond just the size, however, the format has changed, too. First written to keep the reader turning pages, "hiding" major ideas for later chapters, the work's final form is now very straightforward. The "mystery novel" format having been abandoned, readers may now read about issues that interest them and discard those which do not.

This first volume—"What We Can Do"—is designed to outline the mission to Mars in stories. There is little plot in most of these sketches and hardly any suspense, but there is drama as each one describes a small part of the story of transplanting a human society to Mars. This volume is for the dreamers, youngsters, and for those who will intuitively understand what the mission to Mars is all about. This first volume presents a storyboard of ideas, meant less to convince than to show. Read here with that youthful eagerness you once had for space and leave the heavy analysis for later volumes. If the author has been at all successful, what will emerge will be a rapid and enjoyable understanding that *these ideas might just work* and, perhaps, a thirst to read more.

In contrast, the second volume—"Why We Should Go (And Why You Should Not Wait For NASA)"—will appeal to a much more sophisticated audience. This is not to discourage the dreamers, youngsters, and the intuitive as much as it is to give everyone a choice to pursue ideas only as far as they are interested. With three volumes, no one is forced—as in initial drafts—to wade through material that may not hold their attention. The second volume is essential to the mission to Mars, however, because it answers questions about *why* we should go to Mars. Nor could this work avoid presenting the dismal obverse of the mission to Mars: that NASA offers Marsophiles little or no hope for anything other than public relations, toys, and glossy pictures for the internet. And, to be convincing, we must answer that great mantra of the modern western world—"Let the Government Do It"—as it regards Mars.

The Third Volume—"How We Can Do It"—is for the politician and the economist. It contains significant detail about the programs and ideas proffered in the earlier volumes and how they will operate. It answers questions that those with an open mind should accept, even if it puts off most of the Nay Sayers. You know the type: the ones who will never accept any ideas but their own. The Nay Sayers' chorus will be that the mission to Mars will never work. Now, no work of significance can hope to persuade *everyone*. Many materials in the final volume dance the fine line of overworking an issue, but as the third volume attempts to be convincing to a graduate-school-level audience, it was deemed important to include, as a matter of completeness, any issue whose absence might otherwise seem an ink spot on the mission to Mars' parchment.

In a word, the mission to Mars is about revolution. A revolution of ideas, a revolution of programs, a revolution of opportunity, and a revolution as regards human society. The final reorganization of this work may leave some readers of this volume with a false impression about the dramatic *political* message underlying the mission to Mars. Indeed, save for the "Introduction" (coming up next), political ideas are almost entirely missing from this volume. That cannot be helped in a volume devoted to painting word pictures rather than arguing concepts. Still, the reader intent on taking the long road through all three volumes needs to understand that the pleasant notions and programs presented will not be achieved without a dramatic political struggle against the forces of the *status quo*.

Did that last paragraph startle you a little? That was its purpose, because threaded throughout these three volumes is the idea that *everything about human society and life* will change once normal human beings have the following choice. Live as you and your ancestors have lived or decide to leave the earth and build a brave, new life and society in space. The mission to Mars will lead the charge and help create the infrastructure to allow just such a choice.

Next, a word of advice. People who want to dream about weird new technologies or Sci-Fi Channel space colonies with *no* chance to enter the real world should leave off with this volume. The political and economic ideas developed in Volumes II and III, which may seem "unrelated" to space are, in fact, necessary to create a proper framework to understand the mission to Mars and its prospects for success. It is fair to be surprised to discover the heavy political and economic aspects of these three volumes, but all three take the position that you cannot hope to achieve results in the real world if you do not provide a political and economic structure for your program. Whereas the dreamer books simply *assume* people or programs for Mars and proceed to set them off on all sorts of fantastic adventures in fiction; or *assume* a capability to build weird new science and technology for their never to be realized fantasies, the mission to Mars *explains* how to stitch together common political and economic structures toward the end of settling Mars. Not *if*, but *how*. Hopefully, the reader will recognize most of these ideas and come to appreciate how they might be used to create something real, in a real world.

Finally, I counsel the reader to read all three volumes before becoming a skeptic. What I trust will happen is that even the greatest skeptic will come to the conclusion that "maybe this mission to Mars stuff will work". The issues have been addressed. The ideas are not radical. The political and economic institutions have been proven to work. The work has been written to be a logical construct that even critics might admire; appreciate the logic, even if you disagree with one or more premises. Indeed, it would be a victory if the reader is left saying "These ideas would work, except I don't believe that the mission to Mars could ever "_____"—you fill in the blanks.

In any case, it is time to starting introducing ideas.

Introduction

"Didn't it strike you as a crazy plan?" Art asked, his eyes round. "Yes it did!" Coyote laughed. "But all the good plans are crazy, aren't they …?"

—*Green Mars*, by Kim Stanley Robinson

The three volumes of *Mission to Mars: 2025* intend to usher in a new era of mankind. As such, they portend a revolution. Most genuine revolutions open doors not previously considered and lead along paths not previously trod. So it is here. This volume may not contain material most people would expect from its title. This will become more and more true in Volumes II and III. For those who only want entertainment between video game marathons, I recommend the shelves of your local library, which will likely contain at least one or two titles about travel to Mars. There you can revel in titles of quasi-science fiction; complete with modern lamentations suggesting that the author's ideas are nothing more than impossible dreams. You might even find, among the formulas and complex diagrams of futuristic space vehicles, some smattering of an agenda to get to Mars. But fear not; the strong focus upon things technical places virtually every one of those books on the far side of the plausible.

And that is where most of you want to be, isn't it?

This book is not for the faint of heart. Now comes the millennium and I intend to change *everything*. This book announces the revolution—such an interesting term for those of you in the pocket protector crowd—so put it down if you are not ready for revolution.

But if you are strong enough and humble enough, I invite you to read. You, too, may find yourself atop the barricades.

The three volumes in this series are highly political. But as politics is the art of the possible, they were written to be practical, too. They describe a workable plan for you, dear reader, to get to Mars, accompanying thousands of other settlers who will self-select to make a political and personal commitment to the cause. Go back to dusty bookshelves if you want science fiction. This book describes certain events that will unfold over the next few decades, events that will occur in a reasonably charged atmosphere, but nonetheless before your eyes. People take risks in revolutions to make their world a better place. Some people die. So it will be with the mission to Mars. But the reward will be an excursion to Mars, first by a few grimy-faced pilots, then by a few skilled workers, and finally by thousands and thousands of people, who will scream for six space-bound months for their kids to shut up when they inquire with all the tact of youth: "Are we there yet?"

I am by training and experience a soldier, a lawyer, and a businessman. I have had an oscillating interest in space, with most cycles on the downside. Like many, I wanted to become "an astronaut" when I was six or seven, but eventually I lost interest in space. I continued, however, "to check into the net" during my life, grabbing the occasional book about space. I cried when the first space shuttle blew up, but only for a day; for I had no sense at all that anything other than lives had been lost[1]. I knew, like most of you know, that human beings have not been serious about space since the 1960s. The post-Challenger hiatus in the US/human space program delayed nothing because we were doing nothing.

If there is a deity, perhaps my own oft-manic life has been a kind of cosmic basic training for a divine purpose. For on a hot day in June, 2001, I went into the library in Rio Rancho, New Mexico, determined to do work on something important—I have forgotten what—but instead grabbed *The Case for Mars* by Robert Zubrin. Occasionally, the procrastination bug bites and on that day a short trip to Mars was intended to postpone real world work for an hour or two. The book, for those of you who have not read it, is quite good though, respectfully, mostly of the sort I described above. The book laments the lack of funding and paints proposals on a canvas of certain failure, knowing that funds will only be appropriated upon an extraordinary convergence of circumstances. What was interesting to me, however, was that I saw the missing pieces to Zubrin's puzzle. He had the engineering ideas. I saw the economic and political solutions. I promptly began to describe these pieces to him in a letter.

Three days and a dozen pages later, I realized that Zubrin would likely throw my letter away. (Or worse, send me form mail to sign over a check to join his "Mars Society".) Who was I, after all, to tell him that I saw things he had not seen

---

[1]   This introduction was written before the Columbia disaster. I have few feelings about that event. Part of it was, no doubt, that I was called to military service in Iraq almost immediately after the event and missed most of the press coverage. The greater part of this disinterest, however, relates to the utter frustration with an inside-the-beltway bureaucracy that seems irrelevant to the colonization of other worlds.

in a lifetime of distinguished professional achievement? Moreover, my ideas were radical in the sense that it excluded NASA and a "business as usual" approach to Mars. I decided the best choice was to prescribe my own medication for the Martian flu that so many suffer. You see, as a lawyer and someone who has done deals and seen deals done, what was missing in Zubrin's book, as in so many other books of that ilk, was a legal and financial structure to manifest good ideas. Zubrin has a decent car to sell, but has set up his car lot in the desert. In contrast, the mission to Mars intends to close a really big deal. Like any deal concerning a radically new idea, one takes the entrepreneur as one finds him. The legal and financial structure I envisioned was not Zubrin's idea and it would be impossible to remold his ideas into the radical, new mold I saw. His ideas must stand or fall on their own merit, as the ideas of the mission to Mars need to stand and fall on their own merit.

Out of the blue something—or some deity—had led me to a place where my life's purpose had suddenly become very clear. My knowledge of Mars and rockets and such before June, 2001 was threadbare; but that is irrelevant, because this book is not about engineering rockets to Mars. The plan in this book has nothing to do with tweaking a rocket fuel to get 3% more thrust from an engine. Nor is it much interested in the kinds of strange new technologies that only an engineer could love. Rather, this book is about creating political and economic structures to get rockets *built* and to get a *human society* to Mars. Dealing with real problems in the real world is something I have done for many years. As I have been mysteriously led to this place and see clearly the job before me, I intend to get this job done. If you let me do that job, maybe *you* can build that engine with the 3% improved thrust.

So engineers, you have been forewarned. There are no rocket formulas in these books. Instead, I offer here what you hate more than anything else in the world: an economically and politically feasible—I did not say easy—plan that has gathered a select few of your ideas for tethering to the real world while, alas, excluding many others. The result will not be your daydream but, as at your high-paying job, something that somebody might actually want and something that makes economic sense.

Here is the big question: How many of you *really* want to go to Mars or to get into space? This book holds to the premise that a great many people just might. Do *you* want to go to Mars? If you do, then hop aboard.

The only misapprehension about this whole project is that everyone on the planet might answer: *Marsraketen? Nein, danke.* This book is premised on the idea that there exists a few thousand people who want to go into space and to Mars. I could be wrong. I may be so out of touch with the MTV crowd as to fail to realize that "The Right Stuff" went out with the 1960s. This is not, however, what *I* believe. Rather, I so much believe there are people who would choose to go to Mars that, once the mission to Mars gains momentum, I believe people will be turned away at the gates. If I am wrong, so be it. I intend nevertheless to make the effort. Perhaps the pocket-protected-video-game crowd really has decided that the virtual reality of the *X-Box* is superior to anything offered in the real world. Or that Americans and others have so lost their spirit that instead of blazing new frontiers, they simply whimper at Uncle Sam's door, awaiting the next bone. Still, this would also mean that there are no adventurers left in the world; and it is an awfully big planet. I believe the tough-minded and the spirited still await a real challenge, but only time will tell.

These three volumes attempt to be reasonably thorough, in an "*I'm gonna' keep writing 'til you believe it*" sort of way. As such, there are sections or even whole volumes of the project that may not interest you. This is to be expected, especially as some of you have already begun to wonder if you can get a refund on this book. For instance, if you are a reader who only wants to dream then you may not want to read about economics in a book about Mars. The economics and politics will work, so I spend lots of time demonstrating this. What I spend much less time on is "proving" that you can terraform Mars or calculating exact schedules for Mars voyages. While I believe that being thorough is the best way to get several thousand of you to sign on the dotted line and join the mission to Mars, I cannot emphasize enough that the three volumes tackle the *real* issues keeping us from Mars and ignore "trivial" issues. Afterward, the trekkies can argue about non-issues like what is the best design for Martian spacesuits.

Many of the pages in Volumes II and III will admittedly not be an easy read. If you want something easy, find a Mars book with pretty pictures. Rather, the three volumes constitute a *political statement*, laying out a *feasible plan* for events that are just ahead of us. If it pains the reader to think hard about issues that are intended to grapple with reality, then there are other books for him. The thoughtful reader, on the other hand, is invited to understand these volumes in the broad and mostly controversial context in which they were intended.

Mars Socialites will likely find much of the work distasteful. They will want rockets or weird "Martian commune ideas". They will rebel at reading the philosophical and analytical background for the ideas in these volumes. Such readers are invited to search out these volumes' strawman technical discussions, but cautioned not to confuse these discussions with its core political and economic materials. There are no apologies here that the audience is more the plumber from California, the electrician from Arkansas, and the dreamer from Russia than the smarmy know-it-all whose *hobby* is Mars. As political books, written to outline a real plan, the target is those who will be pulled first by the political philosophy and secondly by the possibilities on Mars. The Mars Socialite, as someone who will have long ago surrendered his own dreams for space, is not required for the mission to Mars and is not expected to become an early member. If, after some years, some of the Mars Socialites find that they can no longer stomach NASA, and begin to follow with interest our small community arching toward that red speck in the sky, growing stage-by-stage in some desert somewhere, so much the better. In the meantime, the mission to Mars is content to move forward without them.

I also argue that the options for a Martian society are not wide-open. Engineers can choose to substitute one kind of rocket for another or one type of mission design for something different. Humans cannot choose, however, to discard proven forms of social organization willy-nilly in favor of their own unproven pet project and expect to get to Mars. *I believe that what is presented here is the only realistic option for a human society on Mars in the 21st-century.* I also believe that what is described not only might happen, but that it *will* happen in the next few decades, because it is grounded deeply in ideas that people can understand *and trust.*

That having been said, you must understand that the mission to Mars' ideas are best understood as *process* rather than *substance.* This is especially true as it regards technological discussions. *Process-oriented* ideas seek to create a faithful mechanism to ensure a desired result, without trying to specify in advance every possible detail. Substance-oriented implies a design for a clock radio. To be substance-oriented in the case of a human society going to Mars, that clock radio would have to be so complicated as to be beyond the capacity of even a long series of books. *Getting a human society to Mars needs a process, not a recipe.* A quick example can illustrate the point. Consider the extraction of water from the Martian regolith (Mars' surface, consisting of crushed rock). The process-oriented approach of these volumes states that since humans have been extracting water from the ground on earth for many thousands of years, and since there are no technological or economic reasons to doubt our ability to extract water from the Martian regolith, one may assume the extraction of water on Mars. A substance-oriented process, of course, would wax eloquent upon the vagaries of drills, permafrost, etc.

So process is at the heart of these volumes about the mission to Mars, processes that, of necessity, will not be endlessly faithful to the few specific technical ideas that have been presented; processes which will keep the reader's head above the waterline of endless numbers of technical issues which would otherwise divert attention from the *real* issues of colonization. These processes will become readily apparent as pages are turned.

The underlying political and economic discussions are process-oriented and are not subject to great revision. If you want a socialist Mars where "there is no marriage", then look elsewhere. Now, the social and political ideas of the book might surely be improved upon, but only as plastic surgeons can "improve" the looks of some Hollywood goddess. A nip here and a tuck there around the edges; as long as *that face* remains *that face*. Fortunately for you techies, all the technical stuff is *substance-oriented* and thus amenable to your better ideas. Just don't try to sell your idea that all criminals on Mars are to be tried before a judge-computer, so that no thought need be given to a legal system on Mars, except how to program your judge-computer. The technical aspects of the mission to Mars—the stuff that fills most daydreamer books—serves on these pages as a simple strawman, to be improved upon by eager young minds found in the bodies of people of all ages and from all parts of the world.

In general then, the programs offered here are not meant as *approved solution,* as becomes the format for most Mars books. What is presented here is a political statement about what can be and what will be. These volumes were written to be magisterial, as that term implies a gentle authority. The preceding paragraph is one aspect of this authority.

The disinterest in footnoting minor and subsidiary points is another. This book is not gospel, but a roadmap. Read it as such. Find your own route, but use what is presented here as a tool to reach our common objective: Mars. None of the volumes are formatted to be a college-text, with 4 footnotes on every page and a 300-volume bibliography. Some might use this lack of documentation as evidence of the lack of merit, choosing to opt for form over substance. Note however, that the Bible is not a "footnoted" work, nor Plato's dialogues, the American Declaration of Independence or Marx's *Das Kapital.* Great ideas are not always packaged to be acceptable to bureaucratic scholars. *Mission to Mars: 2025* is written as a manifesto, hoping to add a new brick to the house of human achievement. The modern trend for "scholars", usually working for large corporations or huge universities, anticipating generous salaries for *scholarship* sometimes allows form to overshadow substance. Thus, it is often better to spend time footnoting a "safe" book, than to develop a revolutionary idea that may not be well received within the hierarchy. The mission to Mars proposes some *big* ideas, something you may not have seen for a while. It asks you to look up what you don't trust.

The second point about footnotes is that this book expects to have a lifespan of several decades. Footnoting that Mars has a 38%-as-much gravity as the earth is trivial and requires no footnotes. Footnotes about water on Mars are similarly of limited utility to the main political and economic ideas of the mission to Mars and, of course, much of this data will quickly be supplanted by better information. Argue this technical point or that in meetings at your local Denny's; it matters not to the central message here. It has been hard enough trying to keep up with the minutia of "Are we going back to the moon or not?"[2] Remember that these volumes are not directed to the self-important trekkie who, in any case, will be less receptive to the ideas in this book than the average, thoughtful person. Footnotes appear on many pages, but they offer sidebars into otherwise small points. If your mind cannot stomach an ideas book without footnotes because you are used to format rather than thoughtful consideration, then I ask you to consider how open your mind really is.

Ultimately, the mission to Mars hopes to convince the world, and especially the US, of two things. First, that there are valid reasons to get to Mars immediately and second, that the mission to Mars *should be allowed to try*. Once the national and international consciousness is raised in these two matters, all else will be easy. The end result will be that the reader's, and ultimately mankind's, little switch, the one that powers hope and optimism and a strong belief in the future, will again turn on. *In many ways, that is what this book is all about.* Turn that switch; the switch that has been off for so long; turn it back on.

There is a great deal of exciting material here, but because these are serious books, there is also much that must be absorbed. Have fun with the exciting stuff. Hiss at things that you do not like, but hiss gently, because you may find your ideas changing as you later come to realize why X had to precede Y in order for Z to be explained. Those of you who do not want to mess up your tidy little world with political and economic realities are free to skip those sections or even those volumes that ask you to think hard about fundamental issues. You are invited to take the easy road up to the vista. Those who want the full experience, however, should take the rocky, unsafe roads through all three volumes, even if you need to make several detours. The final view might be the same for either reader, but the experience will be much fuller and more satisfying for those taking the long way around. Most of all, be patient. These three volumes are long and they address serious and difficult topics. You will not finish any one of them in one sitting. If you do, you should consult your mental health professional immediately. And for the young, read even if you do not understand it all. It will be those young in age and young in heart who will make the mission to Mars work. They are the ones who will usher in a New Era for mankind. Neil Armstrong was wrong when he took that "one small step". It will be the mission to Mars that will make "one giant leap for mankind".

Daniel F. Goerke

---

2   A reference to President George W. Bush's "interesting" proposal to go back to the moon *after* his second term is run. Sounds like another "to-be-continued" episode in the NASA soap opera.

How to Think About Mars

Our ideas are only intellectual instruments which we use to break into phenomena; we must change them when they have served their purpose, as we change a blunt lancet that we have used long enough.

—Claude Bernard

The most difficult part of this three-volume format is that a basic methodology for this work—the need for the reader to *rethink* his ideas about space and "astronauts"—is not presented until Volume II. There, an entire "Part" of the book, five chapters about re-grooving mental patterns to allow better ideas to take root, is devoted to the task. As the methodology in this volume is to *paint pictures* about the mission to Mars, it is easy to dismiss many of the ideas here with an effete flick of the mental wrist, declaring that there is nothing of worth between these pages, when in fact all that has been truly revealed is the existence of yet another NASA indoctrinated brain.

If the skeptic and those with an analytical bent can endure Volume I's 350 pages of sometimes-lively examples and stories, they can begin a direct assault upon the more fundamental ideas of the mission to Mars in Volume II. Unfortunately, we are still left with the idea that *the way the reader thinks about Mars* may prejudice her ability to grasp the materials presented in this volume or in Volume III.

The purpose of this chapter then is to get the reader to "re"-open a mind as she reads these pages. This work's view of Mars is probably miles—light-years— from how you currently conceive it. Remember, this volume is purposefully not intended to *convince* readers of the validity of the mission to Mars' ideas, but attempts to communicate a vision, as an artist attempts to express an idea using his particular medium. Many will understand where the mission to Mars will go in Volumes II and III and, indeed, some may find the ideas in this volume so intuitive that a detailed explanation of those ideas is unnecessary. In any case, the full story can only be painted on a canvas where Mars is seen not as some distant and forlorn planet, but as an earthlike planet just a little ways off, that offers literally to double humanity's prospects and resources.

In that light, the words from the back cover are reprised:

> *Mars is a remarkably earthlike world of vast potential, separated from us by as little as the strings of hanging beads in some homes separate one room from another. And, as with supernaturally abandoned space in other homes, only the ghosts of bad ideas keep us off a Mars we could make our own.*
>
> *The purpose of this three-volume work is to allow the reader to understand how humans can step through those beads, exorcise those ghosts, to double the size of the human plantation and, at least as importantly, to usher in a New Era of mankind.*

From The Simple ...

To digest this word-paint, the reader is asked to ponder the following elaborations:

Mars Is The Earth's Twin In Almost Every Matter of Importance

There are five attributes of the earth that help define human possibilities. The first is that earth contains life. Second is that humans are deeply tied to rich supplies of water. Third, is that volcanism and other geological phenomenon have created resources that humans may exploit to advantage. Fourth, is that humans breathe oxygen. Finally, a multitude of attributes on earth coincide to allow plant life to flourish.

Only two celestial bodies in the solar system have all five attributes. Consider Figure 1:

	Life	Water	Resources	Oxygen	Plants
Earth	X	X	X	X	X
Moon				X	
Venus	?	X	X	X	
Europa[1]	?	X	X?	X	
Mars	X	X	X	X	X

Figure 1: A Snapshot of Our Solar System

---

[1]   A moon of Jupiter, often heralded as containing many attributes important to life.

Assumptions:

  a. Venus is too hot to sustain terran ecosystems.

  b. It is not unreasonable to question whether Venus and Europa have life and to posit that Mars *has* life.

  c. Oxygen is loosely bound with carbon on Venus and Mars and with hydrogen on Europa. It is tightly bound with silicates and other minerals on the moon.

There are also a few other factors which make Mars the earth's twin. First, consider that Mars' day is almost the same length as is the earth's or that the Martian axis of rotation is also almost exactly that of earth's. Mars and earth receive relatively similar amounts of solar radiation and there is no reason to believe that terraforming could not create climates on Mars similar to those found on earth. Mars probably once had surface waters and its total land area is almost the same as the earth's.

The earth shares attributes with many other celestial bodies in the solar system, but it shares most of her attributes only with Mars. Humans will one day be able to live on many other planets and moons of this solar system. Only on Mars, however, will they ever be able to live much as they do on earth, with a breathable atmosphere and comfortable climate, grass and trees outdoors, and outdoor insects and mammals keeping the exterminator busy.

Vast Potential

Not only is the surface area of Mars as great as that of earth—effectively doubling humanity's living space—but humans have not spent the last million years harvesting its gold, iron, and other resources. True, there are currently no lakes or streams on Mars to catch fish or forests to harvest wood or shoot deer, but all of that can change. There is no reason that human settlers could not transform the face of Mars to become a home for countless terrain species, to set into motion the kinds of ecological forces that have created abundant *living* resources humans use to better their lives. Living, mineral-based, or otherwise, Mars' potential to improve the fabric of human existence is almost beyond comprehension.

Little Separation From Earth

Most people do not understand how close Mars is to earth or about the relative ease of designing a mission to reach it. Initially, trips will have to be measured in months and not in the weeks those readers old enough to remember Apollo might recall. Still, the problems are virtually the same. Get into orbit, take a trip across space, land on a new celestial world, and then return to earth. Many of the 1000 pages in these three volumes explain how this can be done and how—especially as we choose to settle Mars instead of "exploring" it—costs can be driven orders of magnitude down from what NASA would need.

Bad Ideas Have Kept Us From Mars

Few would argue the utility of a free market in ideas or the power and efficacy of new ideas to transform human society. It was only after European humans amended their ideas about a "flat world" that Europeans were able to "discover" the New World, to transplant colonies there, and to set into motion powerful ideas of human freedom and national identity. The cement-headed mentality of the European traditionalists (or those nuclear physicists who refused to accept Bohr's theories about the atom, or those today who argue that dinosaur fossils are 6000 years old, etc.) is found in great measure today at NASA. The idea that some big, expensive government program is the *only* possibility for space is just waiting to be exploded in favor of better ideas. *We have had the technology for 35 years.* The answer to the question: "Why have we not gotten to Mars?" is simply that humans have not yet thought properly "outside of the box" for a solution. Once NASA's bad ideas are replaced, we will see footsteps on Mars.

Little Physical And Economic Effort Is Needed To Get To Mars

As indicated, *for 35 years we have had the technology* to get to Mars. The primary excuse over the years has been that a Mars program would "cost too much". Most of these three volumes stress the point that it is political and economic questions that stand between us and Mars. With due respect to the possibility that the author is a windbag, most of the 1000 pages are devoted to exploring these political and economic questions and to developing solutions for them. The surprising result is not only that human colonies can be established on Mars quickly, but that the cost per person and in sum are reasonable.

A New Generation Will No Longer Be Seduced By The Dead Hand Of NASA-Think

Those of us old enough to remember Apollo can recall the excitement we held about the future and the confidence we all felt in NASA. A mortal sin of the Apollo era is that this feeling has not been passed down to succeeding generations. This work does not so much judge NASA as it identifies its condition. NASA is moribund and must be replaced if humans are to regain their hopes for space. Many remain wedded to NASA, but this is a terrible mistake. It is one of the old ideas that must be cleared away, like so much scrub brush if something new is to be planted. Many older people may no longer trust NASA to takes us into space, but—far worse—young people no longer *care* about NASA.

The Human Plantation Will Double

It will double in size and more than double in potential. It is easy to understand Mars' physical dimensions or that a human society on Mars implies twice as much land available to humans. What is probably less clear is that the potential for new businesses, new opportunities, new societies, and new ways of doing things will be more than doubled when the creative, the disgruntled, the flighty, and the dreamer have so much undiscovered country on the potter's wheel.

The idea of "twice as much land" is actually a very interesting question, as it requires not only a prediction of how much land on Mars might be suitable for human habitation in future years (not occupied by Martian lakes, oceans, polar regions, etc.), but also involves the question of how much land humans actually use on earth. Besides Antarctica, vast sub-arctic regions, deserts, and other remote areas only marginally suited for habitation constitute much of the earth's landmass. No serious proposal has ever been made to "terraform" earth to maximize the quantity of land available for human use. On Mars, however, not only will terraforming be a real question, but it will be real important. As will be discussed in greater detail in volumes II and III, the author believes that Martians will ultimately decide to create a cool world, one that will likely result in a small "Arctic Ocean", but that will also have temperate and a few sub-tropical climates near Mars' equator. The result will be far more land available on Mars for human habitation and cultivation than on the earth, perhaps even *twice* as much.

The potential for human society on Mars is the classic example of an idea that the intuitive few will grasp immediately, but will require considerable explanation for most. The main idea here is that *Mars as frontier* will ignite and fuel human creativity for at least 500 years, before Mars starts to "fill up".

Humans In Space Means A New Era For Mankind

The final, least intuitive part of the mission to Mars will be its importance in historical terms. The "discovery" and colonization of the New World in the Americas was, by most accounts, a watershed event in human history. The creative forces unleashed not only revolutionized ideas about economics and politics, but the example of a free America and a laissez-faire economy forced the rest of the world to double-time to catch up.

Settling Mars will be another watershed event, but one of even greater force than the New World. The US, the greatest nation in the history of the world, will be found wanting in many ways, as compared with fifty-year old city-states on

Mars. The US will accordingly be forced to re-evaluate its headlong plunge into the nanny-state experiment that it and the western world have indulged for nearly a century. Freedom on Mars will challenge America and all the earth to its core, leaving in its wake new boundaries of prosperity and human dignity.

... To The More Advanced

Now, a few more ideas, most of which follow from the last section.

Twinkle, Twinkle Little Mars, How I Wish An Entrepreneur I Was ...

We begin by considering what it would have been like to have invested in America before it was settled. The answer, in a word, is that it has been history's most amazing investment opportunity.

An investment in Mars will be ten times as valuable, since there is room for ten Americas on Mars. Any business started on Mars will continue to grow at 10–20% per year, for a thousand years. How would you like to be the first Subway sandwich franchisee on Mars? How rich will that person become? The frontier life and the lack of heavy-handed government regulation will allow for an explosion of new businesses and industries. Life and the vibrant, new society on the Red Planet will be enviously viewed as being better than even the best places on earth, bringing in money and easy credit.

Most importantly, Mars will offer opportunities that will not be available on earth. Not only will there be a demand for new ideas and new inventions to cope with the new conditions on Mars, but the old industries of earth will be demanded in full volume, with Martians stepping up to satisfy the demands. Pick the industry: steel, automobiles, insurance, shoes, construction. Everything will be required and it will be a new settler on Mars—and only rarely some hundred-year-old business on earth—that will get paid to deliver.

In Less Than One Hundred Years, Earthlings Will Vacation on Mars and Vice Versa

Two hundred years ago, it was unthinkable for the average European to travel to America on a whim. One hundred years ago, the same was true about Australia. Unthinkable today, one hundred years from now, humans from the two planets will routinely travel back and forth—on business or for pleasure—creating huge pools of wealth and possibilities for human enrichment.

NASA will never deliver this reality. NASA would still be studying Mars in one hundred years. The mission to Mars will pioneer these interplanetary bridges that today seem impossible to create but which, in the near future, will have seemed almost inevitable.

More New Era

One of the great lines in all of science fiction was written by Robert Heinlein when he said that once you are in orbit around the earth, you are halfway to anywhere. By way of explanation, the idea is that the force of the earth's gravity is so immense that once an expenditure of fuel is made to get into orbit around the earth, the amount of fuel needed to reach most other destinations of interest becomes almost trivial. For the flight to Mars, the fuel required to put human settlers and their equipment into orbit around earth will dwarf the requirement to ease them into a trajectory toward Mars. Thus, once there is a *human society* on the other side of this great cosmic divide, humans will be halfway to anywhere.

In the spirit of staying firmly tethered to reality, however, let us understand this idea to mean that humans will be able to build several new societies in our solar system once there is even one spacefaring society. The presence of such societies will challenge existing earthbound societies to keep up with the guaranteed human progress that space societies will make. Some of this progress will be measured in new forms of social organization and bold new freedoms, but most of it, of course, will simply mean vast opportunities for McDonald's, General Motors, entrepreneurs taking advantage of new circumstances and, most of all, for poor slobs who simply want a better life for themselves and their families.

This New Era of how humans understand themselves, as members of an interplanetary rather than international community, will create a new syntax and grammar for how humans relate to one another, to government, and to other organizations in their lives. For some it will be as important as rethinking what it means to be human. For others, it will simply be figuring out how to face Mecca when they pray on Mars.

Of course, far beyond Mars and this solar system are the stars. Humans may not have the ability, once in space, to easily reach for that next star, but most will agree that lots of time will be spent thinking about how an interplanetary community might one day become interstellar.

The Hardest Think To Do Will Be To Change How One Thinks About Space Travel

Humans have been brought up to believe in space as the domain of huge government bureaucracies, with arm-patched astronauts, working for an earth-based hierarchy that may or may not have a great deal of genuine interest in space, as compared to the politics of their own bureaucracies. As for *humans* in space, the nerds have got it both right and wrong. They have it right in thinking that space will end up being a workplace, employing a sometimes-dissatisfied staff, mostly going about the business of making money. They have it wrong to imagine that these businesses will create goods and services demanded by government or huge, earthbound cartels. An infrastructure for space will be built, but most will work, as on earth, to service the needs of workaday people and commonplace lives. Think of space, then, in terms of Wal-Mart, not the Pentagon. Even with a future filled with genetic engineering and computer downlinks directly into the human brain, humans will continue to care most about families, their relationship with others, a little excitement, and making more money in the coming years. These values will be carried into space and will form the basis of the mission to Mars' society, as NASA-bots and bureaucrats aboard the International Space Station never will.

The importance of imagining life on Mars and in space as a life *we already know in the western world,* is an important first step in understanding the ideas in these three volumes. It is important to think of Mars not as a place to gather rocks, but as a future home to millions and even billions of people, living a privileged life, just slightly better than the lives led in the most advanced nations on earth. If Mars can be viewed as a *home* and not a place that in the early 22nd-century might house a few NASA scientists, most every idea falls directly into place. This is not an easy idea for most people and hundreds of ideas and pages are devoted to the task of painting this picture. This is the hardest first step: think of Mars as a home and all else will come easy. Continue to buy into NASA-think and the ideas here are difficult.

What The Mission To Mars Is

Follow the Yellowbrick Road

These three volumes are meant to serve as roadmaps for the colonization of Mars. As with Dorothy, these three volumes are not suppose to solve all problems. Rather, they are a Yellowbrick Road to Mars. There will be uncounted numbers of problems—large and small—that will have to be solved along the way, but the mission to Mars argues that the ideas in these volumes can be counted on to lead to Mars. Remember, it was not the Yellowbrick Road or even the Great Wizard that held the answers. It was faith in themselves that provided Dorothy and her pals the courage, the brains, the emotional depth, and the ticket home.

In a sense, this chapter proposes two main ideas. First, that human beings can achieve whatever they set their minds to achieve and, second, that there are a great many weighty and valid reasons to go to Mars. Thus, the better question is not *why* we should go to Mars, but rather why we are not there already.

Freedom

In a very real sense, the mission to Mars is not so much about establishing colonies on Mars as it is about re-establishing an archway to human freedom. The lack of real frontiers in modern human society means that the options before many humans who desire to improve their lives are few. A good—and mostly tragic—example of the limited options available to us today is the ocean of people coming (illegally) across the American-Mexican border. Mexicans cross illegally into the US not to go to New York Yankee baseball games, but because their life options in Mexico are so dismal. Unlike most of human history, frontiers have disappeared on earth[2]. Today, there are very few choices available for those who hope to improve one's life. The great irony of the Information Age, therefore, is that people come to know that freedom and riches exist, yet (Third Worlders) often have no option to explore possibilities for themselves.

The Third Rail

Freedom is the theme music. Creativity is the counterpoint.

In many subway systems, two rails are for the train's wheels and the third rail is electrified to supply the vast amounts of power necessary to run the trains. Touch that third rail and you are going to die.

At the mission to Mars, touching the third rail means you are going to live.

The mission to Mars' third rail is creativity. Creativity is the flame—arising from sparks of freedom—that both transforms societies and consumes individuals. Almost everyone is born with some measure of creativity. Almost everyone who is allowed to express their creativity becomes alive in a way that those who trade their labor for wages can only imagine.

The real reason that NASA will never reach Mars is that it employs 50,000 people and allows very few to be creative. The mission to Mars, in contrast, will attract 2,000 people, demand that they be creative, and will get to Mars in a few decades. Creativity will flow through the mission to Mars like the Mississippi drains the Midwest; it will dominate every manner of activity and will strengthen every program. Creativity will ignite so many new ways to solve old problems that it will astonish the world.

Most importantly will be the way creativity will enrich the lives of individuals at the mission to Mars. Not only will better and more efficient means of space travel be developed, but a new society will be in a position to develop new *everythings*.

---

2   If only there *were* a frontier for Americans to cross, much of this "Mars" discussion would be moot.

Better ways to deliver health care, better ways to educate children, better ways to deal with marriage and divorce, better everythings. And at the mission to Mars anyone who is willing to devote the time and the effort to try something new will find a place for their new ideas. And everyone who discovers that other people value their new idea will live as richly as the kings and queens of old; and will live far better than the drones who inhabit the hive of our modern western suburbs.

The mission to Mars' third rail is creativity, a rail that will bind together its tiny society and wraps it up in a bow of success.

What The Mission To Mars Asks of Readers

The three volumes of this work will have been successful, and the mission to Mars will succeed, if two simple ideas can be communicated to people across the planet. These ideas relate to political ideas, not technological issues and, once pondered, will be easy to nod to agreement.

There Are Reasons To Go To Mars

The most fundamental question of all is whether there are *good* reasons to go to Mars.

You Should Let Us Try

This one is harder but, honestly, unless you are a dyed-in-the-Spandex hrrrrr-umppphhh machine, is it really so difficult to concede that people should have the freedom to live their dreams, even if the dream is crazy?

Part II

On Earth and Beyond

"This is Abraham," I said. Then I put two rocks underneath the first one in the shape of a family tree. "Here are Ishmael and Isaac. The question the world has been trying to answer for centuries is, 'Which direction does Abraham's lineage go?'" ...

"If you're looking at the realm of ideas ... it doesn't matter." ... He took his hand and swiped away ... [the] rocks, leaving only the rock of Abraham. "Abraham changed the world because he brought one idea to the world."

"So what's the idea?"

"The idea is that what's important is the power of ideas—human ideas. Not rivers. Not idols. Not stones. Not land. Abraham went into the desert, a place of nothing, and created something entirely new. And that something new was based on something invisible. He collected technology and knowhow from all the places he visited. He mixed them with this big, unknowable, untouchable God, and he passed that down to both of his sons. And that's what changed the world. If we're fighting over stones, we're missing the point. Abraham was about a single idea, and that idea he gave to us all".

—From *Abraham*, by Bruce Feiler

The Basic Ideas

What we anticipate seldom happens. What we least expected generally happens.

—Benjamin Disraeli

The plan laid out by this book is simple: bring together in one place all people who desire to go into space and to Mars, harness their talents using proven theories of democracy and capitalism, and set them to the task of creating an independent, non-NASA community. This "earth Embassy" will build a self-sustaining economy whose single-minded goal will be to transplant itself to Mars. This plan avoids the endless tar pits of the rocket daydreamers who talk strictly of science and engineering, first acknowledging and then avoiding the BGBS[1]-crafted obstacles to human settlement of Mars: political will and economic cost.

The plan envisions creating an economy where the talents and labor of people who want to go to Mars will be captured to achieve the colonization of a new world. What the mission to Mars proposes is neither dependent upon anyone's permission nor dependent upon some unrealistic structure. It is rather a workable plan to transport large numbers of people to Mars. Once on the Red Planet, not only will Mars be terraformed to sustain familiar eco-systems, but transplanted industry from the mission to Mars' earth Embassy will help to create a Martian economy capable of harvesting the riches of Mars and of nearby asteroids. There will be a land boom on Mars, offering limitless vistas for entrepreneurs to grow new businesses and to expand those they started at the earth Embassy. Over the course of a few hundred years, Mars will become *Homo sapiens'* second home.

---

[1]   Big Government, Big Science, a major point of discussion in Volumes II and III.

The instrument of this Red Planet Revolution will be the creation of a government[2], the Confederation of Martian Republics, which will establish itself "in exile" at its embassy on earth. Not only will the "government form" of organization be able to harness the political will necessary to get to Mars, but it can also take rational political positions about its power to launch spacecraft, license doctors, emit pollutants into the air, etc. It can rule a Mars of *Martians,* rather than endure generations of colonial rule by "foreign" know-it-alls. And, as will discussed in detail in Volume III, a government will be far more able to harness resources as compared to a corporate or other form of organization. A government will virtually guarantee success if enough people are willing to trade their labor for a chance to go into space, to get to Mars, and to establish a viable society there. Most importantly, however, will be that a "trial" government on earth will be able to work out issues and procedures to ensure that, from the beginning on Mars, freedom and democracy will operate as well and *even better* than they do on earth.

The mission to Mars rests upon three main legs. The first has already been mentioned: the earth Embassy. The second is the "ticket" program, whereby residents of this earth Embassy purchase tickets to Mars. The ticket system serves to yoke the energies and talents of those participating in the program and is the foundation upon which the economy of the earth Embassy will grow. The last leg is the Fleet. This quasi-military organization will socialize and train people for their roles in space and on Mars; and it will harness at below-market rates some of the labor required to achieve the mission to Mars' objectives.

Unfortunately for the impatient, the sub-title of this volume is *"What" We Can Do*, not *why* we are doing it. A more developed discussion of the positive attributes of the earth Embassy, "tickets to Mars", and the Fleet will have to await volumes II and III. For now, the reader must be satisfied with a brief description of these three legs here and a single chapter devoted to each in this volume.

The earth Embassy will seek to create an increasingly self-sufficient economy. This economy will serve the needs of the residents, plan for its own transplantation to Mars, and will build the ships and infrastructure needed to achieve the political objective of settling Mars. The earth Embassy will also serve as a proving ground for government on Mars, after a fashion that will work out Martian solutions to the dilemma of "too many egos" that plagues both current "private" plans for Mars and human society in general.

---

[2]   How to create this "government" and how to accommodate its "embassy" in a "host nation" are major aspects of this work.

Tickets to Mars will be the bedrock of the earth Embassy's economy. Everyone at the earth Embassy *must* purchase a ticket to Mars, payments on which will fund projects at the earth Embassy. The efficacy and the practicality of this system to pave an economic boulevard toward Mars will occupy many pages in all three volumes. The ticket program allows for the unique idea that *anyone* can go into space and to Mars. Thus, those who *self-select* to dedicate their lives to the mission to Mars can avail themselves of this opportunity, not just a few bureaucrat-astronauts lobbying for a seat on a mission to collect rocks.

The Fleet is the last major leg of the mission to Mars. It will consist of everyone over the age of 18, requiring 31 days of annual service from each of its members. Five to ten percent of the Fleet's membership will consist of relatively lowly paid full-time members, who will help units perform missions while its part-timers are away at their civilian employment and who will help advance administrative missions. Even most of the "full-time" Fleet members will not be able to make a career of the Fleet, but will consist of those who decide to enlist for a few years, pilots and other leaders in training, and a few high ranking leaders who will serve short full-time tours, to allow for consistent decision making on the most important programs.

The power of the idea of settling Mars rather than *exploring* it cannot be overstated. This recurrent theme will wind itself again and again into the pages of all volumes. Not only will settling Mars be easier than "exploring" it, settlement will dramatically reduce costs. Part of the difference can be understood as one of having settlers in Boston and in Mexico City, versus petitions to the queen every time you need to get to the New World. This idea is admittedly much different from the NASA paradigm of "exploring" space and most readers are expected at least initially to be skeptical of it. As each brick of the mission to Mars is layered upon a previous row, however, "settlement" will become—self-evidently—a good idea.

The Long View

We now present a satellite's view of the Confederation's plan to take human society into space. While the mission to Mars' first steps will be critical to eventual success and probably of great interest to the Nay Sayer, the purpose here is to paint a much broader picture—to imagine what can be. Consider the campfire now. Worry about the matches later.

The Confederation will have three main incarnations, one provisional and two "permanent". During the first, provisional phase, the "Era of the Earth Embassy", the Confederation will organize resources and a society to transplant to Mars.

This phase will end in the year 2100, when the Confederation's provisional constitution will be revised and replaced with a permanent constitution. This era will have several major sub-components. Figures 2 and 3 show some details about this "era". In very general terms, there will be 15 years of unmanned rocket flights and trials, followed by 10 years of pre-Mars manned space flight, another 25 years to transport 50,000 humans from the earth Embassy to Mars, and finally, 40 years of expanding Martian settlement. The creation of self-sustaining Martian communities will have become a well-known process by the end of this era. And, fifty years after the earth Embassy opens for business, there will be a population of 50,000–75,000 people on Mars, capable of sustaining themselves and working hard to uplift their standard of living from "comfortable subsistence", to "quasi-western equivalent", and finally to the highest human standard of living anywhere. By the year 2100, there should be about 700,000 people on Mars, in four or five independent republics, living an envious lifestyle. Trees and plants will grow abundantly on a warming planet, where water cycles in the environment like it does in a terran tundra. Humans will not be able to walk outdoors without oxygen-assistance, but otherwise we will feel very much at home.

The second era of the Confederation will begin with the adoption of its permanent constitution. As Mars develops, the organization that took humans to Mars will transform itself into a Martian-style Commonwealth of Nations. During the 400 years of this "Era of the Republics", a new republic—a new, Martian nation—will be granted independence about every six years. Just as important, a permanent home for the Confederation government will be built on Olympus Mons, the tallest volcano in the solar system. Immigration will keep the planet's population doubling about every fifteen years. It will not be long before the independent republics on Mars become recognized as among the most advanced of the human nations and before the societies on Mars become the envy of much of the earth. Indeed, a continuation of recent human history on earth may very early on allow Mars to supplant the earth as the most progressive human society, leaving behind troubles on the Indian sub-continent, the Middle East, and elsewhere on the water planet. Life from earth will fill infant Martian ecosystems as busy terraformers tinker with the planet's temperature and fiddle with the hydrosphere[3]. These same scientists and engineers will shake their head at "political" decisions not to release wolves or bears in a certain area or that the planet's average temperature is to be increased another two degrees Celsius in deference to the desires of the powerful "Republic of X". For much of this era, humans will be able to breathe the Martian atmosphere short-term, even though there will be dangerous levels of carbon dioxide that will require some pill or technology to

---

[3]   Circulation of water on a planetary scale.

avoid difficulties. By the end of this era, however, the atmosphere will be fully breathable, rendering any extra-oxygen-devices unnecessary.

The last of the Confederation eras, beginning about the year 2500, will find the Confederation in its permanent state. Almost all of the planet's major political and social issues will be argued and resolved by leaders in 70 independent "republics", each of which will have populations of 5–20 million people. The common defense of the planet will continue to be in the hands of the Confederation Fleet, as will be the terraforming maintenance operations. During this era, the last of the frontiers on Mars will begin to disappear. Of course, by then Martians will have led the way to the stars, where they may form the backbone of an even grander political organization, perhaps enlisting non-human sentients as citizens and allies. It may take another 500 years for Mars to fill up, but the best and most adventurous humans will start to think of Mars as too full and will away to these other parts of the galaxy. The real question at that point in human history will be how Martians view the earth. Will humans on the home world have finally sorted out 5,000 years of historical non-sense or will the true center of human society have shifted to Mars, where the rich people live? Will the earth in 2525 be the proud ancestral home? Or will it be the slum from which some few humans have escaped? Remember that human civilization started in an Iraq that was, until very recently, controlled by one of history's most infamous dictators. The formerly great Mesopotamia certainly is, and may remain for many decades, a slum land where Saddam's thugs and now self-righteous terrorists exploit the Iraqi people, obscuring the grand story of Mesopotamian culture. As human society on Mars moves forward, many human societies on earth could appear to Martians as backward as the Third World appears today to westerners.

Event	Begun Years After Starting Earth Embassy
Unmanned Rockets	10
SSTOs	15
Explorer Transporters	20
Shuttle Transporters	24
Caravan Transporters	28
Ark Transporters	35

Figure 2: Era of the Earth Embassy

Event	Beginning/End (Landing Year)
Create 49 bases	1–30
Terraforming	5–500
Planet Surveys	3–250
Settlement of 2nd Hemisphere	75
Settle Olympus Mons	20–150
First Independent Republic	40
Second Independent Republic	45
All Republics With One Settlement	150
Martians Can Breath Atmosphere	150

Figure 3: Era of the Confederation

Event	Beginning/End (Landing Year)
Oxygen Content Above 5%	200
Large Bodies of Water Form	250
Planet's Final Temperature	250
Higher Animals Released	200
Mammals Released	300
Independent Republic Every 5 Years	200–500

Figure 4: Era of the Martian Republics

The Cost

The "cost" for these plans will consume hundreds of pages in all three volumes. Know for now that "cost" is a misleading term. *The mission to Mars does not need money to get to Mars, it needs capital.* This capital will be created at the earth Embassy. As a point of reference, however, consider that George Bush I had a proposal for Mars that would have cost about $500 billion. Robert Zubrin's "Mars Direct" proposal[4] might cost 50 billion NASA dollars. We believe that this ten-fold reduction in cost bodes well for

---

4   Mars Direct is a plan to shoot astronauts directly from a launching pad on earth to Mars, by-passing ships in orbit, etc.

the mission to Mars. It may not get to Mars for $5 billion, but the ideas in the three volumes will demonstrate to many readers that it will be able to muster the financial wherewithal to implement its program.

The difference between creating a budget—which NASA must do for each of its programs—and building a society—the mission to Mars' goal—is that cost is everything for the former and almost trivial for the latter. This statement might seem radical, but evaporates with just a moment's contemplation and, in many ways, the 1000 pages of this work have been written to demonstrate this point. Still, numbers are useful to understand the dimensions of the task. For that reason, $500 billion for a NASA program and $50 billion for Mars Direct will serve as starting points for analysis throughout these volumes.

To Space and Beyond!

This section discusses the mission to Mars' strawman "space program". As indicated before, this program has been developed not as a blueprint for space, but merely as a skeleton upon which to layer the fat and muscle of more important ideas. The author, in fact, *expects* that the sometimes trivial technical ideas presented here will inspire some readers to come up with cheaper, better, and faster programs that will ultimately be chosen to transplant a society to Mars.

Single-Stage-To-Orbit

The most challenging physical problem for the mission to Mars—as it is for NASA—is to get into a low earth orbit. This work adopts as its own the traditional solution to this problem. Thus, SSTO—single-stage-to-orbit—vehicles which can be used to transport people and equipment into orbit both cheaply and routinely become a paramount first step.

Inner Asteroids

The mission to Mars may decide that sending teams to one of the inner asteroids makes a great deal of sense. Not only might they be able to test some of their ideas for Mars under a program to the inner asteroids, but many of the raw materials that the mission to Mars may need might be obtained from them. If so, the cost for materials to transfer people to Mars might be reduced by orders of magnitude, as water, metals, plastics, etc. are moved from one point in space to another, avoiding the great cost to "lift" these things into orbit. Now, mining space itself for resources implies manufacture and assembly in space and these are not trivial matters. What might be done and how the mission to Mars can create incentives to establish such businesses and capabilities consume not a small part of later chapters of this work.

The Moon

Alternatively, the mission to Mars could send a party to the moon. The vaunted "lunar ice" might serve as a resource to make rocket fuel or be used as drinking water[5]. Like an inner asteroid, the moon could be used as a test-bed for habitation technology, etc. Indeed, the moon would be, in most respects, simply the largest of these inner asteroids. The (relatively) strong lunar gravity, the great cost to obtain resources on the moon, and its proximity to where humans are coming from and not going to, however, all argue in favor of using an inner asteroid and not the moon, if way- or science-stations are to be built along the road to Mars.

Transporters

Four separate series of transporter vessels will be built to transport 50,000 people to Mars. Each series of transporter vessel will have a slightly different mission, culminating with the Ark Transporters, which will be built to move the majority of people. Figure 5 summarizes the information presented in this section

After four years of sending four people per trip to Mars using "Explorer Transporters", the "Shuttle Transporter" series will begin operations. These vessels will carry six settlers per trip, plus its crew of two, landing twice a year. Four years of flights, augmented and accompanied by continuing flights with the Explorer Transporters, will build Mars' population to nearly 100 after eight years of settlement. It will then be time to begin the large-scale movement of people from the earth Embassy to Mars.

Transporter Name	Explorer	Shuttle	Caravan	Ark
Settlers Carried Per Mission	2	6	100	1000
# To Be Built	10	6	9	18
Generations of Vehicles To Be Built	2	2	3	4
# of Missions	70	60	50	40+
Total Settlers Transported	140	360	5000	40,000+

Figure 5: Basic Information About Transporter Vessels

---

[5]   The ethics of denuding the moon of its limited supply of water are strikes one and two against the "Moon-Base" batter. Removing water and other resources from an asteroid is one thing. Taking much or all of the water off the home world's only moon is something that probably should not be done, at least not without some sort of well-conceived remediation plan.

"Caravan Transporters", with a small crew and room for 100 settlers per voyage, will run for five to ten years before the Ark Transporters, to work out the issues of moving large numbers of people to Mars. Fifteen years after the first humans land, there will be two thousand people on Mars and the Martian economy will begin to take on elements of self-sufficiency. This self-sufficiency will be assisted by the organized movement of corporations and businesses to Mars on each Caravan Transporter mission.

Mass movement of people to Mars will be the job of the "Ark Transporters". Each Ark Transporter will transport 1,000 settlers to Mars. There will be 35–45 Ark Transporter missions to transport residents of the earth Embassy to Mars. The first Ark flight should convince any remaining skeptic of the ability of the Confederation to fulfill its obligations under the ticket contracts. Once all ticket contracts have been fulfilled—and 50,000 people have been transplanted to Mars—these Ark Transporters will be sold to commercial transport companies.

As each new, larger type of transporter vessel is brought on line, it will begin to fly missions in convoy with transporters of earlier series. Thus, the first Caravan Transporter mission to Mars will convoy with one Explorer and one Shuttle transporter, carrying a total of 108 new settlers for Mars[6].

Deimos[7]

The first steps "on Mars" may not be on the Red Planet at all. Rather, the first landing will probably be on Deimos, one of Mars' moons, which will become a natural spaceport for those traveling between Mars and the earth. There will also be much ink about why this should be so, but the main idea is to minimize the complexity and cost of the spacecraft to be built. Thus, specialized (simple) spacecraft will lift humans and cargo into orbit around the earth, other specialized spacecraft—transporters—will carry them to Deimos, and a final series of specialized spacecraft will shuttle people and cargo between Mars and Deimos. Using simple, specialized spacecraft and natural celestial bodies like the inner asteroids and Deimos to assist with mission to Mars' objectives will go a long way to reducing overall costs.

"Patches"

The secondary purpose of the Ark Transporter system will be to speed planetary development by inhabiting a bite-sized slice of Mars. After the initial focus on

---

[6]   There will be two flights planned per year. Volume III discusses how this can make sense theoretically (averaging two flights per year) or actually (making administrative and technical improvements that actually allow two flights to launch per year.)

[7]   Mars has two moons, Phobos and Deimos. Phobos is the larger and Deimos the smaller.

one or two settlements, the mission to Mars will refocus on settling a new world. Thus, each Ark Transporter mission will be structured to create an entirely new settlement on Mars. Figure 6 shows a Confederation plan to establish 49 separate bases on Mars, laid out in a 7 x 7 grid, to root humans onto a Texas-sized Martian plantation. This grid is divided into a number of different "patches", or settlement regions. Each new settlement will be about 100 miles from the next. As each patch of Mars becomes reasonably well known, its limits and potential understood, the future of the patch can be factored into the larger plan of Martian development. Each Ark Transporter mission will be directed to a new settlement, to fill out its part of a new patch of Mars.

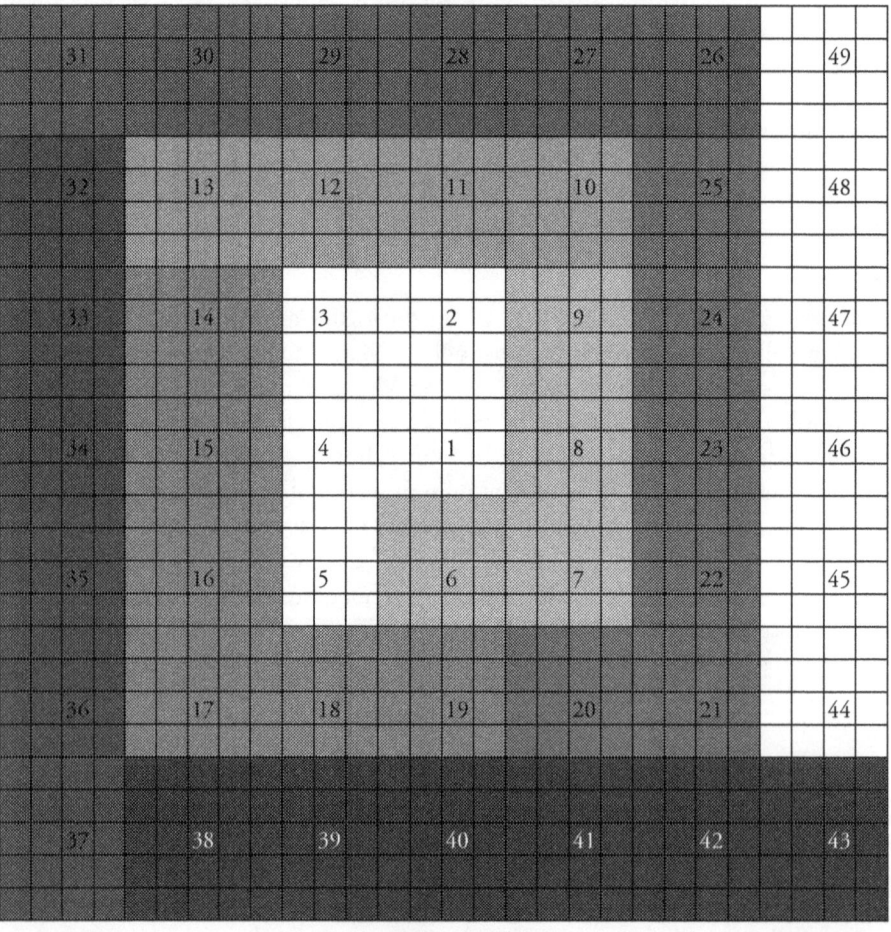

Figure 6: Bases and Patches

Figure 6 shows how this patch system might work. Each number represents a base. The different shading represents the patches of Mars to be settled. While the bases will be settled in approximately the order represented here, the necessity to harvest vital resources will likely force a few settlements out of order, especially in the early years.

The end result of this "patch" settlement idea will be to transplant a society from the earth Embassy onto 1% of Mars, covering an area 600 x 600 miles. Forty-nine communities, each with about 1,000 residents, will have been created. Mars will get "explored" by its residents, living and working on the planet. The patches shown in Figure 6 do not represent political entities, though that aspect of Martian development has also received some attention in Volume III. Rather, each patch will be designed to assist with the development of a freestanding economy on Mars. In a very short time, other communities will be created, some of the original communities will grow rich, others may begin to wither, all of this dictated by the desires of the people on Mars, the economic opportunities each settlement presents, and the resources readily available at each settlement.

Second-Inning Stretch

At this point, the ideas and goals in this chapter may have begun to sound like something from a bad movie on the Sci-Fi Channel. But remember, you have just started the first volume of a three-volume project. It will take three volumes to explain how these ideas can be brought to fruition and why they should. And even more importantly, these volumes explain that what is being proposed is not outrageous at all, but rather a logical step—arguably the only *possible* step—to get a human society into space and planted on Mars. The ideas in these three volumes are not easy. Unlike daydreamer books, they examine political and economic ideas in great detail, with a mind to persuading the reader that what is being proposed can be done. As said before, however, not everyone needs to read every page and review every spreadsheet to be persuaded by or interested in the mission to Mars.

Now that you have read these "Basic Ideas", you can pick and choose to read what you wish. Of course, the three volumes have been laid out to be as logically progressive as is possible. Skip some materials and you may find unfamiliar terms or ideas or—even worse—misinterpret these terms. Really good books, however, are often entertaining even when read out of order or, after the first read, reread in snatches and swatches. Maybe your experience of reading *Mission to Mars: 2025* will be something like that.

Down the Path to Understanding

"The rebellious war ... is manifestly carried on for the purpose of establishing an independent empire ... I need not dwell upon the fatal effects of the success of such a plan. The object is too important, the spirit of the British nation too high, the resources with which God hath blessed her too numerous, to give up so many colonies which she has planted with great industry, nursed with great tenderness, encouraged with many commercial advantages, and protected and defended at much expense of blood and treasure."

—King George III, opening Parliament in October, 1775

"They planted by your Care? No! your oppressions planted em in America. They fled from your Tyranny to a then uncultivated and unhospitable Country—where they exposed themselves to almost all the hardships to which human Nature is liable ... And yet, actuated by Principles of true English Lyberty, they met all these hardships with pleasure, compared with those they suffered in their own Country, from the hands of those who should have been their Friends.

"They nourished up by *your* indulgence? They grew by your neglect of Em: as soon as you began to care about Em, that Care was Exercised in sending persons to rule over Em ... sent to Spy out their Lyberty, to misrepresent their Actions and to prey upon Em ...

"They protected by *your* Arms? They have nobly taken up Arms in your Defence, have Exerted a Valour amidst their constant and Laborious industry for the defence of a Country, whose frontier, while drench'd in blood, its interior Parts have yielded all its little Savings to your Emolument ...

—Colonel Isaac Barré,
member of Parliament and British soldier,
disfigured at the Battle of Québec

We begin with some consideration of what it would be like if Washington decided to send humans to the Red Planet. Here we discover that the momentous task of sending humans to Mars becomes first a comedy and then a tragedy.

Mars According to Garp

A semi-prominent engineer, Robert Zubrin, proposed that NASA shoot manned rockets directly to Mars. His "Mars Direct" program was supposed to cost $50 billion to organize and $5 billion for each mission.

NASA approved the plan, but the director wanted to put his imprint on it. Accordingly, six years after Mars Direct, NASA convinced Congress to approve its "Mars Indirect" program. The price went up to $65 billion, with $7 billion budgeted for each mission. The program anticipated twenty years to prepare the first manned mission, with five subsequent missions occurring at three-year intervals.

Thirty-six years and $130 billion[1] later, the first manned mission was launched. Unfortunately, Zubrin could not be present, as he had passed on during the long wait, missing the launch date by two months. He died a happy, but unfulfilled man, knowing that mankind would finally walk on Mars, but never having gotten his own chance. The mission was a spectacular success, but public enthusiasm for more Mars programs never materialized. The six-month window from launch to arrival at Mars prevented much sustained interest. The fact that the crew spent only two weeks working on Mars before they blasted back to earth transformed the magical into the ridiculous. A second mission was prepared, but instead of the estimated $15 billion for the launch, a major redesign of the mission meant that it would cost $20 billion. The second mission departed for Mars one year to the day after the first crew arrived back on earth. Again, the world watched, then quickly returned to its soap operas and wars in the Middle East. Already, $150 billion had been spent to put seven scientists and one senator on Mars, this time for two years. That senator was an important member of an appropriations sub-committee, however, so NASA's Mars program was assured of continued support.

Smiling and glad-handing for the cameras, but in private most reluctantly, the President signed a bill committing the US to another twenty-year program, this one costing $500 billion, to lay a foundation to settle Mars. As with the International Space Station and the Lunar Colony project, most of those geeking for the cameras knew that the actual program would be pared down over the years. Still, the next launch was only two weeks away and the next ten missions, several per year during favorable launch windows, were virtually assured. Deep in the bill, however, was a provision that paid off the President's anti-nuke supporters. There

---

[1]   This section assumes constant dollars.

would be no further construction of nuclear power plants on Mars or in space. Instead, NASA planned to transport solar panels to Mars, raising Martian energy costs five-fold. The senator from New York remarked to a British reporter, "It was a small price to pay to keep space free from the dangers of nuclear weapons." The Brit gave the senator a puzzled look, not knowing if the senator realized that a nuclear power plant was not a bomb.

Only one "virtually assured" launch was made before war broke out. The war against the Norwegians was popular at first, but the call for massive amounts of new military spending meant that fifteen Mars launches were cancelled or delayed. Six missions were actually launched during the war, but mission parameters were scaled back to keep total costs below $300 billion. After taking Oslo's alpine ski resorts in bitter fighting, the Americans banned the export of lutefisk and proclaimed victory. Nasty guerilla action continued in the north, however, as tough Norwegian sailors, put out of work by the lutefisk embargo, took to the hills.

The day after Washington declared victory, sixty-three years and $450 billion after Zubrin had proposed his Mars Direct program, a direct CNN link from Mars showed an American scientist holding a slimy rock up to the camera. She smiled when she said that she had discovered life on Mars. The world was stunned. Immediately, another $500 billion was appropriated for ten more missions. Equally as important, NASA announced several mining contracts with private mining firms. The last paragraph in the news release revealed that the mining consortium planned to build its own ships and to haul its own equipment, rather than risk the vicissitudes of NASA's programs. The cynics in the world wondered if the timing of both announcements, so soon after the end of the war against the Norwegians, had been canned. Certainly, howls from the UN about violations of "international law" were muted in the celebrations. Still, the Leftists and the UN began to lobby hard against the spread of "Big Business" and "American imperialism" to Mars and into outer space.

A month passed and NASA made another major announcement concerning Mars. Fearing that humans could "kill Martian molds and bacteria" if they fiddled with the Martian environment, efforts to terraform the planet were scaled back significantly. A few experiments continued, but winning the day were arguments that raising Mars' temperature (daytime summer temperatures in Mars' Northern Hemisphere had already risen 15° centigrade) might affect Martian life, even cause its extinction. Those scientists saying that Martian life was very simple—hence hardy—and was mostly protected in the deep cracks and crevices where it was found, were ignored. Though the scientific consensus doubted that human terraforming would cause massive extinctions, the policy stood. Years of terraforming work was undone and the billions of dollars in terraforming equipment NASA had brought to Mars over the years would have started to rust, had there been more time to turn underground ice into water vapor.

For a short time, the announcement also silenced critics who wanted to minimize costs related to Mars. Ending the terraforming program would save many billions of dollars. NASA had always realized how vulnerable its Mars program was to the budget-cutters and, to the extent possible, NASA had followed the political winds and kept the program low profile. Entrenched interests would keep the money flowing, even if NASA's bad ideas and mismanagement prevented anything better than new money chasing bad programs. NASA kept up its Mars-shuttle program with overhead costs of nearly $15 billion per year, kept up its $10,000 per pound supply chain for the growing number of scientists who spent two to four years on Mars, and spent billions to launch new equipment for them.

There was no lack of ideas to get free enterprise into space, but NASA had boxed itself into a corner. Bad laws, no political support, and little physical infrastructure meant that although private mining and other commercial operations could experiment with commerce on Mars, the lack of an economic infrastructure prevented any real development. NASA's old programs to explore Mars and to provide scientific opportunities for its scientists had no money for a Martian *economy* that might lower some costs. Granted, they grew some food on Mars and harvested some water, but these were science projects, too, not efforts to build an economy. In the end, NASA gave seminars and conferences about establishing farms, electric companies, a Wal-Mart, and manufacturing capacity on the Red Planet; internally these matters were discussed in meeting after meeting but, of course, nothing ever happened. The way things were organized, the way international legal structures had been created, nothing *could* happen.

On the 100th anniversary of Mars Direct, a small ceremony was held in Houston. Zubrin's grand achievement was celebrated with cake and news that the first twenty civilian settlers would travel to Mars aboard a NASA rocket. Mars Station, as the scientists had named their base, had over 75 quasi-permanent residents and had seen three births. The twenty settlers went to Mars as caretakers of the base, to relieve the science and engineering staff of some of their more wearisome duties. Mars Station was especially happy about Mrs. Chen's decision to go, as she had announced that she would open a Chinese-style take-out stand on Mars, if she could receive the proper provisions[2]. Zubrin's great-grandson read the news release with bitterness. Like his ancestor, he had dedicated his life to Mars, but because of a sour disposition and the controversy over an article he had written early in his engineering career, he had been blacklisted from ever going to Mars himself.

On the 150th anniversary of Mars Direct, after spending almost $4 trillion for Mars (about $25 billion per year, constant dollars), the US announced it could no longer support Mars financially. The mining and small commercial routes created

---

[2]   Mrs. Chen never received her provisions and the take out stand never materialized.

over the years would continue as best they could, but there would be no more NASA missions or financial support for Mars. The 1,100 Martian natives[3] greeted the news with mixed emotions. Having "American citizenship" never meant much to them, as few of them had ever been to earth. Over the years, the Martians had grown weary of the "other Americans" lording over them from so far away. The US had always bowed to UN pressures not to claim any territory on Mars and had mostly bowed to the UN's demands to limit property rights there. The long-awaited decision that it made no sense for the US to spend so much of its money on a place so far from its shores, for little or no return, had finally been taken.

Mixed emotions on Mars gave way to resentment, however, when a United Nations' of Earth ("UN") delegation arrived, intent upon asserting political control over Mars. The residents greeted the Governor-General, his staff, and the Pakistani "peacekeepers" very coldly. The Americans took down their flag, to be replaced by the flag of the UN. Resentment subsided, however, when Martians learned that the UN regime would be both ineffective and mostly disinterested in certain black market trading. NASA had brought to Mars the US' law-loving and reasonably efficient government and had thereby suppressed most black-market activities. Forced off the dole and forced to earn their own keep, Martians began to grow rich under UN mismanagement.

Fifteen years later, riots and skirmishing broke out on Mars. The twenty-two Pakistani "peacekeepers" were no match for the Martians, especially as the Pakistanis had no stomach for their role as colonial enforcers. During the fighting, one man was killed on each side. Heime Zambudi, the last UN Governor-General, left Mars disguised as a woman. The next day, the Republic of Mars was proclaimed. The first order of business, according to the new president, was to restart the old American nuclear reactors and to resume terraforming. "It's too damn cold here", she said, "and things don't have to be this way".

Midwestern Space Clubs

While "Mars According to Garp" was partly tongue-in-cheek, it was meant to convey the kinds of difficulties that any earthbound government would face on Mars. Even assuming that the American government would make its best effort toward Mars, there are many reasons why the Americans will never achieve much there[4]. The mission to Mars proposes that its own organization not only will beat the Americans (and everybody else) to Mars, but that it can quickly build a society that will, in short order, surpass human societies on the home world.

---

[3]   There were another 200–300 visiting scientists and politically connected tourists.

[4]   This will be a major topic of discussion in Volume II.

To make this point, consider three hypothetical Midwestern Space Clubs and their own ideas for Mars …

The purpose of these clubs is to design and build life-size models of spacecraft for manned missions to Mars. The clubs intend to allow the public to tour the model spacecraft and to explain how such vehicles could be used to send a human to Mars and back.

The first club is in Warren, Michigan, where the club wanted to build a model SSTO, a ship designed to go into orbit, return to earth, and be available for re-use within a few days. The next club is in Bryan, Ohio, about 100 miles south of Warren. The club there also built an SSTO, but this one was for Mars. The Bryan club could have built a model identical to the SSTO for earth, but it wanted to build an improved vehicle to reflect Mars' lower gravity and to allow their vehicle to have greater mission capabilities than the Warren SSTO. In addition, their vehicle was designed to be modular, to be easily assembled and disassembled, for transport to Mars. The last club is in Freemont, Nebraska. There, the club would build a full-scale model of a "transporter ship" to carry settlers to Mars. Of course, reality has a way of intruding upon the best-laid plans, so the original plan to build a ship capable of carrying 100 settlers to Mars was scaled back to a smaller ship carrying 25.

It took the Bryan club only two years to open their model to the public. The Michigan club, jealous at the quick results in Ohio, made a dramatic effort and finished its own SSTO a few months later. The clubs were thrilled to discover how much interest there was in their models. They decided to continue building, especially since there was every indication that the Freemont model, being much larger than the two SSTOs, would attract huge crowds. The two SSTO clubs held a joint meeting and decided to keep working, this time creating "prototypes" of their SSTO vehicles, using materials as realistic as possible.

A few years later, the Freemont club finished its transporter ship. Predictions about the success of this enormous model were correct. The SSTOs, after all, were much smaller than the transporter. The Nebraska ship allowed people to walk around inside and to feel as if they were actually on board a space ship. Although all three clubs crafted many details into their models[5], no club spent more than a few thousand dollars for construction. Charging a small admission not only allowed each club to hire guides for its models, it also created a fund to pay for the prototypes. The Freemont transporter model alone was making a profit that averaged $25,000 per year.

---

[5]   Installing old desktop computers as consoles and writing simple software to make attractive displays, stringing wires and lights, simulating fuel cells, etc.

The opening of the first prototype SSTO in Bryan, Ohio, just a few months after the opening of the Freemont transporter, made the three sites into nationally recognized tourist attractions. Many people commented that these three spacecraft reminded them of the movie about the Iowa cornfield turned baseball diamond. Once built, they indeed did come. The Michigan club hurried to finish its prototype.

The continuing success caused the clubs to dream the impossible: if we can build models and prototypes, could the three clubs, working together, build something that actually works? With the assistance of professors at the Universities of Nebraska and Michigan, real engineering and real testing of systems began. The clubs grew greatly in membership, as large numbers of people volunteered to assist with the creation of a real "space ship". The club members, in fact, grew impatient with the professors, who always seemed to be three steps behind. Club volunteers installed engines while designs were still being made at one of the schools; they installed computers while software was still being written; etc., but in the true spirit of tinkering, everything got put together and when something did not work, someone was always able to figure out how to fix the problem. Usually, it was a college student. Sometimes it was a volunteer, who suggested something that the college-boys never would have considered.

The working SSTO vehicle, being built in Michigan by that club's crews and crews busing in from Ohio and Nebraska, soon generated plans for a real "launch". The test envisioned the crude ship shooting 100 feet straight up and then landing safely. Volunteers to fly the quasi-death trap were turned away. Years of successful effort had people believing that, even if something went wrong, the ship could land safely. So, in a cornfield outside of Ann Arbor, the teams assembled. First a dozen, then a few hundred people showed up to watch. The pilot entered the "space craft" and the countdown began. Everything was ready. The people watching were thrilled. For less than $100,000, ten years, and countless hours of donated work, a "working spacecraft" was about to be tested in flight.... 5 ... 4 ... 3 ...

A team of FBI agents arrived to halt the launch. Department of Transportation (DOT) and Federal Aviation Administration (FAA) regulations prohibited the test under several sections of administrative rules. The ground control team and the crowd were stunned, but they complied with the agents' orders. They knew that if the federal government was there, it was there to help.

The pilot emerged from the SSTO bitterly disappointed, but her husband was much relieved. He did not have as much faith in the club's handiwork as she did. The director of NASA, watching the proceedings from a concealed vehicle parked a few hundred feet away, breathed a sigh of relief when he heard that the agents had successfully shut down the launch.

The Ted and Maude Show

The mission to Mars is not a pitch for "private enterprise for Mars". Rather, it will be a government and not simply "clubs" like those described in the last section that will get people to Mars. Why "private enterprise" alone is insufficient for the task will begin to be explained here.

In this episode of the Ted and Maude show, Ted, the space engineer, has come up with a great new idea for "going it alone" into space. He will build rockets to launch space tourists into orbit, at a cost of $10 million per ride. Maude looks around at her four-bedroom house in the suburbs, with two SUVs in the drive, swallows hard, and nods as she tells Ted that she will risk these things for his dream. After all, Maude is a good wife. She is unwilling, however, for Ted to risk his engineering license, credibility within the small community of NASA engineers, or jail, and asks Ted about these other risks.

Ted replies that there may be some untested international agreements about space travel, licensing, etc. with ambiguous language that could possibly get him into trouble, but that these things should not pose a great risk. He then tells Maude that his engineering license cannot be taken away from him, unless he is convicted of a crime or has a civil judgment against him for grossly negligent engineering. Maude's ears perk up, knowing that Ted usually thinks things through pretty well, but is occasionally blind to what others might see. She asks what would happen if his new enterprise were found guilty of criminally violating some law that had never been tested in court. After all, she says, a lawyer can only speculate about these kinds of risks, not articulate them with specificity. Ted has no good answer. Maude further asks why this, or some other disaster, might not further cause civil judgments against him for grossly negligent engineering. Ted says weakly, "Everything in life has risks". Maude goes on to say that she also believes Ted will be finished in the aerospace industry if the business fails, especially if these other contingencies manifest themselves[6]. Ted sheepishly disagrees, but admits that she may have a point. Maude concludes that Ted is risking not only their money, but their entire future and social status. She does not think Ted's great idea is worth the risk.

---

[6]   This point clearly rests on the definition of "finished" and "business failure", but the point remains. If a person "bucks" the government, as Ted is proposing, and that business fails, not only will the professional credibility of the entrepreneur be questioned, but legal troubles may continue for some time after. This former mid-level manager might get hired to teach at a small university or might get a dead-end job in the industry, but it is unlikely that a person who has set up some "alternative to NASA" will ever again get hired by NASA or be put into a high-visibility position in a company trying to get NASA contracts.

Suppose, however, that Ted finally persuades his wife of the merits of his idea and forms Renegade Space Orbitals. Now, mindful of his wife and his lawyer, who has warned him of the risks of being asked by some smarmy, nasal-toned lawyer, "So, Mr. Ted, before your rocket blew up and hurt all those people, you had never actually *built* a rocket before. Is that what you are saying?" Ted must also avoid being co-opted by the existing BGBS community. After all, Ted may be very attracted by a contract to build a rocket for NASA, bringing in tens of millions of dollars, which looks remarkably like the orbitals on his drawing board. Just a little government financing to back the research. Or so he thinks, until ten years downstream, with four slips of the launch schedule, Ted realizes that he must choose between his lucrative work for NASA and his dream of launching millionaires into space. Then one day, as Maude looks up at him from her fruity, mid-afternoon cocktail, sitting under an umbrella blocking the blistering heat of the Bermuda sun, Ted realizes that it is NASA and not Renegade Space Orbitals that has won.

Millionaires To The Moon

Even if some entrepreneurial projects begin to move forward in the next few decades, there are still immense structural reasons[7] why these projects will be walking up hill in both directions. Consider a rather simple and potentially profitable business: sending millionaires to the moon. Such an enterprise seems like it may be possible given the success of the recent SpaceShipOne flight, where a private organization flew into space twice in two weeks, with a single aircraft, to return safely both trips. This was a *private* venture, but not yet a *commercial* one. Let us, therefore, briefly examine some of the circumstances that the SpaceShipOne organization—or any would-be space entrepreneur—operating out of the US would face.

The US government, a "competitor" whose space agency does not appreciate competition, must license your project. The standards set by its BGBS-bureaucracy increase costs many-fold through government-developed safety precautions, DOT and FAA operating procedures, etc., where cost is rarely an object. The entrepreneur will find it hard to compete for the best engineers, who understand that the one organization may have a few dozen engineers (lower career options) and the other thousands of engineers (infinite career options). The entrepreneur will likely need to rely on private funding, rather than traditional capital markets, because his project is not only unusual, not only will require significant funding,

---

[7]   A more complete analysis of why private enterprise alone is insufficient for Mars is found in Volume II.

but must germinate in legal limbo, where laws are unclear and where there are no precedents for banks and other investors to follow[8]. While the entrepreneur will likely raise some money, there is no marketplace to calm the bankers' fears, so their loans are small. Nor can the entrepreneur answer the bankers and critics when they ask: "You charge $1 million per ride. What will you do if NASA arbitrarily decides to give rides for $100,000?" The entrepreneur limps along, hoping that his business can develop in an international political climate that may view his efforts with suspicion. It would not be surprising if the entrepreneur eventually gives up or spends most of his time working on NASA contracts that have the effect of co-opting his ideas and efforts.

It's Just Too Easy ...

Beating up the government is easy and this book has much bigger fish to fillet. As will be emphasized in all three volumes, the purpose here is to launch a revolution, not to try to "reform" NASA or to get some federal money appropriated. Now, the mission to Mars revolution does not mean to rally people to the hills with their blunderbusses. Instead we intend to convince a small number of people that they can get to Mars and to convince a larger group that we should be allowed to try.

Space is not and it cannot be business as usual, run by beltway insiders plotting election victories against unfunded challengers. There are people who will no longer accept a space opera conducted by such maestros. And, as is the right of any free person—you are *free*, aren't you?—these people will take action into their own hands. This book gives them a roadmap.

---

[8]   A commercial operation *will* get sued. Its investors must understand the financial risks they run, unless they are willing to lose *all* of their investment and even greater losses, in some cases.

Economics 101

This chapter will allow the reader to glimpse the framework of the mission to Mars. While preceding chapters have presented castles in the air, this one reaches down to bedrock ideas. At chapter's end, the intuitive reader might glimpse a clear outline of the mission to Mars on the horizon, no longer to be surprised by whatever she might later see.

The Mars Community

The mission to Mars proposes not a business, but a society for Mars. The creation of an *independent community* for Mars allows the distillation of the political and economic interest that exists for manned space flight.

Let us begin by positing that the US has the capability to send someone to Mars, a capability that clearly exists. It then follows that some *subset* of the US would also have this capability. In this section, we call this subset the "Mars Community" and investigate how large it would need to be—or how small. We start with the idea that this community requires all 300 million Americans but, as you will see, this community can shrink rapidly with a few assumptions.

Two "cost" figures have been used thus far in this volume. A NASA-designed US mission to Mars might cost $500 billion, whereas Robert Zubrin's "Mars Direct" might cost $50 billion. Mars Direct proposes to use straightforward engineering, off-the-shelf technologies, and refuses to create huge, budget-absorbing bureaucracies that soak up money. In addition, its clearly stated objectives will avoid politically motivated redesign. Assuming this ten-fold reduction in costs would also be possible for the mission to Mars, our Mars Community shrinks from 300 million people to 30 million. This is not simply "efficiency", a concept that will be addressed in a few pages. Rather, this ten-fold reduction is created by simplifying design and using a "If-you-want-to-take-Vienna-take-Vienna" approach[1].

For the sake of clarity, "shrinkage" means that whereas the initial 300 million people could afford a $500 billion project, a project one-tenth the size would require only one-tenth the number of people to support it. Thus, the Mars Community now requires only 30 million people. Of course, we will now consider other factors that will reduce this number significantly.

In fact, we are now going to reduce the size of the community by a factor of 1,000, from a Mars Community of 30 million people to one of 30,000. Here, we assume that the people at the Mars Community agree to support this mission to Mars with 75% of the community's economic wherewithal. This contrasts with a strictly BGBS-mission, which would likely only be able to obtain funding equal to 0.075% (less than one-tenth of one percent) of total economic activity in the US. In the Mars Community, the total economic output devoted to Mars can be 75%, because everyone at the Mars Community is completely (or at least 75%!) devoted to sending manned missions to Mars. Reducing the Mars Community by a factor of 1,000 gives us that Mars Community of 30,000 devoted people.

At this point, the reader should consider that a community of 30,000 is not infeasible, even if it would admittedly remain difficult to create. These 30,000 would be invited, of course, from everywhere on the earth, not simply from Boston and the Jet Propulsion Lab. Fortunately, there is another factor that will further reduce the required size of the Mars Community.

That final factor is called "Efficiency". Efficiency is created by the employment of market-oriented systems and non-BGBS ideas to build the necessary infrastructure for Mars. Regarding this idea of Efficiency, there are two general and two mission-to-Mars specific points to be made.

The most obvious point is that a market-oriented program can be much more efficient than a government bureaucracy. Little more will be said. If you do not

---

[1]   Straightforward engineering implies several things. First, it means getting to Mars for Mars' sake, without building fleets of extraneous vehicles, space stations, or moon bases. Thus, simplicity of design and function, a rare commodity in NASA's manned programs. Buying off-the-shelf is also unknown to NASA, but can greatly reduce costs. Finally, straightforward engineering implies that "good" designs are accepted and steps forward are made, rather than seeking endlessly and expensively for the "perfect" design. Maybe it would be prettier to have your cavalry lead you into Vienna, but if your infantry are in front, the simplest thing to do is simply to march your dog-face soldiers forward to the Emperor's palace.

accept this idea then little of what is written here will be persuasive. The second simple point is that if a traditional NASA mission to Mars would cost $500 billion and if Zubrin's Mars Direct (also as NASA program) would cost $50 billion, the mission to Mars (non-BGBS) *might* achieve its objectives with $5 billion. This additional factor of "10" has not yet been added to this analysis[2].

There are two other ideas to add to the mix, both of them highly specific to the mission to Mars. First, is the idea that *a settlement program is much, much cheaper than an exploration program.* There will be literally dozens of examples presented in these three volumes as to why this should be so. Finally, we must address the poverty of the word "cost" to account for the mission to Mars program. If, theoretically, *everything* could be manufactured or grown at an entirely self-sufficient "commune", then the "cost" for the mission to Mars might, in some respect, correctly be said to be "zero", as no "US dollars" at all would have been spent outside of this self-contained community.

All of these points will receive extensive discussion in this work, hopefully to cement these points for the open-minded reader.

For now, we assume that "Efficiency" can be a factor of "10" and see that the Mars Community needs a mere 3,000 people, working together, to achieve a goal of sending a manned mission to Mars. Figure 7 summarizes the discussion to this point.

Insert here a reality check of sorts. Is it really so difficult to believe that a dedicated group of 3,000 people, working full-time for years to accomplish their goals, could not build spacecraft, manufacture fuel, create the software, etc. and so craft a program to get people to Mars? This book takes the position that it is not.

We have started with the assumption that the US could undertake manned missions to Mars and that we could take a "theoretical slice" of the US to distill the necessary political will and economic wherewithal required to achieve this goal. We have determined that a well-organized, focused community of 3,000 people *might* accomplish what the mission to Mars argues BGBS never will. We suggest that a Mars Community of 30,000 might be possible and that a disciplined community as small as 3,000 people can achieve efficiencies of which NASA can only dream. This analysis has probably been intuitive to many and even tedious for those who have an innate understanding of the human potential to achieve great goals. Nonetheless, there are other reasons to believe that this model for a Mars Community has indeed been a *conservative* analysis. For simplicity, we will simply say that the "Efficiency" factor we used may, in fact, have been too low. We present five ideas to support this statement.

---

[2]   The ten-fold reduction for straightforward engineering from the last page is not the same as the Efficiency idea. "Efficiency" here is a combination of factors on this page, combining to produce another ten-fold reduction.

	Factor	US	Mars Community
Start	- - -	300,000	300,000
Straightforward Engineering	10	"	30,000
Economic Commitment	1,000	"	30
"Efficiency"	10	"	3

Figure 7: Population Required (Thousands of People)

First, is the probability that more than 3,000 people would be attracted to the Mars Community. If the idea of going to Mars attracts 9,000 instead 3,000—from a world of six billion—then the first blanket of safety is fluffy indeed.

Second, those who truly understood the meaning of "The Third Rail" section in the chapter called "How To Think About Mars" will also understand that the cost-savings achieved by relying on a creative *society* rather than a managed *bureaucracy* by itself may be greater than ten to one. As there are no limits to what the human mind can conceive, there are no limits to what may be achieved by subsidizing rather than suppressing creativity.

Third, as the efficiencies to be garnered from a *settlement*[3] rather than exploration approach are difficult to quantify, both because they are so numerous and because efficiencies relate to possibilities not found under NASA exploration scenarios, the value of settlement over exploration has likely been understated. To cite just one example, the ability to create banks on Mars will jumpstart an economy there, as Martian businesses fight for money. Every business on Mars will have a "built in" market for many decades, as everything is demanded by the arriving settlers. Instead of asking for a government hand-out, Martian banks will find sources of capital, loan money to those building the Martian infrastructure, and make money themselves[4].

---

[3]   Settlement will cost more than exploration in the sense that it will cost more to send 1,000 settlers to Mars than to send five scientists there. Once *any settler* has reached Mars, however, it must be understood that cost issues will begin to disappear, as people from across the globe clamor to get aboard the next ship for Mars and help fund the remainder of the project. If this seems to be poetical sleight of hand, the reader is invited to revisit the issue after having read all three volumes.

[4]   It is a virtual certainty that large sums will become available for investment once the first mission to Mars' humans arrive on the Red Planet. Have one *settler* plant one foot on the Martian surface and money will be available for anything. Such economic certainty will allow for far lower costs of capital than would exist in circumstances

Next, consider that Efficiency can be improved by passing the costs to get to Mars to individuals and especially to private businesses working at the Mars Community. This does not eliminate costs, but if a chemicals company moving to Mars to manufacture fuels there must pay to move itself and its equipment to Mars, one should assume that great efficiencies can result.

Finally, consider that no timeline was included in this analysis. A *community* is not a budget. If the plan was to send a manned mission to Mars after 15 years, but the community was willing to work 30 or more years if necessary, an additional "safety" factor is available for the model.

The ideas presented here about the Mars Community are not meant to *end* the discussion about finances for Mars, but to begin them. The author compliments those who remain skeptical and suggests that it will be those capable of *constructive* skepticism who will turn these ideas into reality. This three-volume argument has hardly begun. Trust for now that the major concepts just introduced will be developed in much greater detail in these three volumes. The figure below summarizes the discussion about conservative assumptions.

"Conservative Assumptions"
Many more than merely 3,000 might be attracted to the Mars Community
"Creativity" will be harnessed as never before
Settlement is much cheaper than "exploration"
Pass costs on to settlers and businesses moving to Mars
If there are good reasons for delay, the first landings need not occur in "2025".

Figure 8: Efficiency

A realistic, non-BGBS recipe for Mars can create a different kind of Titusville-NASA community, free the residents from bureaucratic red-tape, empower them by making a wondrous dream their daily reality, and let them work aggressively toward reaching the Red Planet. One must hold a PhD in cynicism to believe that humans could not get themselves to Mars under these circumstances.

---

where a NASA future would depend entirely upon who was the President and who was in Congress. Indeed, funds will be difficult to find after a NASA landing on Mars. "Why should we spend more money on Mars? I thought we already put a woman up there", asked the tight-fisted senator.

"But wait!" The Skeptic Will Say ...

"... even if it only costs $5 billion, how will fund your project?" To be sure, the BGBS mind will think almost entirely in terms of *funding*, as befits a psyche glued to Washington. But even *some* BGBSers understand that economic power begins not in Washington, but outside of Washington with *people* and *ideas*. One of the great divides between readers will be between those who can and those who cannot understand this important point. Indeed, one of the great tests for readers who make this admission about economic power is for them then to stipulate that the mission to Mars *might* harness the relatively small sums it would require for its project. In other words, if you really believe that economic power is found outside of Washington, the key question about the mission to Mars is not whether there might exist sufficient economic energy for its ends, but whether it proposes methods efficacious to them.

The next chapter will discuss "Tickets to Mars". These tickets will provide the economic fuel to "fund" the mission to Mars. Those who still do not understand where "funding" will come from might understand then how the earth Embassy can create financial wherewithal from payments on these tickets to Mars.

These volumes seek to provide answers to the hardest questions about Mars: economic means and political will. The answers offered will not please everyone, but answers will be given. In the case of generating sufficient economic wherewithal to empower the mission to Mars, the answers will be provided over the course of many pages in all three volumes. Most of the detailed economic analysis[5] related to this issue is not found in this volume, as we paint here with a very broad brush. Since there has been so much groupthink regarding NASA and its supposed singular path into space, this chapter hopes only to introduce some major economic ideas, to allow the reader to rethink space "outside of the box". But consider this: if a *society* reaches for Mars, rather than a bureaucracy, how to fund your project is a much smaller question than asking how much that society wants to get there.

The Capital of Gilligan's Island

The idea of human capital is one that is glibly used, but seldom understood or analyzed outside of the small community of economists. This concept, of course, refers to the skills of the people in a given group. A group of four or five people

---

[5]   Much of what awaits the reader in Volumes II and especially Volume III are economic ideas and analysis. These discussions offer details designed to convince people that the ideas are *reasonable*, not that they are definitive.

may seem an unlikely start for any grand enterprise, but consider how many very large corporations once started with an idea in a garage.

Now suppose that there are not four or five people, but twenty highly skilled and devoted people. The project that is proposed is not futuristic in that it will not require vast new kinds of technology; it is only grand in its *reach*. The real question, as hinted at in the last section, is whether the mission to Mars can begin to interest a few hundred, and then a few thousand, to live a life of dedication, with near-western standards of living, to make manifest an idea that has enthralled many tens of millions of humans in the past one hundred years. This section discusses two basic economic ideas that most readers will readily concede, that will allow some to see their way ahead, and that will point a few all the way to Mars.

Suppose that three people were stranded on a deserted, tropical island with no shelter, no tools, and no supplies. These people might initially despair of their own deaths, but before they did, they would do well to take inventory of "the capital" on the island. First, they would have land and access to the oceans and the streams of the island. The earth itself would yield up a bounty of fruits and fish. Next, consider the human capital found in the skills of the three people. If one castaway is a farmer and knows how to fish, he can offer to feed the group. If another is a carpenter and handyman, he can offer to build shelter for the group. Finally, if one person is a craftsman who can build many things and is resourceful to boot, she can find supplies when others cannot. Suddenly, the future of the three does not appear bleak at all.

The group argues for a day or two about duties. Then, one of them notices the small supply of clam shells on the beach. This person collects all fifty shells and suggests that the group use the clam shells as a medium of exchange. Money. The others agree and the tiny economy is off.

Farmer charges one clam per day for the food he gathers and prepares. Housebuilder will build a home for Farmer. They agree that the Farmer's small house should cost 100 clams. Craftsman, on the other hand, wants a really nice house. She offers to pay Housebuilder 200 clams to build this nice house. She also begins to create some tools for Housebuilder and for Farmer. She sells these tools for ten clams each.

The group lives on the island for several months. Housebuilder stays busy building the houses. Craftsman is busy making nets and tools and pouring off oils from coconuts. Farmer supplies the food. They trade, using clams when necessary. Some days it seems that Craftsman is the richest, some days it seems the Housebuilder is richest. No one complains, because what all three really want is to stay fed and to be comfortable until they can figure a way home.

One day, Sailor washes ashore on the other end of the island. She is also stranded and, knowing nothing of the other three, fears she will die. Sailor knows how to build and sail boats, but little else. She is in despair and on the verge of starvation when Craftsman finds her and brings her to their small village. Sailor tells the others she can build a boat to get them all back to England, but estimates that it will take six months to get everything ready. The others realize that they can simply incorporate Sailor into their society. They agree to pay her two clams per day to build and outfit the ship and sail them home. Sailor readily agrees. She uses one clam per day to buy food and promises to pay Housebuilder 100 clams to build her a shelter. She uses the balance of her clams to buy tools from Craftsman.

Six months after Sailor arrives in the village, the four board the vessel and sail home for England. As they depart the island, they all remember how fearful they had been of dying. Little did they realize that they could cooperate to survive, to eat, to find a way home, and even to live well.

The obvious lesson of Gilligan's Island is that human capital can create goods, services, and a reasonable economy, even in circumstances where such things seem impossible to the suburban mind. The other lesson, perhaps less obvious, is that hard currency is not always necessary to create a money economy. Where there is agreement, it does not require US dollars to make an economy work. Clam shells can work just as well.

An economy can be created *anywhere* there are people. There is no reason that an "enclosed" or "self-contained" economy cannot be built at a community whose goal it is to get to Mars.

I'm Keeping My Gold

One of the great myths created by space "explorers" is that off-world settlements need to trade raw materials with the earth in order to survive. The idea is made absurd by the oft-quoted statement: if there were diamonds lying on the surface of the moon, it would not pay to ship them back to earth. This diamond idea is not only false absolutely—the cost to ship the diamonds would likely be very low compared to their value—but it is also false in another, more profound way. This statement assumes that the only value for a diamond would be *on earth* and ignores the idea that diamonds can be valuable *wherever humans are found*. If there were a commodity such as diamonds on the moon, those diamonds could be used to increase the capital of the human society living on the moon. Then houses and farms and factories could be built, creating far less dependence on the earth. How would this work on Mars?

The banks on Mars will have branches on earth or will interface with earth banks. Someone who has an account at a bank on Mars can make a purchase from a company on earth. Computers make this task routine. If you want to purchase a "Baby Cries Endlessly" doll for your Martian Christmas and you can only buy this doll on earth, you email your order to earth (we ignore shipping costs and time), the doll company presents your payment at the earth branch of the Martian bank, and ships your doll. The earthbound doll company gets its money and you get your doll. Here's the big question: why would the earthbound doll company trust the Mars bank to make the payment? That's where the diamonds come in.

There are likely to be many, many things on Mars that will make people rich. Gold, silver, a furniture factory, etc. If a Martian bank has $50 billion in gold bullion in its coffers, businesses on earth are not going to worry about using its branches on earth. When Billy Joe Walterruziak, high-school dropout who has somehow made it to Mars only to stumble over a $1 billion silver deposit when he got off the plane, puts his money into a Mars bank account, that bank's reserves go up accordingly. This, or some other bank, will find the earth-to-Mars-and-back financial trade lucrative and will create ways to facilitate the trade. Merchants will deal with this bank's branch on earth because, ultimately, they can sue the Mars bank if payments are not received. If a judgment is entered against the Mars bank, this bank will be forced to pay, even if it has to ship the gold at its own expense back to earth[6]. With sufficient assets in its coffers, anything can be bought through the Martian bank. It does not matter whether these reserves are physically located on earth or on Mars. Thus, there is no reason to ship your gold or diamonds or iron ore back to earth. Just show the earthers how big your bank account is and they will come a-runnin'.

Much more importantly for the mission to Mars, however, is that if this Martian bank has accounts worth many billions of dollars, then it can provide money to help grow the Martian economy. Consider the circumstances of a Martian colony with one thousand inhabitants, many of whom want to build or improve their homes. In Case One, banks on Mars have only $1 million to lend out. In Case Two, there is $100 million to lend. In both cases, we examine desired loans of $100,000.

---

[6]   Operating under the same principles under which foreign banks operate in the US. Yes, they could simply flee the country (or the earth!) to avoid paying a $150 judgment. Being businesses, however, foreign banks are scrupulous about complying with all financial judgments, to keep operating and to keep making money.

	Case One	Case Two
Capital To Be Lent ($)	1 million	100 million
# of $100,000 Loans Made	5	200
Interest Rates High or Low	High	Low
Money Remaining For Business Loans	Little	Lots

Figure 9: A Savings and Loan on Mars

Figure 9 suggests that the economic activity possible on Mars under Case One will be relatively low, while the economic activity possible under Case Two will be relatively high. As the ability to create a viable economy on Mars is one of the keys to the mission to Mars, the advantages of Case Two speak for themselves.

A Martian bank that has holdings representing large accounts "backed" by Martian gold or diamonds (or a highly successful hog farm) can help to fuel a powerful Martian economy. Martian banks with little cash will struggle to fund all of the development that will need to occur on Mars.

Said another, less polite way, is that if Martian settlers can pay to develop their own economy, albeit in a pioneering setting, the only thing standing in their way will be internal or external enemies, the kind that face any society. If Martian settlers must constantly look to a "government" or a "NASA" to pay for their programs, little or nothing will ever get done.

So, I suggest that all Martians keep their gold on good, old mother Mars. The more gold a Martian bank has, the greater its ability to facilitate financial transactions. If there is a vibrant economy on Mars, one that is building up the assets of its Martian banks, there will be no need to ship products off-world. The local economy can be used to pay for projects on Mars. And when there is a need to ship something to or from the earth, a strong Martian bank will have the financial wherewithal to facilitate this trade.

Once the mission to Mars begins to transplant its people to the Red Planet, it will seek to build up the capital available to its colonies there and to use this capital to help develop the Martian economy, an economy that will have had its planning and operational roots deeply planted at the earth Embassy.

Guarding the Mexada Border

There are those who believe that only government programs for space can become reality. In many cases such people have gotten so used to huge government apparatuses and their accompanying bureaucracy that they fail to see the advantages of non-governmental solutions to problems. This section addresses the issue.

Now that terrorism has become a threat to America and its citizens, it is likely that the US will decide it needs to patrol the Mexada border to prevent illegals from crossing into the country. Two options to accomplish this mission are presented to the President for review.

The first option is offered by General Walterruziak, father of the soon-to-be silver magnate on Mars. General Walterruziak has briefed the President that he wants to open up ten square miles of National Forest land astride the border and build an army base there. General Walterruziak estimates the total cost to be $15 billion. The base will house 10,000 soldiers, who will be used to patrol the border. The $15 billion will be used to build the base, procure its equipment, and install supporting technology. The general estimates that it will take five years to get the base fully operational and that, once built, annual operating costs, including salaries and expenses, will be $2 billion per year.

The President listens carefully. He is concerned that these costs, though not difficult to get through Congress, will eliminate for the remainder of his term even his scaled-back domestic agenda. He eagerly awaits hearing his second option.

Larry Smith, a fellow from the Plato Institute, a Washington think-tank dedicated to limited government and free-market ideas, proposes an entirely different solution. Instead of stationing troops along the border, Smith wants to sell the ten square miles of National Forest land to developers and to provide incentives "to grow" a town of 10,000 people. The plan would net the government $20 million proceeds from the sale of the land. This money would be put into a fund to pay "bounties" on each illegal caught crossing the border. The bounty system would be open to all citizens of this new town, who would be paid $10,000 for each illegal they caught. Not only would the bounty system bring residents rapidly into the new town, but it would also provide a great incentive for people actually to watch the border[7]. Developers would have incentives to seek private capital

---

[7]   This is not a minor point. Grumbling soldiers, lead by sometimes less-than-competent officers may not achieve the same kinds of success that "border entrepreneurs" could achieve. If somebody crossed the border while the army was on watch, the incident gets reported to the general in some morning briefing and would soon be forgotten. If somebody crosses the border under Smith's plan, somebody just lost out on a $10,000 bounty. Incentives *fuel* one system while bureaucracy *operates* the other.

and investment to create this new town and the new businesses to support the residents. These developers project a town of 1,000 residents (including construction workers) within six months, 5,000 within a year, and a population of ten thousand two years after the sale of the land. There will be no "federal operating costs" other than to create a federal office to process paperwork and pay claims, and to replenish the fund as it gets used to pay these claims. Once the people of Mexada realize how extremely difficult it is to cross the border near the town, illegal immigration is expected to stop completely, keeping out people without legal access to the US and much more importantly, *keeping out potential terrorists.* At that point, the town's economy would be self-supporting and would not need the bounty income to survive. The town will generate a significant new tax base for state and federal income taxes, adding tens and hundreds of millions of dollars in additional tax revenues to their coffers. Smith outlines a few corruption- and violence-prevention ideas (prohibition against carrying firearms, registering and deputizing bounty hunters, a program to prevent fraud, etc.) all of which seem to provide common-sense solutions to the concerns of a 21st-century, media-crazed society.

The President favors Smith's plan[8]. He cannot help but wonder how the use of government powers to structure non-government solutions might resolve other issues facing the nation. The case of the Mexada border illustrates to him the differences between a Big Government solution to a problem and the application of incentives, infrastructure, and societal investment to that same problem.

Funding Star Fleet

There is an interesting premise in the *Star Trek* television series that human beings no longer work for money, but instead work to better themselves and to improve the human condition. This premise is not well developed, but the producers

---

[8]   Lamentably, Congress would not pass legislation approving the plan. In a bloated government that thrives on BGBS ideas, the Pentagon, "Big Labor", and a few big defense contractors successfully lobby key Congressmen to support the expensive, new military base. The Pentagon wants the mission and 10,000 new troops. Big Labor wants a thousand new federal bureaucrats (civilian employees at the base) it can solicit for union representation, and defense contractors want the new contracts that the base represents. The base becomes reasonably successful, but only at an estimated cost of $895,000 per illegal immigrant caught. This is almost 90 times the cost "per illegal alien" of Mr. Smith's plan. More importantly, the border remained porous, as it is difficult but not impossible to cross the border guarded by the army.

constantly make fun of species that are businessmen[9]; make almost no provision for "money"; provide the *Enterprise* and *Voyager* crews with "replicators", not stores or corporations to obtain food and other supplies; and hint strongly at sources of virtually unlimited power, like "warp" engines.

The purpose of this section is not to geek out about a TV show, but to consider how the "Star Fleet" premise might serve as a model for the mission to Mars. Specifically, how would Star Fleet build its ships and provide for its needs? Our purpose, of course, is to provide additional insight into the mission to Mars. The reader is encouraged, therefore, to digest this section fully, as it forms a foundation for several ideas important to these volumes[10].

The Star Fleet problem is actually quite trivial in a circumstance where there is limitless energy and energy-driven replicators exist to manifest every material whim. Each component of the starship would simply be created in a replicator. Being unlimited, there would be no cost for the energy and presumably a replicator could create anything, limited perhaps only by size. Thus, when a part is needed, it would simply be replicated. Money might disappear, as there would be little incentive to exchange anything, if a replicator could create everything.

The free-energy scenario sheds no light on problems that might be faced by the mission to Mars, other than the thought that the abolition of money might actually be a step *backwards* in sophistication. Without money, there would be nothing to motivate the slothful and much less incentive for humans to stretch themselves through competitive efforts. In a moneyless society, the policeman cannot trade security for the baker's bread and one cannot trade the manufacture of sixty light bulbs for a ticket to see the Spurs play basketball. If there were unlimited energy and replicators, people might be convinced to be bakers and basketball players and starship captains simply for the love of the job, but the basic notion of incentive to eradicate laziness would be gone. Humans may *someday* reach a point where money is not carried in a wallet, but it will likely always be credited and debited, perhaps by using a plastic ID card. In many ways, modern western culture is almost there. Still, it remains *money* that is being credited and debited. It is the complete abolition of money, where no medium of exchange exists, that is difficult to imagine. Rather than try to contemplate

---

[9]   The Ferengi, who are driven by a profit motive and guided by their "Rules of Acquisition".

[10]   Some of these important ideas include: a "two-currency" monetary system, how the labor of the community's members might be harnessed to provide most goods and services, the economic reality that some things must be obtained from the outside the Mars community, etc.

these matters and fall off the edge of this book's flat world, this scenario will not be further addressed.

We now consider the Star Fleet problem in the context of reasonably priced energy. Since energy is not unlimited, it must be bought and familiar economic issues and principles return[11]. The Star Fleet assembly line would still replicate parts, but now economic decisions must be made. Prices, money, and "an economy" would exist. Perhaps larger parts of the hull would be bought from traditional markets, while small, complex parts would be replicated. Much more importantly, Star Fleet would need funds to operate. It would need to buy space parts and would need to pay its crewmen. Star Fleet's circumstance would resemble that of the mission to Mars, where there is a workforce willing to be organized and motivated, even if participants view money as a secondary or tertiary concern. We divide Star Fleet's needs into three main categories and see how it might operate in this universe of replicators and limited energy.

The first issue here will be how Star Fleet repairs its vessels. Each captain would be given a budget for repairs. This budget will allow for rational decisions about whether to "fix" a problem, "replicate" a component, or "purchase" something from outside Star Fleet (see Figure 10). Many real circumstances could involve all three but, for simplicity, we separate them. If there is a problem with some of the communications equipment, the solution might be to replicate some complicated part. If there are engine problems, the solution might be to have the ship's engineers recalibrate or otherwise fix the problem. Finally, if new "dilithium crystals" are needed for the "warp" engines, it might be necessary to use part of the ship's budget to purchase new crystals from a private source.

	Communications Problem	Engine Problems	"Dilithium Crystals"
Fix		Repair the Engine	
Replicate	Communications Equipment		
Purchase			Purchase crystals

Figure 10: Fix, Replace, or Purchase in a Limited Energy Universe

---

[11] Limited resources must be paid for. Paying for things implies budgets. Budgets imply husbanding of resources.

Now imagine a mission to Mars with ships and its own maintenance issues. The Mars Community will essentially have the same issues as Star Fleet. When maintenance problems develop, the mission to Mars will either need to make repairs using its own personnel, where diagnosis and reconfiguration alone will suffice, replace a part that can be manufactured at the Mars Community, or replace the part by purchasing the component from the outside using hard currency.

	Communications Problem	Engine Problems	Fuel For Rocket Engines
Fix		Repair the Engine	
Inside Purchase	Manufacture Communications Parts within the Mars Community		
Outside Purchase			Purchase fuel from outside of Mars Community

Figure 11: Fix, Replace, or Purchase: Partly-Developed Mars Community

Figure 11 analyzes three Fleet issues when the Mars Community is only partly developed. Where a communications problem has developed that can be repaired with a replacement part manufactured at the Mars Community, this part is purchased from within. Where a major engine overhaul is needed, Fleet labor will be mobilized to repair the engine. Neither matter anticipates the need for any hard currency, but relies strictly upon the economic forces developed within the Mars Community to solve a problem. Finally, fuel for the engines is not yet manufactured within the community, so this fuel must be purchased from the outside, using some of the community's hard currency.

Using Figure 12, we revisit these problems with a moderately sized Mars Community. To fix a communications problem, a circuit board is again purchased from a manufacturer at the Mars Community. And again, problems with rocket engines can be repaired within the community. Fuel for the rockets, however, can now be purchased from within the community, saving precious hard currency for other matters.

	Communications Problem	Engine Problems	Fuel For Rocket Engines
Fix		Repair the Engine	
Inside Purchase	Manufacture Communications Equipment within the Mars Community		Fuel Now Manufactured Within Mars Community
Outside Purchase			

Figure 12: Fix Replace or Purchase: A More Developed Mars Community

The second issue before Star Fleet in the reasonably-priced-energy universe will be the crew's access to replicators to create items for personal use. Endless access to these replicators would not be possible, since the replicators consume energy, a limited resource on the starship. Even the *Star Trek* programs envision this issue. The show's solution is to provide replicator rations and to pay "credits" for use during shore leave. Each crewman could decide what made most sense to replicate and what she should buy during shore leave. A newly assigned ensign might want to replicate a stereo for his quarters, at a cost of 500 replicator credits. The ship's main computer would track each crewman's balance of replicator credits and when the balance was zero, the crewman could not use the replicator until more credits were issued. Another crewman might want a personal computer, but decides that it would cost too much to replicate and so decides to buy one at a market during her next shore leave. A judicious budget of replicator credits and "hard currency" for shore leave would maximize the crewman's purchasing power.

The mission to Mars will institute a system similar to that just described. Instead of credits and replicators, however, the earth Embassy will have its own internal money system—Confederation Script—to allow for the purchase of items from *within* the community and "hard currency" when there is a need to make an outside purchase. As it is the intention of the mission to Mars to create a happy, fun-loving society, where the snap of beer cans is heard after a long day's work—nothing could develop relying only on the slave labor of mindless drones—this system would be important. It would not be terribly much different from Star Fleet, if certain items were available from within the community (corresponding to replicated items), while other items had to be purchased from without (shore leave purchases). Whether the beer consumed at night was purchased from a brewery using Confederation Script or with "hard currency" would depend upon whether someone had opened a brewery at the Mars Community. A system using two forms for payment might seem cumbersome but, as with the Star Fleet crewmen, this need not necessarily be so in the age of credit cards and computers.

The last issue before Star Fleet is how to provide the crew the necessities of life. Here, we consider food, clothing, and medical care. Housing is not an issue, as each person will be assigned quarters. Food and clothing will likely be replicated. Health care will be provided by trained personnel aboard the ship, who provide this service as part of their duties, much as the engineers keep the ship's engines running. Medicines and medical equipment will also be replicated, as it would doubtless be cheaper than purchasing such sophisticated products. It is noteworthy to see how little money would actually be needed to keep the ship's crew in the staples of life.

As with Star Fleet, food and other necessities could be provided as compensation for the labor of Mars Community members. Properly structured, the mission to Mars could have its farmers and food processors trade their wares for medical treatment; with people building rockets trading a place on their rockets for what they eat and wear. Like the Star Fleet crew, there may not be a great need for hard currency except to purchase the odd item that cannot be supplied from within the community. While these ideas about Star Fleet and replicators are fictional, the reader is invited to consider that a community like this Star Fleet might provide for most of its own needs, if it is properly organized. There is no theoretical impediment to the creation of such a self-sustaining community, just as the Star Fleet economy does not generally tax the credulity of it audiences. The question thus becomes how to organize this community to get to Mars, not whether such a thing is possible.

Scroungers

Recycling operations will be very important to the mission to Mars. Many details of these efforts will be described in Volume III. For now, understand that the mission to Mars will attempt to recycle everything and what cannot be reused will be burned to generate electricity. Sophisticated systems to recycle metals will be created, far beyond what is done today. Broken glass will be reprocessed, as will paper and plastics. Wooden furniture will be restored, the wood reused, or burned for energy.

For many years, recycling will also imply refurbishment. Most everyone in the US has a basement with grandma's old stove and a dozen other odds and ends that might easily be returned to use. Though the stove is old, someone or some business with the mission to Mars can refurbish and resell it. The $100 deposit required when you buy a *new* stove at the earth Embassy will provide a great incentive to buy from a refurbisher to avoid some or all of this deposit. It also means that when you are finished with the new stove, you will take it back to a refurbisher to recover your deposit. Those things that might easily be recycled could fill ten more pages. Hopefully, the reader has already heard of many of these ideas. If not, go ask the Sierra Club or the Natural Resources Defense Fund[12].

A unique industry at the earth Embassy will follow from these recycling initiatives. These businessmen will be called, by those delicate of heart, "the low-cost resource acquisition managers". The rest of us will call them "scroungers". If there are several dozen businesses at the earth Embassy "refurbishing" everything, there will be a market for people to sell these refurbishers "neat stuff" from outside the earth Embassy which otherwise would go into the trash heap. Go to the oil patch and you will see rusting pipes. Go to a university and you will find basements filled with old laboratory equipment just waiting for the day it no longer hurts "to throw it all away". Go to any government or military facility and there are so many things wasted or "stored" it will make your stomach hurt thinking how much it all costs you on April 15. Business has the same story. As the mission to Mars gains credibility, "scroungers" will find virtual gold mines of possibility. Would a non-digital gas chromatograph be useful? Why not? Even if it must be "refurbished", the mission to Mars can get a piece of equipment at no cost, requiring merely the labor of a qualified "refurbisher" to create a valuable piece of equipment for the mission to Mars. Not only will an entire "scrounger" industry be created, but refurbishers will lay the foundation for manufacturing at the earth Embassy and later on Mars. Thus, recycling, according to flexible but intrusive "standards" set by the Confederation, will help to create the kind of society that can first make itself self-sufficient and then get itself to Mars. Scroungers will get equipment and supplies at a fraction of the cost that would be required if these items were purchased on the open market and 1% of the cost NASA would pay.

Money and "The Search for the Soup"

We now indulge in a sidebar that should tickle the ironic in most hardy Martian explorers. More, it should help frame an understanding how a capitalist mission to Mars might harness ambient economic forces to achieve its goals. The ideas here are not meant as major sources of economic power for the mission to Mars,

---

[12]   These organizations may be unreasonable about their recycling ideas for the developed western economy, but many of their ideas will be incorporated into the larger plan of the mission to Mars.

but to be a jolly example of what might be done at a properly organized Martian community.

The "Search for the Soup", a hot, steamy bowl of Campbell's enriched Martian bacteria, can be harnessed by the mission to Mars to speed its success in at least two ways. First, and admittedly most mercenary, is to document this search by Martian settlers for television and movies. Likely, the funds raised by having exclusive rights to these documents will amount to tens and possibly hundreds of millions of dollars. This money will not be available at first, but will flow in rivers once "the soup" has been discovered. Less mercenary, though still first-cousin to the dastardly, will be to limit the search for the soup to those associated with the mission to Mars. This may seem most un-science-like, knowledge supposedly being free and not restricted for economic gain, but in fact is quite commonplace in the modern west. Arguments from the Left notwithstanding, drug companies cannot invest billions of dollars and then give away their miracle drugs. If they were required to do so, the line of new and wondrous drugs would cease. There is no reason that the mission to Mars should not adopt a similar position. As intellectual property rights such as copyrights and patents give artists and inventors economic protection for their work and incentives to produce more, so the mission to Mars will claim property rights to help fund its efforts toward Mars. These economic protections will be somewhat controversial, as they will stem from a Confederation that will long struggle for "recognition", but they *will* be honored[13]. Some scientists, agencies, and governments will join the mission to Mars, or at least cooperate with it, based upon these kinds of ideas and intellectual property rights. The mission to Mars will conduct the research on-site, allowing access only to those individuals who are willing to help to settle Mars or who acquire a license for the information. The Martian scientists who work on the project will be free to publish and this published information will become part of the public and scientific domain, free from special property rights. The actual rocks, "soup", fossils, sites, etc., however, will remain the property of the Martian government or the persons or companies that have found and developed this

---

[13]   The only sheriffs or military on Mars to enforce property rights will be those of the Confederation, operating under Confederation rules. Mars will thus be claimed by the Confederation, which will resist the effort of earther "invaders" to operate there, except by permission. Obviously, this will not be a real problem for many years, as earth governments will likely remain reticent about spending the huge sums needed to get to Mars and unwilling to spend such sums simply to bring "rocks" back to earth. In addition, the idea that scientists and engineers from earth will be welcome to work at mission to Mars colonies, at the cost of only a few million dollars per person per year, will further erode efforts on earth to duplicate the infrastructure the mission to Mars will have spent so many years and so much effort to create.

property. Thus, those persons who want to become famous for Martian research will need to decide whether this desire balances the sacrifices requested of them by the mission to Mars.

These policies will offer the mission to Mars significant leverage to raise funds and to recruit top-notch persons to settle the planet. In addition, these policies will allow the just protection of the efforts of those who desire to settle Mars, against the sometimes rapacious, political, and bureaucratic activities of "Big Science" that may have little respect for the sacrifices and labors of a young PhD researcher working in relative obscurity at a mission to Mars base.

As indicated at the beginning of the chapter, the ideas here offer a first glimpse of the mission to Mars. Of course, this first of three volumes has just begun, so we have only touched the surface of a dozen major ideas. The rest of the pages will work in the invariable questions about "space tourists" and Martian domes and how to transport people to Mars. Still, those who can now envision our "Mars Community" should have little trouble understanding—and helping to improve upon—the ideas found on the path ahead.

Tickets to Mars

Have you been half asleep and have you heard voices?
I've heard them calling my name.
Is this the sweet sound that calls the young sailors?
The voice might be one and the same.

I've heard it too many times to ignore it;
it's something that I'm supposed to be.
Someday we'll find it, the rainbow connection,
the lovers, the dreamers, and me.

—Kermit the Frog in *The Rainbow Connection*

Who gets to go? Everyone. Anyone. Whoever is willing to put in the effort required to help the mission to Mars achieve its goals will some day be able to walk upon the Red Planet. The mission to Mars will, of course, face certain realities and its ability to absorb political immigrants and those with significant health problems will force some limits, but these will be very low hurdles.

We take up here three ideas basic to the creation of economic wherewithal at the Mars Community: the ticket contract, life under this "ticket contract" at the earth Embassy, and translating a ticket into a seat for Mars. This system distills into a real world option the desire of those who would be willing to trade their labor for a ride into space and to Mars. The twin attractions of this system will be that life might be lived as a middle- or upper-class American (an inducement likely to attract great numbers of people from the Third World) and the reality of becoming a Martian pioneer.

The Ticket Contract

The device most central to the financing of the mission to Mars will be the sale of "tickets" to each adult resident of the earth Embassy. Each ticket will actually be a contract with one of the various "Travel Corporations", chartered and regulated by the Confederation, to provide goods and services to the "ticket holder". The most basic aspect of the ticket will be that, upon accumulation of CS[1] $1 million in a special "Travel Account", a "ticket to Mars" will have been purchased. The Confederation will authorize these Travel Corporations to negotiate the right to withhold up to 80% of an employee's paycheck for deposit into this account. The sums in the account will earn a return at a rate set under a negotiated contract, but within guidelines established by the Confederation. As will be discussed in detail in a later volume, the required CS $1 million will take anywhere from 8 to 26 years to accumulate, even for the lowest paid member of the earth Embassy.

The "ticket" contracts for several things. First, it purchases a home at the earth Embassy, which must meet certain Confederation-mandated standards for space and construction. Next, the ticket buys an allotment of food for the ticket-holder during his time at the earth Embassy and during travel to Mars. Naturally, the ticket buys passage into space and to Mars on a Confederation transport ship. Arriving on Mars, the ticket also provides a home there, which must also meet standards for space and construction[2]. The ticket holder will also have the right to claim a parcel of land, up to one thousand acres in size, on any Martian real estate that has not already been claimed. Finally, but only if desired, the ticket buys a round trip from Mars to earth and back. This second "trip to Mars" will allow the individual not to feel stranded on Mars and to provide for a last opportunity to see loved ones on earth, after a real Martian experience.

The ticket system will harness the abilities of prospective Martian settlers. It will exchange talent and labor for a reasonable lifestyle and transportation to Mars. The reader can decide for herself whether this is a fair trade. The mission to Mars assumes that thousands, tens of thousands, and perhaps hundreds of thousands

---

[1] CS refers to "Confederation Script", the currency issued by the Confederation for purposes that will be discussed in great detail in the remainder of this work. The dual monetary system was introduced in the last chapter. The initial exchange rate will be one CS $ for one US $. Confederation policies and the realities of its economic system will dictate whether or not this exchange rate holds.

[2] Not necessarily the same standards as required at the earth Embassy. The homes on Mars may be smaller and will likely be, of necessity, constructed differently from those on earth. A democratic process will determine what minimal standards must be met for housing. Everyone is free to buy, sell, rent, or remodel their home—at the earth Embassy or on Mars—at any time.

of talented individuals will find this exchange to their taste. This system, patterned in part after the system of indentured servitude that helped settle America, has worked historically and can work again in the 21st-century. The voluntary exchange of talents and services for a ticket to Mars will create an earth Embassy with a not-so-small society of strong, determined people whose labor will be harnessed to build the equipment and institutions needed to settle the Red Planet. The Confederation's earth Embassy will house engineers to build propulsion systems, workers to build the spacecraft and other equipment, doctors to establish medical facilities, lawyers to create a (better) system of laws and courts, "Domino's Pizza" franchisees to provide pizzas for the people of the earth Embassy and on Mars, bankers for banking services, etc., all in exchange for a seat on the next bus to Mars. The personal circumstances of individuals at the earth Embassy will be examined several times in later chapters and volumes, but it makes sense to consider now a more specific application of this process to a specific ticket holder.

Not So Plain Jane

Jane arrived at the earth Embassy desiring to become a Martian settler. She had to be eligible for immigration, but as these standards exclude very few persons, immigration was not a problem. Jane had only to prove that she could find employment and enter into a ticket contract before she was admitted as a resident. While there are numerous guest programs for people interested in learning more about the earth Embassy, to become a resident one must join the Fleet and, more importantly, enter into a ticket contract with one of the Travel Corporations. As Jane had been a legal secretary and had obtained a position at the earth Embassy with a solo practitioner[3] entering into a contract for a ticket was not a problem.

Jane's first 31 days at the earth Embassy was in service with the Fleet. The Fleet offers no "basic training". Rather, most training is done within Fleet units; in addition, every year crewmen undergo three or four days of "regimentation" training. Regimentation for the lower ranks ensures that members of the Fleet know how to march, wear their uniforms, know basic military skills, etc. Jane's first tour was with the Academy Regiment[4] where she served as a classroom coordinator for Fleet training classes. In some respects, Jane was lucky. Most Fleet recruits spend their first tour doing construction or other manual labor on projects needed to build Fleet ships and infrastructure.

The reason that a person is introduced to the Confederation through Fleet service is simple: if someone shows that she is unwilling to work as a member of

---

[3]   Shakespeare would be pleased to know that most Confederation lawyers will only work part-time and neither is it likely that there will be large law firms at the earth Embassy.

[4]   The "regiments" will be discussed in the next chapter.

the Fleet when she first enters the earth Embassy, chances are good that this person will not long remain a member of the Confederation. Better for everyone to discover these things sooner rather than later.

Upon successful completion of her first tour with the Fleet, Jane started work for Kwanzaa Blackmun, a lawyer originally from South Africa. Mr. Blackmun has proven himself to be exceptionally skilled at negotiating the red tape standing between potential immigrants from the African continent and the Confederation. In his practice, Mr. Blackmun supervises a process where he helps to recruit people from Africa and processes immigration paperwork needed to get them through US immigration[5] and to the earth Embassy. Jane is paid CS $25,000 per year to process most of the American immigration paperwork for an average 100 African immigrants per year.

Jane was given a tiny house under the Travel Corporation's contract. The house was her first and she did not like living alone, so she rented the 400 square foot house to a family from mainland China who was raising their family in tiny accommodations to save money. Jane met a friendly woman during her first week at the earth Embassy and the two women decided to rent a trailer together. The singlewide trailer was not big, but it did give Jane the feel of living in a working class American neighborhood.

Jane's plastic money pays for all the Sloppy Joes and macaroni and cheese she can eat at the main Confederation buffet. The food there is bland, however, so she often goes to one of the many private homes selling food. Many people, especially Indians and Chinese, have a side-business of selling food from their family trailers. These are legitimate and "licensed" Confederation businesses, required to meet minimal standards for cleanliness and accommodations for their "restaurant" guests. The Fleet also runs several "mess halls". Jane and her friend find that the mess hall run by the 1st Marine Regiment is especially good, so they eat there several times a week.

There are several stores that sell food products at the earth Embassy, but most of these cater to Asians who insist upon eating delicious food. Most of the westerners tolerate the mess halls and the buffets and treat themselves once or twice a month to a real meal outside of the earth Embassy. A bus service runs into town and elsewhere. Several of the most popular stops are at nicer local restaurants. As Jane and her friend have a small refrigerator; they stock it with drinks and a few munchies. They bravely tried the Confederation-made potato chips, but these were not very good. Since "real world" junk food is hard to get (the Confederation buys aluminum and machine tools, not Fritos), "a few" munchies means exactly that: just a few.

---

[5]   Assume that the US is the host nation.

Jane has about $400 per month "to spend". Even with that small sum, she feels lucky. Many four-person families have even less. Jane receives about US $75 per month in hard currency, depending upon the latest Confederation policy. The balance is paid to her in Confederation Script. The women fix up their trailer with nice things, keeping in mind that shopping usually means going to someone else's home to buy used goods or crafts or furniture made during a person's free time.

Jane is thrilled about eventually going to Mars. Her contract calls for her to pay on her ticket for 19 years. She hopes to be selected to travel to Mars just about the time that the ticket is paid off. Jane visits relatives almost every year. Most employers at the earth Embassy try to be lenient about work schedules, as life there can be hard and opportunities for family and friends need to be seized. So, after every 31-day tour with the Fleet, with a CS $2000 paycheck from the Fleet in her pocket, Jane heads back to Georgia for a week or ten days with her parents. Once, she took her crewman's uniform along. The whole family smiled, but her father seemed especially proud, despite the uniform being little more than dark-green work clothes with a couple of crude insignia sewn on.

Jane is sure that she will return to earth after a few years on Mars. She loves her family too much not to come back. She thinks she will return to Mars, though, to finish what she has started. Only a few die-hard Confederation-types talk about never coming back to earth, as no one really knows what life on Mars will be like. Jane plans to accept whatever her life brings, though she mostly believes that she will find love, happiness, and fulfillment living at the earth Embassy and on the Red Planet. In any event, Jane is equally certain that she will have lots of money in the future, whether on Mars or on the earth, and has few concerns about financial insecurity. The Confederation's "forced savings", the opportunities on Mars, the possibility that she could sell her story someday; all these things make her realize that the monthly CS $325 and US $75 she "brings home" is not such a bad salary after all.

Tickets to Ride

Three major hurdles must be passed before a person can board a ship to Mars. First, that person must have paid off his ticket[6]. Next, the individual must have completed a basic level of Fleet training, so as to be prepared for life in space and on Mars. While this training will not be extraordinarily extensive or difficult, it must obviously cover basic skills that a pioneer on Mars will need. Finally, there

---

[6] A Travel Corporation will be the entity at risk if the ticket is not paid. Before it releases a ticket to Mars to settlers whose accounts are not paid off, the Travel Corporation must be reasonably certain that the balance will eventually be paid.

must be a "need" for that person on Mars. In most cases, this "need" will be determined by a system similar to the one described in this section. Note that part of the "need" will be to ensure the integrity of families and of the ticket system. Long-time members of the earth Embassy must get sent to Mars, not just the youngest and most virile men or the best-educated women. Seniority of ticket will be an important aspect of this system.

Once routine movement of settlers to Mars with the Caravan Transporters begins (eight years after the first landing), the Seat Selection System for seats on these Caravan Transporters might have eight categories of seats.

Categories of Seats	Percentage of Seats
Needs of Confederation, family emergencies	5
Needs of the Martian Economy	15
Needs of Specific Businesses on Mars	15
Transportation of Children and Younger Citizens	15
Transportation of Older Citizens	12
Date of Ticket Contract	25
Re-Unite Families	10
Return Settlers to Mars[7]	3

Figure 13: Categories of Seats and % Breakout

Class	Notes
Work Skill	Assigned by the Fleet
Corporate Priority	Assigned by Specific Business and Rolled Up by Fleet
Youth or Age	--
Contract Date	Date Ticket Contract Was Made
Re-Unite Family	As Determined By The Fleet

Figure 14: Travel Classes

---

7   While the opportunity to return to earth will be available to every Martian settler, the reality will be that there will be few return seats back to Mars. Thus, a non-pilot who decides to return to earth may encounter a lengthy wait to get back to Mars.

Categories of Seats	Description	Number of Assigned Seats
Needs of Confederation, family emergencies	One doctor One computer specialist for the Confederation's computers One senator (yes, there will be "pull", even in our little utopia) One wife, to be by the side of a dying Fleet crewman, who has developed cancer while on duty	4
Needs of the Martian Economy	15 different job skills needed on Mars	15
Needs of Specific Businesses on Mars	Marsmovers, Inc. moving to Mars—6 seats Mars Energy moving more people to Mars—4 seats Precision Metals moving to Mars—6 seats	16
Transportation of Children and Younger Citizens		16
Transportation of Older Citizens		12
Date of Ticket Contract		25
Re-Unite Families		9
Return Settlers To Mars		3

Figure 15: Seat Assignments For The 22nd Caravan Transporter Mission

Each individual will be assigned a priority in several different categories. These categories might be those seen in Figure 14.

Figure 15 illustrates this system. It envisions seat assignments for the 22nd Caravan Transporter mission to Mars, at a time when Mars' population has reached 2600. This mission will have a crew of five and 100 seats available to settlers.

The following individuals have been called to claim a seat on this Caravan Transporter:

Mary

Mary contracted relatively early for her ticket. Her husband was sent to Mars eighteen months before. She has worked sporadically at the earth Embassy as a secretary while her children, now six and nine, were young. Mary's codes are:

Class	Transport Codes
Work Skill	2549
Corporate Priority	9
Youth or Age	N/A
Contract Date	248
Re-Unite Family	2

Figure 16: Mary's Transport Codes

Mary and the children will be assigned seats on the transporter to re-unite them with John, since she was second in line in the "Re-unite Family" category. The children will also fly with Mary, now that both she and her husband will be on Mars.

Sam

Sam contracted for his ticket three years ago. He is not married. Sam is a highly skilled nuclear engineer, whose employer, Mars Energy, has been allocated four seats on the next ship to Mars. Sam's transport codes are:

Class	Transport Codes
Work Skill	125
Corporate Priority	4
Youth or Age	N/A
Contract Date	8274
Re-Unite Family	N/A

Figure 17: Sam's Transport Codes

On the Caravan Transporter, Sam will be assigned seats as one of the four people from Mars Energy being sent to Mars to upgrade the Martian power grid. A fellow engineer at Mars Energy will not be on the caravan, because that person's Corporate Priority code was "5" and only four people from Mars Energy get to go on this mission.

Melissa (Who Just Graduated From High School)

Melissa just turned 18. She contracted for her ticket only five months before. She is not married. Melissa plans to attend college and study biology. She does not hold any job where an employer is allowed to assign her a "Corporate Priority" code. Melissa's transport codes are found in Figure 18.

Melissa's Youth Code of 5 was assigned to her at random from a group of people under the age of 23. Since six young people and eight children will be assigned seats on the next caravan transporter, Melissa will get one of those seats.

Class	Transport Codes
Work Skill	N/A
Corporate Priority	N/A
Youth or Age	5
Contract Date	28,399
Re-Unite Family	N/A

Figure 18: Melissa's Transport Codes

Hans

Hans is 62 years old. He has worked at the earth Embassy for 28 years as the manager of a retail store. He contracted for his ticket 31 years ago. Hans was married, but his wife passed away unexpectedly two years before. Hans' transport codes are:

Class	Transport Codes
Work Skill	379
Corporate Priority	18
Youth or Age	9
Contract Date	156
Re-Unite Family	N/A

Figure 19: Hans' Transport Codes

Hans was assigned a seat because of his age. It is important to the Confederation to send people like Hans to Mars to show that it intends to honor all contracts for Mars, even with someone who has reached the beginning of "old age". Hans' Age Code was 9. Twelve people are being assigned seats in this category. They offer Hans a seat and he decides to take it, as Hans cannot be certain how much longer he will be healthy enough to make the trip. Hans chooses to be sedated for the lift-off from earth, which he experiences without incident.

Although Hans' financial circumstances are quite good, he is planning to "start a new career" on Mars. The low gravity there means that a retirement age of 85 is not unreasonable (see below), as long as a person is mentally able to continue to work. As Hans' family has a history of longevity, because his skills as a retailer are in great demand on Mars, and because he is a little lonely now that his wife is gone, Hans hopes to work for another 15–20 years on Mars, to add a rich, new chapter to his already full life.

Once a person is assigned a seat, she must decide to accept the trip to Mars or decline it so that the seat can go to someone else. If a seat is declined, that person retains her Class rankings for the next ship to Mars. If a second seat is also declined, however, then she goes to the end of the line in the Class under which that seat was offered. Note, however, that she retains her rankings in the other Classes. Thus, a person may legitimately delay her trip to Mars and still be able to go within a few years.

Hans and Health

The positive aspects of Mars for the elderly are related to its lower gravity. Not only will Hans' heart not have to work as hard to pump blood, it will make lifting easier, and the lower gravity will reduce the number and severity of injuries related to falls, a major cause of death for older people. On the negative side, the lower gravity will likely increase bone deterioration and reduce muscle mass. Both of these health issues may be amenable to relatively simple treatment (exercise and calcium supplements), but only intensive medical research on *people living on Mars* can provide real information about the health impacts of the Martian environment.

These health issues illustrate another way that mission to Mars' costs will be lower than NASA's costs. NASA would study the "gravity issue" endlessly and expensively, using methods that cannot simulate Mars' gravity[8]. Thus, NASA would spend lots of money but would be able to draw only limited conclusions; it would have to send humans to Mars to obtain real information. This, of course, is what the mission to Mars proposes all along. The mission to Mars intends to send people, study the effects, and then find appropriate solutions. If this approach sounds too "risky" for you, stay at home and watch newsreels about the heroes on Mars.

A great difference between the mission to Mars' program and NASA is that the mission to Mars does not intend to exclude people if they have minor health problems. The exact definition of "minor" can only be set as the program progresses, but the true test will not be how fast someone can run two miles, but whether a person can survive on Mars with the limited infrastructure there. Note the word "survive". The mission to Mars believes that it makes sense to have sick people on Mars, to stimulate the need for doctors and to provide an immediate database of medical information. People can make their own decisions about the risks and rewards of a limited medical infrastructure when their time comes to take a seat on the bus for Mars.

The ticket system tells a good part of the human side of the mission to Mars. The rest of the human side is told by the Fleet, the organization that will link each member of the earth Embassy to training for Mars and an opportunity to affect the direction of Confederation programs. Conveniently, that is our next subject.

---

[8]   On earth there is no way to simulate long-term exposure to low gravity. In space, the problem is nearly the same.

The Fleet

"Cooperation is more important than competition."

—Naomi Wildman, *Star Trek Voyager*

The Fleet is the second major leg of the mission to Mars. The purpose of the Fleet will be to train people for Mars, to serve as a social leveler by providing a common experience for everyone at the mission to Mars, and to serve as the Confederation's engineering general contractor. Several other themes behind the Fleet's organization will be developed in Volume III, but the purpose now is to show how important—indeed essential—the Fleet will be to the mission to Mars. For those who may have skipped the discussion of basic ideas, we review.

The Fleet will consist of everyone over the age of 18. It will require 31 days of annual service from each of its members. Five to ten percent of the Fleet's membership will consist of relatively lowly-paid full-time members, who will help units perform missions while its part-timers are away at their civilian employment. Even those few members on full-time duty cannot expect to make a career of the Fleet, but rather will consist primarily of those who enlist for a few years when they are young, pilots[1], and other leaders in training. A few high ranking leaders will also serve full-time to allow for consistent decision making in the most important programs. Finally, some lower ranking crewmen might be put on active duty when they are between jobs, so the Fleet can also serve a social welfare role.

Theoretical ideas about the need for a military structure may be fine for discussion, but asking people to "soldier" 31 days per year is a very different matter. Many will look askance upon any program that requires universal military service and *purposeless* service in uniform will make it impossible to attract recruits for the mission to Mars. As has been indicated, however, the Fleet forms a pivotal leg

---

[1]   A Confederation "astronaut".

of the mission to Mars' triumvirate. Service in the Fleet will be the opposite of purposeless. Without the Fleet to organize the efforts of those living at the earth Embassy, there will be no mission to Mars. Especially important will be its role as engineering general contractor. As will be discussed in this chapter, the Fleet will have a military character, will emphasize leadership and cooperative efforts toward Mars, but it will not have a military mission at the earth Embassy. It will organize the mission to Mars, guiding and motivating the civilian community toward achievements in space. This chapter's discussion about the engineer regiment will be especially relevant to understanding the relationships being proposed.

Volume III will argue the advantages of the Fleet's "military" organization. The remainder of this chapter discusses how the Fleet might be organized and trained using a notional earth Embassy with a population of 5,000 residents, providing a Fleet of 3,500 crewmen. What is to be discussed here can be scaled up or down according to actual population.

Figure 20 provides an organizational and functional breakdown for this 3,500-member Fleet. A basic assumption of the Fleet is that everyone can and *should* contribute their service to their community, even if this service is as simple as providing support for more active crewmen and projects. Even the infirm will be required to serve, unless someone is in a vegetative state. In this way, even 80-year-old "invalids" and the "handicapped" will be expected to do their part, such as helping with the food, office duties, or maintaining barracks. In turn, even elderly members of the mission to Mars will be important parts of the community and of its ambitious undertaking.

Each of the six regiments in our 3,500-person Fleet has approximately 600 members[2]. Many people will remain in the same regiment for their entire lives, though people may transfer for promotion, as part of their move to Mars, or simply as a matter of personal choice.

The Space Command will consist of two regiments whose missions relate to matters of space. The 1st Space Regiment will provide pilots and direct supporting services for the Confederation's spacecraft. The 1st Space Operations Regiment will run spaceport operations and provide more general support for space operations. In this community of 3,500 adults, about one-third of them will be directly involved in space operations. Thus, a mission to Mars' recruiting slogan "Join the Mission to Mars and Get Into Space" will be more than just a sales pitch. The ease of access to important space functions, as well as importance of these operations, will serve as a powerful incentive to recruit new members to the Confederation. It will stand, too, in the strongest contrast to NASA programs, where it is difficult to get close to actual space operations.

---

[2]   The Academy Regiment may have 800–900 members.

Space Command	Troop Command	Academy Regiment
1st Space Operations Regiment	1st Marine Regiment	Training and Doctrine Battalion
1st Space Regiment	1st Engineer Regiment	Fleet Hospital
	1st Military Police Regiment	Research Battalion

Figure 20: Major Commands and Subordinate Units

The Troop Command will consist of three regiments. These units will be much more traditionally organized than the other regiments in the Fleet. There will be a regiment of marines, who will serve as the Confederation's "infantry"; one military police regiment, which will provide a police capability; and a regiment of engineers, who will be given design and construction responsibilities. As the Confederation and Fleet grow, additional regiments of marines, military police, and engineers can be created. As additional regiments of marines are added, the actual "military" capabilities of one or two hundred of these marines will increase, until the Confederation will find itself with one or two companies of western-style light infantry.

The last major command of the Fleet will be the Academy Regiment. The Academy Regiment will be organized into three unique battalions. The Training and Doctrine Battalion will provide the Fleet a comprehensive set of training courses and programs: academic, professional military development, a medical school, and field training for the units. The Fleet Hospital will provide medical services and otherwise serve as the focus of medical capabilities within the Confederation. Finally, the Research Battalion will be organized to sponsor research for the Fleet in its role as the engineering general contractor.

1st Space Regiment

The 1st Space Regiment will consist of two battalions. These two units will train pilots for the SSTOs and the transporter vessels. This regiment will also command "military" missions transplanting the earth Embassy to Mars.

The 1st Space Battalion will have the mission to plan and execute the manned missions to Mars. As such, much of its activity will focus on pilots and pilot skills, although significant cross training in civilian and quasi-civilian skills will also occur. During those first missions to Mars, every pilot in the 1st Space Battalion will be expected to have at least two and possibly three near-professional skill sets.

This way, if the doctor is killed in an accident, a medic can serve as her replacement until a new doctor can be packed off for Mars.

The 1st Space Battalion will also plan and execute missions (if any) to the inner asteroids. The two to four pilots who fly these missions will learn to operate in non-terran space and to build small bases on new worlds.

At the very start of the mission to Mars, only a few members of the Fleet will be assigned to this battalion and most of them will be involved in planning, rather than conducting operations. As the time for manned missions to Mars (or the inner asteroids) nears, however, more and more people will be assigned to the 1st Space Battalion. Many of these people will come from the 2nd Space Battalion, which will have a more immediate mission assigned to it. In all, both battalions will have a total of 15 trained pilots.[3]

The 2nd Space Battalion will have the mission to train and execute missions into low earth orbit. Increasing with the scope of actual SSTO operations, there will be ten pilots assigned to this battalion. The mostly robotic, computer-controlled SSTOs will not need much human assistance to launch, orbit, re-enter the atmosphere, and land. Pilots will be highly intelligent and well-trained people, but with limited roles aboard the SSTOs. The pilots will be on the ship primarily to take control in the event of an emergency. The 2nd Space Battalion will fill the ranks of the 1st Space Battalion as the mission to Mars moves from earth and low earth orbit activities to the creation of settlements on Mars. Of course, the need to continue SSTO flights will remain even after flights to Mars begin and the 2nd Space Battalion will continue to execute these operations. However, some of these operations will end as equipment is sold or otherwise transferred to civilian control. At some point, the 2nd Space Battalion will be transferred to Mars and assigned a new mission: to develop weapons, tactics, and units to operate as a true military force, to protect Mars from potential aggressors.

The Confederation will likely purchase one or two small aircraft for the earth Embassy to facilitate travel for members of the Confederation and to serve as transport for residents of the earth Embassy. The pilots and support crew of these airplanes will be assigned to the 1st Space Regiment.

The mission to Mars will be a much more egalitarian organization than NASA and will try to minimize the distinction between elites and "others". Restricting

---

[3]   In this notional community of 3500 adults and 15 pilots, there will be one pilot for every 250 adults. In addition to the ten pilots initially assigned to the 2nd Space Battalion, there will be three pilots assigned to the 1st Space Battalion, one to the 1st Space Operations Regiment headquarters, and one pilot to the Space Command Headquarters.

pilot slots to just a few "astronauts" will be resisted in favor of a program where eventually hundreds of pilots will be trained, assigned, and rotated on missions. Three hundred or more people will qualify as pilots during the earth Embassy years of the mission to Mars, providing a broad base of experience in space. The Fleet will also provide the necessary organization and infrastructure for a community brimming with space operations and training, to facilitate the enormous task of moving 50,000 people from earth to Mars.

1st Space Operations Regiment

The 1st Space Operations Regiment will operate the spaceports and provide other support necessary for Space Command missions. There will only be one pilot in this regiment, probably serving as the regimental commander. His flight experience should provide some insight about the needs of the pilot-dominated 1st Space Regiment. The 1st Space Operations Regiment will concern itself with refueling operations, communications, maintenance, fire fighting capability, etc. Seemingly mundane, from the beginning this regiment will have actual operational responsibilities. Unlike the marines (see below) or the pilots, the members of the 1st Space Regiment will not spend their time drilling or training, but actually planning and controlling space operations for the Confederation.

The most visible aspect of the 1st Space Operations Regiment will be to run the various spaceports at the earth Embassy and on new worlds. There will be a spaceport on Deimos and one on Mars. In addition, one or more of the inner asteroids may have some need for a small spaceport, just as Phobos will need a spaceport when operations begin there. The 1st Space Operations Regiment will operate in each of these locations, establishing landing, fueling, and communications capabilities.

The men and women of the 1st Space Operations Regiment will also provide an important community function: a modern and effective fire fighting capability. Most of the training will revolve around space- and Mars-based fire fighting. Most fires during the first 50 years of Confederation activities, however, will not occur on Mars or in space, but at the earth Embassy. An efficient capability to fight fires at earth Embassy factories and in residential areas will be important. It also makes sense that the Confederation's fire marshals come from this regiment.

1st Marine Regiment

The 1st Marine Regiment will be the Confederation's "infantry". Its primary mission will be to "train the trainers" for people in the rest of the Fleet. Thus, the regiment will specialize in military- and space-related tasks. It will also provide a pool of labor for a variety of construction and economic activities at the earth Embassy. The 1st Marines will set the military example for all the other Fleet units, as they strive to become an "elite" organization. On Mars, marines will be used as scouts and escorts for surveying and other missions, where their ability to navigate, live in rough country, and assist in the transportation of equipment will be highly valued.

There is very little probability that the earth Embassy would ever attempt "to defend its borders" in a military sense.[4] However, to begin a tough military tradition, infantry skills will be taught and once on Mars, marine units will be expected to operate as traditional infantry. If it can generate favorable publicity, the regiment might offer to participate in military exercises with friendly governments[5].

The regiment will also seek to create at least one highly capable company-sized unit (100 marines) who will be prepared to serve in peacekeeping roles around the world. Not only would service as peacekeepers further the Confederation's goals for the New Era, but peacekeeping operations would raise morale at the earth Embassy. And, in a slightly mercenary vein, it would present an additional opportunity to bring in hard currency, since mission to Mars peacekeepers would only be employed if paid more than the expenses they incurred.[6]

This potential role as peacekeepers should drive much of the training in the regiment. Mobility, deception, camouflage skills, and hand-to-hand combat will comprise the most important military-style training. These skills will be useful to the marines whether they serve as peacekeepers or in more traditional infantry roles. Marine training will not be like infantry training in the US military, where

---

[4]   It would likely be political and military suicide to "defend the perimeter" of the earth Embassy. Thus, resistance to a host nation intrusion must be met politically. If this is problematic, the Fleet will act to ensure that members of the mission to Mars *do not* resist arrest or other interference by host nation authorities. The only viable option to a military or police action against the earth Embassy will be to find a more congenial host nation. Intrusions by non-governmental organizations (Greenpeace decides to invade the embassy to protest the "pollution of space" by the human virus, etc.) might be met by a non-violent Confederation "police action", organized by the Fleet, but this is an entirely different matter from a "defense of the perimeter" against a host nation's military or police forces.

[5]   Provided, of course, there is no net cost to the Confederation (the friendly government agrees to reimburse all costs associated with this training). This could prove to be a source of hard currency for the mission to Mars.

[6]   By the US, the UN, the European Union, etc. Much more on peacekeeping operations in Volumes II and III.

there has long been a tendency to amass overwhelming numbers and overwhelming firepower against enemy units. Marines will use deception, camouflage, and hand-to-hand fighting abilities to best effect in their military training missions.

The regiment is called a *marine* regiment rather than an infantry regiment because its eventual role will be to fight in support of the Confederation's air and space forces guarding Mars. Moreover, as the Confederation transforms itself into a commonwealth of independent republics, the Confederation's need for traditional infantry will disappear, since infantry regiments from the independent republics can be called into service. What will be very useful to the Fleet in the future will be troops who can operate on air and spacecraft, to seize small objectives that are not to be destroyed by aerial or space bombardment.

One might call the marine regiment special forces, though this currently overwrought term implies a measure of elitism that will be foreign to the Confederation. The ability of some small numbers of troops to conduct military operations with great skill may develop in this regiment. Much more important, however, will be the unit's drive and determination to succeed in whatever mission it is given, even those less-than-glamorous construction and labor missions that will be common early on.

The mission to Mars has only a tangential military character. Creating intense and capable marine regiments for service as infantry will not be a priority. This does not mean, however, that excellence will not be demanded by the commanders of the marine regiment. Moreover, what may be asked of Company A, 1st Marines, a group of 18-year-olds with a few "professional" soldiers to lead them, will be different from what may be asked of Company B, 1st Marines, a mixed group of 18-year-olds, one-month-a-year troops, and a small cadre of full time professionals. Both will certainly be different from what will be asked of company D, 1st Marines, a unit comprised exclusively of 20–30 year-old one-month-a-year troops. The capabilities of the other companies of the regiment will vary, too. Each marine will be skilled, though the skills will differ from one company to the next.

The labor aspect of the marine regiment will be the least glamorous part of its mission. In essence, the marine regiments will provide labor for Confederation and other projects; not to the detriment of other missions, but as a planned part of the marine training cycle. Thus, if there are trucks to be unloaded or supplies to be moved, it will be marine units assigned to the task. When crops need to be harvested and there are not enough laborers, marine units will be assigned. Remember, one of the main reasons so much emphasis will be placed upon the creation of the Fleet will be to help organize the labor of the earth Embassy residents. Not everyone at the earth Embassy will be a craftsman or a researcher. Just as important for ultimate success will be those who help build roads and buildings, often using manual labor. The Confederation's effort can lunge forward if it always has a pool of inexpensive labor to complete important projects. Since it may not be cost effective to buy large amounts of expensive construction equipment, hard-body marines can help to fill the gap.

Finally, marine regiments will serve as test beds for everything related to space and Mars. They will test food, proposed quarters, equipment, and living routines. They will urinate into tubes that will go into wastewater treatment facilities testing for space and will practice setting up Martian shelters for the researchers and engineers. "If it needs to be tested, give it to the marines." If the marines don't break it, the design can be considered a success.

1st Military Police Regiment

During the early years, crime should not be a problem either on Mars or at the earth Embassy. Should a criminal be discovered in the Confederation's ranks, the offender would likely be expelled from the earth Embassy. Nonetheless, there will be a need for some police capability, both on Mars and especially as the earth Embassy grows. This capability will come from the 1st Military Police Regiment.

The earth Embassy community of 5,000 people will require a presence of one duty officer for every 1,000 people. This implies shifts of five military policemen, 24 hours a day, 365 days per year. Two hundred military policemen will be adequate to man these shifts (see Figure 21). Given that the military policemen will require continuous training, that support personnel will be needed to man police labs and other police facilities, and that other administrative personnel will also be needed, the regiment's total should likely be closer to 500 people, approximately the strength suggested at the beginning of this chapter.

Per Shift	5
Per Day	15 per day
Per Year	5500 person-days
# of Shifts Per Crewman (per Annual Tour of Duty)	28
Total # of Military Policemen Required	200

Figure 21: Police Shift Requirement

A small criminal lab company and other specialty companies will be formed for specific regimental purposes. While there may not be much street crime at the earth Embassy, as the community grows there may be some white-collar crime, plots to embezzle funds, or to commit fraud. The military police regiment can be alert to these issues and train detectives or investigators appropriate to the community's needs. In addition, there may be Confederation members indicted by outside authorities. Thus, some capability to handle extradition and provide temporary confinement will be needed. Finally, although the Confederation will not likely ban drugs or prostitution[7], a mission to combat these problems will be assigned to the regiment, requiring an education program.

The regiment might have six line companies, each company having a two-month window of responsibility for policing the earth Embassy. If military policemen are activated in staggered call ups, then there will be hardly any noticeable diminution of ability during the transition from one company to the next[8].

---

[7] Even if these practices are not banned because of the libertarian nature of the mission to Mars, they can be combated by working administratively against those who might abuse the freedom enjoyed at the earth Embassy.

[8] Assume 75-person line companies. Each company will have a span of ten weeks of duty. The first and last weeks will concentrate on "right-seat rides" (RSR) and "left-seat rides" (LSR) with the outgoing/incoming military police company. During the RSR, the new military policemen (MPs) accompany the outgoing MPs in their patrols, to observe and assist the outgoing MPs. During the LSR, the new MPs patrol while accompanied by the outgoing MPs who assist and critique the incoming MPs. The 75 MPs are brought to duty to cover the entire period, as indicated below:

Week	1st Rotation	2nd Rotation	3rd Rotation	4th Rotation
Incoming Transition Week	X			
2	X			
3	X	X		
4	X	X		
5		X	X	
6		X	X	
7			X	X
8			X	X
9				X
Outgoing Transition Week				X

Each rotation will include about 15 MPs, the number indicated in Figure 21 as required to patrol this 5,000-person community. The other 15 members (four rotations require 60 MPs) of the unit (company commander, platoon leaders, desk sergeants, etc.) will work and train as needed during these ten weeks. The labs, files units, detectives, etc. in the other companies in the regiment will be called to duty as needed.

Internal crowd control or riots should not be of major concern. If people at the earth Embassy are rioting, political, rather than police actions will be the antidote. External riots and protests (outside the confines of the earth Embassy), however, are matters that must be expected over a decades-long project. In these circumstances, the military police regiment would execute operations designed to counteract the effects of these riots or actually to put the riots down. Marines could assist the military policemen, who could offer their hand-to-hand fighting abilities and man security stations and checkpoints.

1st Engineer Regiment

All military organizations have engineer formations. The Confederation's Fleet will create engineer units to assist with the construction of property owned, operated, or otherwise relevant to the mission to Mars. The engineer regiment will have a construction engineer company composed of various specialties, such as carpenters, electricians, heavy equipment operators, etc. most of whose work will center on space-related construction projects. There will also be some design work done by the engineer regiments, though this will be quite limited for reasons we now address.

The engineer regiment will be the focus of the Fleet's "Engineering General Contractor" mission in the sense that engineer units will unify the engineering efforts of the entire earth Embassy. The Fleet will not do most of the engineering work proper, but will help to coordinate efforts undertaken by civilian businesses. The regiment's commanders will also make decision's that will influence the general direction of the mission to Mars. While most of these commanders will be "part-time" members of the Fleet, who will be paid only for 31 days of duty per year, they will be expected to shoulder great responsibilities. Thus, an added measure of productivity to help reduce overall mission costs for the Confederation. Many commanders may spend many more than 31 days working on Fleet projects, but most will only be paid a relatively low "military" wage for 31 days' work.

These ideas might best be understood by describing the Engineer Regiment in our 5,000-member community. Figures 22 and 23[9] detail several matters that provide insight into how the mission to Mars might work and about its assertion of increased productivity from Fleet members. The complex sets of relationships described in these two figures indicate several ideas fundamental to the mission to Mars, including:

---

9   This explanation of the organization and interaction of the engineer regiment with the rest of the earth Embassy community will be the most in-depth discussion in this volume. It is typical of discussions found in Volume III.

1) That most work toward Mars will be done by civilians;

2) that "part-time" Fleet leaders (people like *you*) will hold many key positions and will make many important decisions relating to the task of getting people to Mars; and

3) that mobilization of resources by the Fleet, especially the skill of highly talented people motivated by a desire for Fleet promotion and the power to influence mission to Mars programs, will avoid huge salary costs that an organization like NASA cannot avoid.

Figure 22 lists examples of the relationship between the engineer regiment and civilians in typical construction and design projects undertaken at the earth Embassy. While most of these projects will be work done for the Confederation itself, there are also two examples where the engineer regiment will work for private contractors, in the role of a subcontractor. We see, too, how the engineer regiment will be involved in almost every aspect of the design for Mars, in most cases, providing project management or liaisons to private contractors. While the relationship of the engineer regiment (indeed the entire Confederation) and private contractors will resemble that of the US' Defense Department and defense contractors in the US, the power structure will mostly be reversed. In the US, the Defense Department decides what it wants and then hires private contractors to build things. At the mission to Mars, private contractors will bring their ideas to the Confederation and the Fleet, which will provide coordination and funding, without the ability to bulldoze the work being done in the private sector. In a word, the Confederation's Fleet will cooperate and coordinate the work being done by private firms to reach Mars. As a coordinator, the Fleet will have limited power. There will be little chance that a new admiral, a new Congress, or a new President can order changes simply from a desire—respectively—to inscribe the program with the new commander's imprimatur, the new Congress' budgetary moods, or to adapt to a new President's non-space agenda.

Figure 22 also helps the reader understand that the engineer regiment (hence, the Fleet) will manage and coordinate progress. It will rarely become involved in the day-to-day management of any project, almost all of which will be run by civilians. Engineer commanders will have liaison responsibilities with these projects, but the engineers will assist these civilians, rather than become an earth Embassy "bureaucracy".

If an engineer company has been assigned a job to complete an outbuilding at the earth Embassy's spaceport, it can requisition labor from the other Fleet regiments. Imagine a construction project that will require one week to complete. The engineer leader in charge believes that having two laborers to assist with simple tasks can facilitate construction. The Engineer Regiment can request that two laborers be assigned for the week, crewmen who will usually come out of the marine regiment. Fortunately for those of us with weak backs, the physical demands of those in the marine regiment will likely be greater than those in the engineer regiment. Those designing spacecraft, for example, may make far larger contributions to the effort toward Mars if they belabor their designs, rather than build outhouses. As in all military organizations, however, the work of a soldier lies in the accomplishment of any task assigned, even if the task is sometimes dreary.

Description of Project	Who Owns It	Typical Engineer Regiment Activity	Typical Civilian Involvement
Ark Transporters	Fleet, 1st Space Regiment	Rewiring heating and cooling systems	Designed and built by Ark Systems, Inc.
SSTOs	Fleet, 1st Space Regiment	Install new engine system for SSTO	Designed and built by Ark Systems, Inc.
Spaceport	Fleet, 1st Space Operations Regiment	Build VIP facilities to observe take offs and landings	Designed by Mars Engineering Corporation
Wastewater Treatment Facility	"Mars Water Treatment Company"	Construct a "polishing" pond for private company. Fleet paid for its work.	Designed by Mars Engineering Corporation
Roads	Confederation	Conduct traffic pattern analysis	Built by ABC Paving
Fleet Warehouse	Fleet, 1st Space Operations Regiment	Build the warehouse	None
Private Apartment Complex	"SMP Enterprises"	Co A constructs an apartment complex for a private company. Fleet paid for its work.	None (a rare circumstance)

Figure 22: Building the Earth Embassy Infrastructure

Figure 23 is a proposed organization for the 1st Engineer Regiment. In this proposal, there are four engineer battalions, each with a different engineering mission. There is also one support battalion, whose mission will be described below. The organization of the engineering regiment will be similar to those of other regiments in the Fleet, though each regiment will have a different mission.

Each of the five battalions conceived in Figure 23 have a different mission. The 1st Battalion is assigned a construction engineering role. The 2nd Battalion is concerned with specialized engineering, such as drilling water wells on earth and Mars (D Company). The 3rd Battalion is assigned a mission to design Mars- and "space-" specific products. The 4th Battalion is concerned with designing spacecraft. Finally, there is a support battalion that will provide first line medical care, supply, maintenance, food, and a company consisting of the retired and infirm.

We end this section with a discussion about the Engineer Support Battalion. Each regiment will have a support battalion, so it makes some sense to detail its mission. This battalion will be responsible for "logistical" support for the regiment. While "providing logistical support" may sound dull, in fact, the support battalion will be one of the most important battalions in any regiment. Any group of 600 or more individuals requires significant logistical support to conduct business. In the case of the engineer regiment, crewmen on duty need to be fed, their families provided medical care, supplies ordered and delivered, and vehicles and equipment maintained and repaired. We see in Figure 23 that the Engineer Support Battalion contains over one third of the regiment's personnel.

On any given day, about 75 members of the engineer regiment will be on duty. We will use the breakdown in Figure 24 to consider what the support battalion mission might look like that day.

The support battalion itself has 25 crewmen on duty. "Today", the battalion commander and his full-time support crewman are working at the battalion headquarters, the battalion commander working an unpaid day to write efficiency reports for four of his leaders. The medical company has five people on duty. From 7 AM to 7 PM, four crewmen man a regimental aid station. This aid station sees anyone from the regiment (on duty or not) who may have a medical problem, as well as any member of their families. The fifth member of the team works at the Academy Regiment's hospital from 7 PM to 7 AM, where all after-hours patients are treated. Moving to R Company, we see that there are three people on duty. These three older people help in the battalion headquarters and also clean up the regimental area, including doing laundry of dirty linens (some of the 75 people on duty in the regiment are required to sleep in the regimental barracks area). Q Company has 15 people on duty. Twelve of these are working ordinary rotating shifts. Three others are working a special maintenance problem. With 600 people, the regiment has more than 1000 pieces of equipment that need maintenance. Some of the engineer regiment's equipment includes backend loaders, trucks, testing equipment, chain saws, etc. Each piece of equipment must receive routine maintenance. Four people per day are usually scheduled to provide this maintenance[10]. The special maintenance problem is the arrival of two old bulldozers than need major overhaul. The three-man crew assigned the job hopes to get the bulldozers rebuilt during their two-week tour. Four people are working in the mess hall. The usual headcount per day is not always easy to estimate, but 1.9 meals per person on duty is usually a good estimate. This includes a component of meals for those civilians who choose to eat their food in the regimental mess hall. The schedule today calls for 35 breakfast meals, 50 lunches, and 60 dinners. Most cooks work together as teams. Today's team has two people working from 4 AM to 2 PM and two working from 10 AM until 8 PM.

Headquarters	Units	Assigned Personnel	Unit Mission	Annual Participation Of Selected Unit Members
Regimental Headquarters, 1st Engineer Regiment		5	Command and Administration	Commander works many unpaid days, 1 or 2 full-time members, 2 or 3 staff officers working many unpaid days
1st Engineer Battalion		3	General Engineering	Commander works many unpaid days, 1 full-time member, 1 staff officer working many unpaid days
	A Company	100	Construction Engineering	31—day tours for whole company or by platoon
	B Company	50	Engineering and Design	Commander works some unpaid days, others work in small teams called to duty as needed
	C Company	50	Project Management	Commander works some unpaid days, project managers work almost every week, often unpaid

---

10   Each person has a specialty and a secondary skill, certifying the necessary training to provide maintenance for a given piece or type of equipment.

Headquarters	Units	Assigned Personnel	Unit Mission	Annual Participation Of Selected Unit Members
2nd Engineer Battalion		3	Specialty Engineering	Commander works many unpaid days, 1 full-time member, 1 staff officer working many unpaid days
	D Company	12	Drilling	Commander works some unpaid days, troops work when a water well must be dug, planning cell for Martian water wells works 31-day annual tour
	E Company	24	Water Engineering	Commander works some unpaid days, troops work as needed on water and wastewater engineering project as needed
	F Company	12	Nuclear Engineering	Commander works some unpaid days, planning and design cells mostly work 31-day annual tour
3rd Engineer Battalion		3	Mars Engineering	Commander works many unpaid days, 1 full-time member, 1 staff officer working many unpaid days
	G Company	12	Resources	Unit works 31-day annual tours to plan procurement of basic resources on Mars
	H Company	12	Design	Unit works 31-day annual tours to plan and oversee design of first Mars bases
	I Company	12	Space Engineering	Unit works 31-day annual tours to plan to plan and oversee design of Deimos, and possible inner asteroid bases

Headquarters	Units	Assigned Personnel	Unit Mission	Annual Participation Of Selected Unit Members
4th Engineer Battalion		3	Space Transportation Engineering	Commander works many unpaid days, 1 full-time member, 1 staff officer working many unpaid days
	K Company	20	Explorer Transporter	Unit works staggered tours to coordinate planning and design of Explorer Transporter
	L Company	20	SSTO	Unit works staggered tours to coordinate planning and design of SSTOs
	M Company	10	Shuttle Transporter	Unit works staggered tours to coordinate planning and design of Shuttle Transporter
	N Company	10	Caravan Transporter	Unit works staggered tours to coordinate planning and design of Caravan Transporter
	O Company	10	Ark Transporter	Unit works staggered tours to coordinate planning and design of Ark Transporter
Engineer Support Battalion		3	Support Battalion	Commander works many unpaid days, 1 full-time member, 1 staff officer working many unpaid days
	P Company	55	Medical Company	Company provides 365-day per year medical coverage for members of regiment, averaging four medics/nurses/physicians assistants/doctors per day

Headquarters	Units	Assigned Personnel	Unit Mission	Annual Participation Of Selected Unit Members
	Q Company	125	Supply, Maintenance, Mess	Company provides 365-day per year coverage for the regiment, averaging 10 personnel per day to provide supply, prepare food, and repair equipment
	R Company	30	Retired and Infirm	Company provides 365-day per year "duty" personnel for the regiment, averaging 2.5 people per day to help with minor administrative or menial tasks

Figure 23: Organization of the Engineer Regiment

Headquarters	Units	On Duty "Today"	Unit Mission	Unit Members On Duty
Regimental Headquarters, 1st Engineer Regiment		3	Command and Administration	Commander and 2 full-time members
1st Engineer Battalion		1	General Engineering	One full-time member
	A Company	1	Construction Engineering	One full-time member
	B Company	5	Engineering and Design	Small team serving the 31-day tour
	C Company	1	Project Management	One full-time member
2nd Engineer Battalion		1	Specialty Engineering	One full-time member
	F Company	2	Nuclear Engineering	Two-crewman design cell working 31-day tour

Headquarters	Units	On Duty "Today"	Unit Mission	Unit Members On Duty
3rd Engineer Battalion		1	Mars Engineering	One full-time member
	I Company	2	Space Engineering	Two-crewman design cell working 3-day tour to coordinate Deimos site plan with civilians
4th Engineer Battalion		1	Space Transportation Engineering	One full-time member
	K Company	20	Explorer Transporter	Entire unit on duty for a conference
	L Company	5	SSTO	Unit works staggered tours to coordinate planning and design of SSTOs
	O Company	7	Ark Transporter	Unit works staggered tours to coordinate planning and design of Ark Transporter
Engineer Support Battalion		2	Support Battalion	Commander and full-time member
	P Company	5	Medical Company	Five medical personnel to provide rotating medical coverage for the regiment

Headquarters	Units	On Duty "Today"	Unit Mission	Unit Members On Duty
	Q Company	15 (4 supply, 4 mess, 7 maintenance)	Supply, Maintenance & Mess	Four supply, four maintenance, and four mess crewmen serving normal 31-day tours and three maintenance people called in for special duty
	R Company	3	Retired and Infirm	Three people to help with minor administrative or menial tasks

Figure 24: On Duty "Today"

With so much equipment in the regiment to care for, so much food to order, not to mention the large number of office supplies and equipment (computers, copiers, etc.), the four supply crewmen needed to order and account for supplies and equipment in the regiment have their hands full.

Finally, consider that the regiment has 50 other crewmen on duty this day. There are likely to be several calls to the Engineer Support Battalion for emergency or other non-planned business. The support battalion is there to "support", so support battalion soldiers must ensure that support is provided, regardless of what may have been put on the day's calendar.

The Academy Regiment

The Academy Regiment will have the mission to create a comprehensive military academy at the earth Embassy. It will have several missions, all to be accomplished within a disciplined collegiate atmosphere[11]. The four battalions in the Academy Regiment will be named after the primary mission of the battalion.

---

[11] Note that the earth Embassy will have three "universities": Ares University, the University of Mars, and the Fleet's Academy Regiment.

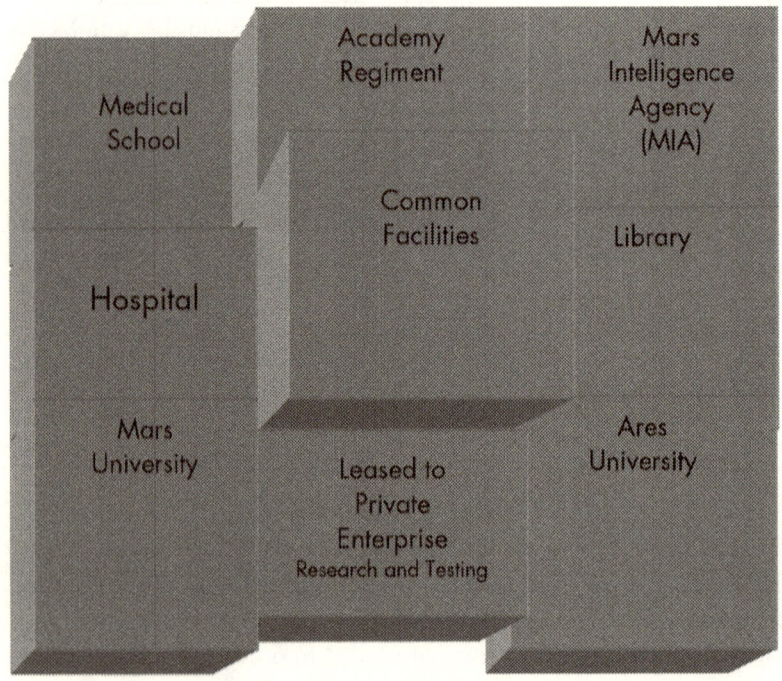

Figure 25: Academy Regiment's Physical Layout

Figure 25 presents a notional physical layout for the Academy Regiment and some adjunct activities. This layout facilitates the Academy Regiment's goals to provide an academic environment and a central intellectual focus for the mission to Mars.

Figure 26 shows the Academy Regiment's administrative organization. The regiment's first five companies will form the Training and Doctrine Battalion. It will organize the Fleet to provide first-rate training regimes for Fleet units and individuals. The Academy Regiment will provide the doctrine and training assistance that allows the units to concentrate on operations, not on "invention-of-the-wheel" training programs. There are two kinds of training activities the Fleet will conduct: individual training and unit training exercises. Individual training will focus on imparting those skills required of a crewman's position and rank. Unit training exercises will provide opportunities for the unit to practice or demonstrate the combined skills of the unit's soldiers, as well as the organizational capabilities of the unit. We next highlight several points to illustrate how the Training and Doctrine Battalion will operate.

Training and Doctrine Battalion		
A Company	Leader Training	
B Company	Recruit and Crewman Training	
C Company	Unit Training	
D Company	Logistical Support Training	Supply, Maintenance, Mess
E Company	Medical School	Trains Medical Personnel
Fleet Hospital		
F Company	Medical Company	Doctors, Dentists, Nurses, etc.
G Company	Medical Staff	Hospital Technical Staff
H Company	Administrative	Administrative Staff
I Company	Medical & Dental Records	Centralized, Computerized Medical Files
Research Battalion		
K Company	Human Research	Physiology, Psychology, Space Related
L Company	Genetic Research	Genetic Mutations and Selective Breeding For Space, Mars, and Terraforming
M Company	Hardware Research	Rockets, Materials
N Company	Firm- and Soft-ware Research	Fuels, Processes, Software
O Company	Martian Intelligence Agency	Compilation of Data Regarding Mars
Support Battalion		
Q Company	Supply, Maintenance, Mess	Academy Support
R Company	Retired and Infirm	Academy Support

Figure 26: The Academy Regiment

B Company will operate a daily reception program. This three-day program introduces new members of the earth Embassy to its programs and to the Fleet. During this three-day session, the recruits will be given baseline physicals, issued Fleet uniforms, given haircuts[12], oriented to the map and layout of the earth Embassy, etc. In addition, there will be lots of paperwork, including ticket contracts and the issuance of a home or apartment.

Both A and B Companies will provide regimentation training. Sometime during their first 31-day tour, recruits will return to the Academy Regiment to spend three days learning to march, etiquette when dealing with senior members of the Fleet, and other basic matters. Fleet leaders must also spend three days each tour at regimentation, learning progressive leadership and management skills and how to handle Fleet administrative matters.

The medical school is listed as part of the Training and Doctrine Battalion. Whether, in fact, the medical school becomes part of the Fleet Hospital is something that can be decided at the appropriate time. For now, since the idea of training people is at the heart of the Training and Doctrine Battalion, this battalion seems the better choice though, no doubt, there will be doctors who will insist that only doctors can administer programs to train other doctors.

The Fleet Hospital will operate as would any hospital. The "administrative" organization presented in Figure 26 will no doubt have little real world meaning as doctors and nurses live their crazy lives, doing what doctors and nurses do. Pray for the people who have to administer the chaos.

The Research Battalion will sponsor research programs, mostly run by civilians, into those areas necessary to achieve the goals of the mission to Mars. While expected to be slightly less chaotic than the Fleet Hospital, no doubt the Research Battalion will give its crewmen long leashes. Of note in this battalion is O Company, the Mars Intelligence Agency. Unlike a library that deposits information, this organization will process information. Volume III details how the Mars Intelligence Agency will differ from a national intelligence agency in the western world. Suffice if for now that this organization may catalogue information about potentials adversaries, but it will also publish this information, not keep it secret. And, as the name indicates, most of the information processed will be about Mars, not Greenpeace.

---

[12]   Fleet regulations about haircuts will be lax. Recruits going to the engineer regiment may only need to have their hair trimmed. Recruits going to the marine or military police regiments, however, may want to get shorter haircuts.

Rank	Typical Responsibilities
Fleet Admiral	Commands Entire Fleet
Admiral	Commander of a Major Command
Captain	Regimental Commander
Commander	Battalion Commander
Lieutenant Commander	Staff "Officer" (at all levels), Pilots
Lieutenant	Company Commander
Ensign	Platoon Leader (30–50 Crewman)
Sergeant	Leader of 10–20 Crewmen
Corporal	Leader of 2–5 Crewmen
Crewman	Worker
Recruit	New Member

Figure 27: Ranks and Responsibilities

Ranks

As a necessary administrative point, we turn to the rank structure of the Fleet. First, there will be no "officers" in the Fleet[13]. Unlike most armies, there will be no distinction between an "officer" and an "enlisted" soldier. Rather, almost everyone will begin as a recruit and work their way up. Even highly educated people will undergo a basic Fleet experience at the lower ranks, though there will be special programs for rapid promotion for those with distinctive talents. Second, Figure 27 allows the reader an understanding of ranks for discussions that occur in this and other volumes.

Commander Tanaka, Pilot

We now introduce Commander Tommy Tanaka, a man hoping to become a pilot, to illustrate how the Fleet will work. We will also contrast his experience in

---

13   Most militaries in the world have a "dual" chain of command, one running through the officer corps and one up a separate "enlisted" chain. While time tested and traditional, the Fleet will adopt the more efficient and businesslike system of having a single chain of command.

the Confederation Fleet with the experiences of soldiers in American active and reserve units.

Tanaka's parents brought him to the earth Embassy when he was sixteen. At eighteen, he decided to remain at the earth Embassy. Moreover, he also decided to serve a few years full-time with the Fleet, not simply to serve as a part-timer. So, Tanaka joined the "elite" 1st Marine Regiment, where he served with distinction in Company A, the most-disciplined and hardest-trained unit in the entire Fleet. About 90% of Company A was in full-time Fleet service, unlike most of Company B and the other line companies in the regiment, which were mostly composed of one-month-per-year crewmen. Tanaka never attended a lengthy basic training course. Rather, each crewman underwent three or four days of "regimentation" training during each training cycle. The initial phase of regimentation taught basic military skills, but Tanaka received most of his early training from his squad and platoon leaders and otherwise by osmosis during his time in his unit. While ten or fifteen days each month were devoted to infantry-style military training[14], the remainder of the time was spent on work details, helping to construct new roads, unload trucks, etc.

A reservist in the US Army structure would spend their first 13 to 52 weeks on active duty, receiving basic military training and then training in their specific military skill, before they started their two-day per month routine. American soldiers in active duty units average five to fifteen days training per month. During most of Tanaka's three-year active duty tour, his military skills would have been comparable to most low-ranking infantrymen in active-duty western armies. Naturally, these skills constituted only a portion of the training he received.

In his last year on active duty, Tanaka was promoted to corporal and made a section leader, in charge of five other marines. During this last year, Tanaka also identified a billet for himself in the 1st Space Operations Regiment. This was a typical path to becoming a pilot and Tanaka wanted very much to become a Confederation "astronaut". Fortunately for him, he could keep his corporal rank in his new Fleet position. So, Tanaka became a one-month-per-year refueling specialist in the Refueling Platoon, Company B, 1st Space Operations Regiment.

After his three-year tour with the marines, Tanaka enrolled at Ares University to study engineering. There, Tanaka spent six years studying full-time for a bachelors and then a masters degree. While in school, he was called to serve six one-month periods of active duty with the refueling platoon. During each one-month call-up, Tanaka helped teach military skills to the other reservists. Former members of the

---

[14]   This "combat" training was almost unique to Company A, 1st Marines. Other marines worked on some military skills, but mostly served on work details and assisted with other, non-combat Confederation missions.

1st Marines were often asked to use their more combat-oriented experiences to assist with map-reading courses, survival skills, etc. In addition, Tanaka trained to conduct refueling operations with mock-up spacecraft, helped to design and build some of the refueling stations and infrastructure for the regiment, and actually fueled one spacecraft for a launch. Unfortunately, that mission was scrubbed when a technical problem developed, whereupon Tanaka's platoon had to unload the fuel from the rocket. Tanaka thought he would get promoted to sergeant in the platoon, but he caught some bad luck there and remained a corporal during his entire six years with the refueling platoon.

An American reservist would spend two days every month drilling with his unit and then spend two weeks every year training at a military base. Tanaka spent 28[15], not 38 days training each year, but his 28 consecutive days were probably more productive than the 38 days American reservists serve, even though his skills deteriorated significantly during his 11-month layover each year. The other difference between an American reservist and Tanaka was that Tanaka derived a good measure of satisfaction from his role with the refueling platoon, especially helping to design and build the refueling facilities. No corporal in the American army has much say in anything his unit does or how it operates.

After school, Tanaka's father wanted him to join the family's new venture[16], making lamps. Tommy knew the value of this venture. Mundane as it was, until the elder Tanaka's business opened, the earth Embassy needed to buy lamps from the outside. Moreover, the skills developed at this business to roll and finish aluminum and other metals was so valuable that Tanaka was often asked to take on special projects beyond making lamps. Tommy decided not to join his father's business. Rather, he accepted a position with a tiny engineering firm that was designing a natural gas pipeline for the earth Embassy. This project seemed to be perfect for Tommy. Not only could he directly employ his engineering skills, but his Fleet work related to it, as the natural gas pipeline would be an important part of the Confederation's capability to make its own rocket fuel. In addition, this work was also partly related to Tommy Tanaka's goal to become a Confederation pilot.

Ironically, Tanaka transferred out of the 1st Space Operations Regiment just two months before his platoon conducted fueling operations for a Confederation SSTO that launched into orbit. A sergeant's slot had opened up in the 1st Space Regiment. It was not a pilot's billet, but it was a job working with pilots. Just as

---

[15]   All members of the Fleet earn three days leave during their annual 31-day tour.

[16]   Though so many at the earth Embassy were in business for themselves, it is difficult to call any of these operations a "family business", since few were more than a few years old.

important, with manned SSTOs beginning to launch regularly, there was a good chance that he would get to fly into space.

Tanaka never got his ride into space. He served for three more years as a crew chief and wrote a manual for crew chief training, but was accepted into the pilot training program before he got his chance to go into space as a crew chief. The three one-month tours he had spent as a crew chief were nonetheless exciting. The SSTOs had begun to fly routinely and he had a full and meaningful schedule during each one-month tour. Writing the Crew Chief Training Manual, which Tanaka viewed as a wonderful challenge and a great opportunity to affect the Confederation's future, was in some respects the least interesting part of these three years. A sergeant in an American reserve unit would never have had the opportunities for service or for excitement that Tanaka experienced. At least as important was the new, personal ethic that Tanaka developed during this time. He had become a Fleet "lifer", meaning that he spent a significant amount of time between Fleet tours taking on-line correspondence courses and volunteering without pay to help out during important regimental operations. Few American reserve sergeants spend anywhere near as much free time on military matters as Tanaka spent on space regiment matters.

Pilot training involved a one-year period of active-duty and three years of two-days-a-month training on top of the one-month annual call-up. Pilots-in-training were also expected to train on their own time by taking on-line correspondence courses. Much of the time was spent training in a simulator, where Tanaka amassed several hundred hours. Once each year, Tanaka flew into space as the "assistant co-pilot" of an SSTO. After the one-year active duty, Tanaka was promoted from Ensign to Lieutenant. Upon graduation from pilot training, Tanaka was promoted to Lieutenant Commander.

Tanaka's pilot training was not at all like pilot training in the American military. The Americans have money, so American pilots have the leisure to concentrate on the complex role they have been assigned and to amass hundreds of flying hours. The mission to Mars has little money. Tanaka was not learning to become a jet pilot. Rather, he was learning to become part of a team that operated the simple spacecraft being designed and built by the Confederation. The SSTOs were not "do-anything", "any weather" aircraft. They were meant to fly only one limited type of mission. The spacecraft were fully computerized and fully automated. Tanaka and the other pilots trained to deal with any emergency that might happen during the SSTO mission. It is difficult to compare Tanaka's experience with pilots of American military aircraft. Tanaka, as a new pilot with limited experience, had almost five years of active military service on his record, but most of this time was not service as a pilot. Still, Tanaka had more launches into space than almost any other human outside of the Confederation. In addition, Tanaka was

34 years old. He was not only a mature individual; he was also a person who had been living close to space operations for almost two decades. In terms of hours flown, Tanaka was inexperienced. In terms of total exposure to his role, however, Tanaka was superbly qualified to help lead humanity into its new era in space.

As a pilot, Tanaka was called to duty for four-ten day periods each year. During each period, Tanaka flew two or three SSTO missions into space. Five days re-training, and then back into space. SSTOs were launching several times a day in this period, as space launches were becoming routine ways to travel the globe.[17]

The mission to Mars was close to launching its first manned mission to Mars. More pilots were needed. Already 50 had been trained and plans for 150 more pilots were moving forward. Once Tanaka had 50 missions into space, he was asked to make a decision: join a new unit with the 1st Space Battalion and get an early slot for Mars or become the commander of the 2nd Space Battalion, the unit that flew the daily SSTO missions. Tanaka decided to go to Mars. On the civilian side, he had finally joined his father's business, which had grown into a complex metals fabrication operation. The Tanakas still made lamps but, in addition, they made metal parts for just about everything at the earth Embassy. Tommy Tanaka was acquainted with most of his father's operations. The ability to operate equipment, to make repairs, and to improvise when something broke was an important aspect in a Mars pilot's job. His experience in his father's business was invaluable to these ends. Because of his military and civilian experience, Tanaka was chosen to be the co-pilot on Explorer Transporter #5, in the third year of mankind's era on Mars. He would assist piloting the transporter ship to Mars, spend two years on Mars using his civilian skills to assist the base commander, and then co-pilot the transporter vessel back to earth.

Tanaka began training for Mars on a two-days-a-month and one-month-per-year basis. Just as important, Tommy Tanaka the civilian received a contract to select equipment to take to Mars and to train himself to operate and supply this equipment. Thus, Tanaka concentrated on his Mars voyage full-time. Even when he wanted to help his father with more mundane chores in the workshop, the elder Tanaka chased him out and had him either relax or—he knew his son well—get back to the Mars mission work.

Tanaka's mission to Mars brought two more settlers to Mars, lots of metal working equipment and supplies, and thereby established a real capability to create metal products at the new base. He began eagerly to anticipate the not-so-distant day when the elder Tanaka would move himself and his business to Mars.

---

17 This idea will be taken up again in the chapter entitled "Single-Stage-To-Orbit" in Volume III, where the possibility of creating a commercially profitable "lift-into-space" program will be discussed.

Tanaka spent three years on active duty during his mission to Mars. Six months out, two years working at the Mars base, and six months back. Nonetheless, Confederation fiscal policy required that Tanaka pay the Confederation CS $2 million for his ticket back and forth from Mars[18]. The idea was that Tanaka, being only the tenth human back from Mars, would now go out into the world to speak to large, eager audiences, speaking engagements that typically paid him US $50–100,000.

Tanaka had had dual American-Japanese citizenship before accepting Confederation citizenship at age 18. He had been born in the US to a father who had come to the US to teach geology at a small university. The family spoke Japanese at home while he was growing up and it was only with great difficulty that the two Tanakas used English during business hours when talking to each other. So, Tanaka's speaking and television tour took him both to the US and to Japan. It was in Japan, however, where there was the greatest excitement. He had become a national hero. He was everywhere on Japanese television for a month after his return. His speaking fees there were presidential.

Tommy Tanaka received over US $10 million from his speaking and television appearances, and from the book he co-authored for the Japanese public. The money was used to pay off the CS $2 million cost of his ticket and to help the family business gear up for its transplantation to Mars[19].

After his role as public hero ended, Tanaka decided to give up his Fleet billet as a pilot. There were more than enough qualified pilots to take his place and Tanaka wanted to spend all his available time preparing his father and the "family" business to be transplanted. No one retires from the Fleet, however, as service obligations remain for one's entire life, so Tanaka transferred to the Academy Regiment where, for one month every year, he taught would-be pilots about spacecraft operations. During the other 11 months, Tanaka devoted all his energy to the business and to its plans for the move to Mars. He could hardly wait. The transfer of his business would be a milestone event on Mars. Though some metal parts were made there now, moving just a dozen key employees to Mars—one of them would be Tommy—would allow for most anything metal to be fabricated on the Red Planet. It was almost more exciting than an SSTO trip into orbit. Just

---

[18]   Tanaka paid a premium for his ticket to Mars specifically because the Confederation knew how much its first pilots stood to earn once they returned to earth.

[19]   Tanaka's net worth when he died was approximately US $85 million. Some of this money came from his books and speaking fees, some came from his share of the family business, but about half came from the singularly attractive investment opportunities Tanaka—like others at the earth Embassy—had in businesses and opportunities at the earth Embassy and on Mars.

fifteen years after first landing on Mars, the Tanaka business would make the Red Planet virtually self-sufficient in small, metal, parts manufacturing.

LT CMDR Thomas Zubrin

The second example of a "Fleet career" will be that of a Fleet "part-timer", to discuss how a person's technical background might be used by the Fleet in its role as engineering general contractor.

Here, we recount the role of Lieutenant Commander Thomas Zubrin, a grandson of Robert Zubrin, who joined the mission to Mars when he was 35 years old. He held two masters degrees: one in electrical engineering and a masters in business administration (MBA). Although he never chose a "business" path for his career and had remained in mostly technical positions, he considered himself to have more than average business acumen. In fact, one reason he was attracted to the mission to Mars was for the opportunity to be involved both in technical aspects of getting to Mars *and* the opportunity to be an entrepreneur. Zubrin's plan was to build solar panels for use in space and on Mars; he was fascinated by the idea of using native Martian materials to this end.

When he first heard about them, Zubrin had been circumspect about the mission to Mars' ideas, programs, and prospects. Still, Zubrin became involved with it over the course of ten years, visiting the earth Embassy twice and contributing a few hundred hours on three projects he worked "on line". Even then, he only decided it was a worthwhile risk after he began negotiating for a large contract to underwrite his new solar panel business at the earth Embassy.

Thus, Zubrin joined the mission to Mars not as a wide-eyed eighteen-year old, but as an accomplished professional who joined only after securing a three-year, CS $1.2 million contract for his solar panel venture. Like everyone else, however, Zubrin was required to spend his first 31 days at the mission to Mars serving a tour with the Fleet. In his case, Zubrin was assigned to O Company in the 1st Engineer Regiment, coordinating the design and construction of Ark Transporters.

Zubrin entered the Fleet as a recruit. Under a special program for accomplished individuals, however, Zubrin would receive one promotion each year, up to the rank of Lieutenant Commander. Nonetheless, it would require six tours of service[20] before he received this rank. During the first five years, Zubrin worked as a member of the engineer company, serving one tour as a squad leader (sergeant) and one year as a platoon leader (ensign). He contributed significantly to the Ark

---

[20]   Zubrin had 228 days of Fleet service before attaining his "prescribed" rank. His six annual tours of duty totaled 186 days plus, as will be shortly described, he served two additional tours where he accrued 42 more days of service.

Transporter project, even as a relatively low-ranking individual, since the emphasis on rank was downplayed in his unit and since his abilities quickly became obvious. In fact, Zubrin was given so much responsibility during his five years working in O Company that he often groused about how little he was paid for the services he contributed[21]. He was especially laconic about the two extra three-week tours he was ordered to serve during important phases of the Ark design.

In his sixth year, Zubrin was promoted to Lieutenant, equivalent to the rank of the engineer company's commander. Although Zubrin was offered the job as the unit commander, part of the deal would be that he must retain this rank and serve as company commander for at least three years. The opportunity to become the Fleet's chief designer and construction liaison for the Ark Transporters tempted him greatly. Nonetheless, he decided against accepting this command, both because he wanted to be promoted and because he did not want to be burdened with the administrative duties of a commander. Thus, in his sixth tour of duty with the Fleet, Zubrin left O Company to become a Project Officer in C Company to develop and acquire certain equipment for Fleet marine units.

Zubrin knew very little about the needs of the marines when he first started working in this role, but he was a quick study. Over the course of five years, one as a Lieutenant and four as a Lieutenant Commander, Zubrin contributed to the design and acquisition of ten new systems and pieces of equipment for the marines. Included were simple items such as individual mess equipment for use in a variety of environments, but there was also more complex equipment like portable, temporary shelters for the marines and simplified drill rigs, allowing marines to drill for water on Mars at temporary camps. As with his work on the Ark, Zubrin was both excited by his ability to make contributions to the complex goals of the earth Embassy and the power to affect these ends directly, with a minimum (as is possible in complex undertakings) of bureaucratic nonsense and outside meddling.

In his middle-forties, and with the Ark Transporters starting to run twice a year to Mars, Zubrin was offered a promotion to Commander and an early trip to Mars. His job on Mars would be to supervise the Fleet's preparation of new Ark bases for the arriving settlers. He eagerly accepted, so for three years Zubrin supervised the construction of six new bases on Mars, creating sufficient infrastructure at each base to allow it to be handed off to the 1000 arriving Ark settlers.

---

21   One of the purposes of creating "the Fleet" is to obtain the services of qualified individuals at "below market" cost, lowering the overall costs of the mission to Mars program. The two reasons this system will work is because people will *appreciate* the opportunity to have an annual "change of pace" to work at their "Fleet jobs" and because people will make modest sacrifices to contribute to the lofty goal of reaching Mars.

Zubrin was on duty full-time in this role and he needed to delegate day-to-day supervision of his now highly successful solar panel construction company. But Zubrin wanted the Fleet promotion and he believed that his Vice-President for Operations at Zubrin Solar Panels was more than qualified to handle day-to-day business for three years[22]. So, he accepted his promotion and was off for Mars.

Zubrin worked on Mars for three years and extended for a fourth year before leaving full-time Fleet service to return to his solar panels business. By then, Zubrin Solar Panels had three different manufacturing operations on Mars and was supplying more than half of all solar panels used on Mars and in space. Back at work in his business, Zubrin started to pursue a huge terraforming contract, whereby a small "space mirror" was to be built to help terraform Mars. Far smaller than the 100,000 square kilometer monstrosities envisioned by science fiction writers, this space mirror would only be as large as tiny Confederation budgets would allow. Still, the materials for the space mirror were to come from the asteroids and Zubrin believed that his solar panel technology could be adapted to build a very cheap space mirror. It was the kind of challenge Zubrin had come to love at the mission to Mars. And although the pressures of his civilian work forced Zubrin to take a Fleet demotion back to the rank of Lieutenant Commander, he did not mind[23]. His new Fleet role put him into the new Military Academy getting started on Mars, where he held various research and teaching positions for another twenty years.

---

[22]   This vice-president had operated the business as its chief before, one month each year, during Zubrin's tours of duty with the Fleet.

[23]   Such demotions were common in the Fleet. A later chapter discusses the mechanics of rapid promotion and subsequent demotion to retain a vibrant leadership for the Fleet.

The Earth Embassy

[Ethan Allen burst into the commander's quarters and demanded, from the still reposing officer, the fort's immediate surrender.] Captain Delaplace gazed at Allen in bewildered astonishment. "By whose authority do you act?" exclaimed he. "In the name of the great Jehovah and the Continental Congress!" replied Allen.

—Washington Irving in *Life of Washington*

The third leg of the mission to Mars has already been introduced: a community striving to become economically self-sufficient, while planning its move to Mars.

We start by taking a step back from the assumed existence of an "earth Embassy", in order to see how such a community might come into existence. Indeed, the reader is asked to consider that the greatest difficulty will not be any particular step forward, as any one can be small indeed, but rather how to get the project started. This next section discusses the prelude to the earth Embassy, places where the mission to Mars' message can be brought to large numbers of people, and where the beginnings of a self-sufficient economy might be staked.

Farms and Visitors Centers

As a precursor to the earth Embassy, the mission to Mars will build a series of recruiting and information stations around the world. These recruiting and information stations will have two basic forms: the farm and the Visitors Center. The difference from one to the other, however, will not be so much substantive as it will be point of emphasis.

Figure 28 lists the various farms and Visitors Centers that might be built as feeder organizations for the earth Embassy. The worldwide reach of these kinds of facilities will provide every continent and many cultures a place where people can learn hands-on about the mission to Mars. In addition, as the farms and the Visitors Centers will, in effect, be mini-earth Embassies, they will serve as potential sites for the ultimate earth Embassy.

These relatively small Visitors Centers will not be difficult to establish. The mission to Mars assumes that the requirement for each Visitor Center to be $100,000 (in hard currency), 300 donated or low cost acres, a few sources of donated supplies from retailers (who would receive acknowledgement and advertising at the Visitors Center), and the commitment of ten "person-years" of labor. With materials to begin building according to a centrally-prepared plan, it will not be difficult over the course of a year to build a small but expandable museum, a fast food establishment for visitors, a gift shop, create agricultural operations to become partly self-sufficient in food, and living quarters for visitors and staff.

After a year, the newborn Visitors Center could be opened to the public, to be operated as a non-profit organization whose goal would be to return a "profit" of hard currency for the next Visitors Center. The public would be attracted by a stylish "museum" of displays, activities simulating Martian and space operations, video presentations of mission to Mars activities, the ability to buy things in a gift shop, and to eat in a franchised fast food restaurant. Year-to-year, each operation would expand, increasing the number of live-in residents who "worked" there. Also increasing would be food production, the size of the farm and research facilities open to the public, and the availability of housing for overnight or other temporary residents. With these increases, an increasing number of visitors can be expected to visit the center who, in turn, might consider spending additional time to explore more fully the mission to Mars' program. With proper management, these facilities would not only grow of their own accord, but would set the table to build the next Visitors Center.

Country	Location (Tentative)	Type	Emphasis
USA	New Mexico	Visitors Center	Near Tentative Earth Embassy
USA	North Carolina	Visitors Center	Regional center
USA	Illinois	Farm	Corn, hogs, cattle
USA	N. California	Visitors Center	Regional center
Canada	Toronto	Visitors Center	Regional center

Country	Location (Tentative)	Type	Emphasis
Canada	Alberta	Farm	Wheat
Mexico	Near Tourist center such as Acapulco	Visitors Center	Link to Spanish-speaking New World and to tourists
Brazil	Rio de Janeiro	Visitors Center	Possible host nation
France	South of France	Visitors Center	Regional center
Germany	Anywhere	Visitors Center	German technological prowess
UK	Near London	Visitors Center	British and Irish legacy of colonizing a world
Hungary	Near Budapest	Visitors Center	Outreach to former eastern Bloc
South Africa	Near Johannesburg	Farm	Southern hemisphere and outreach to Africa
Iraq	?	Visitors Center	Cheap costs and outreach to Arab world
India	Near urban center	Visitors Center	Regional center
China	Peking	Visitors Center	Regional center
Japan	Mid-sized city	Visitors Center	Regional center
Russia	Moscow	Visitors Center	Regional center
Ukraine	Kiev	Farm	Wheat
Australia	Sydney	Visitors Center	Possible Earth Embassy Site

Figure 28: Visitors Centers

The key to the Visitors Center idea is the same as the key to the earth Embassy: to grow an economy, offering jobs and a mission, not to create a roundtable committee to argue about the design of spacecraft to reach Mars. Even though the Confederation will not *be* a business—it will be a "government"—it will be *run* like a business. After the first few difficult years, more and more people will decide to work at the Visitors Center because they will find it is in their personal economic or professional interest to do so. Those who come because of a personal interest in Mars and space may, in fact, never be the majority of those who choose to participate at a Visitors Center. That's OK, however, since every society needs workers and, besides, the point is to create a society, not a debating club about Mars. Those who are at the earth Embassy purely out of self-interest can make their own decisions about the ticket contracts, payments on those contracts, etc.

Figure 29 shows the many categories into which people working at the Visitors Center would fall. Some of the people employed at the operation would be simple employees, disinterested in the mission to Mars, working in the Subway sandwich franchise, farm laborers for the adjoining fields, or as contractors performing an otherwise unfilled role. Some would be college students or "part-timers" who had decided to devote a semester as an intern with the mission to Mars[1]. These people may have research projects they work on year-round, come to the Visitors Center to demonstrate their work, or to complete aspects of their research. Other "part-timers" might be carpenters or plumbers who spend a week's "vacation" with their family living at the Visitors Center. While they plumb a new building, their children participate in a program learning about (and testing) a new shelter design for Mars, their spouses go shopping at nearby malls or attend a sci-fi convention. Visitors and others will use the Visitors Center as a hotel, where they pay hard currency or trade their labor for a place to stay. If the Visitors Center is in New Mexico, perhaps they will arrange to stay at the Visitors Center for a week, send their kids to the programs there and take a day trip to Santa Fe to check out the art and architecture. "Ticket holders" would begin working off their ticket to Mars by working full-time at the Visitors Center, living there with their families. As they would be trading their labor for their tickets to Mars, these people would not be paid much "hard currency", but would live, eat, and send their children to the Visitors Center school, as part of their contract for Mars.

---

[1]   Gaining credibility in the academic community will allow Visitors Centers to make permanent liaisons with colleges and universities. Thus, internships at a Visitors Center could become part of the normal curriculum of these schools.

Category	Notes	Activities Assigned
"Employees"	Used to fill roles otherwise unfilled	Hourly-employees, specialized contractors
"Ticket holders"	Persons working for the mission to Mars, either until they reach Mars or until they decide to leave the program	Managers or any appropriate job
College Students	A fascinating place to earn money and to spend a summer or a semester earning college credits	Hourly workers, researchers, program leaders
"Part Timers"	People "testing the waters" to see if the mission to Mars makes any sense for them	Laborers, researchers, program leaders
Visitors	Tourists, people wanting to know what the mission to Mars is all about	Laborers

Figure 29: Categories of Those Participating in the Mission to Mars

Two final comments about Visitors Centers. First, it should not be too difficult to convince some economically-strapped (or progressive!) community to donate 300 acres of land for a Visitors Center, especially in the western parts of the US. The potential pay-off for the community would be enormous, at the cost of transferring otherwise vacant scrubland. Second, the reader must understand that to build a human society on Mars requires human realities to become part of the social structure. If the Visitor's Center is run as a Saturday flea market, it will never work. On the other hand, if it is managed by a highly dedicated and professional staff, is able to partner with universities and work with their college students and professors; if the Visitors Center presents an attractive and modern appearance, with DOT-highway signs announcing the way; and if the displays and activities at the Visitors Center are first-rate and not merely cardboard and narratives with misspellings, then the Visitor Center will not only achieve its immediate goals, but will also serve as a powerful tool to attract people and supporters to the mission to Mars.

Taking Advantage of Opportunities: The Iraqi Visitors Center

As this book is being written, there are large tracts of un- or under-utilized land in Iraq, a nation with a reasonably well-educated population and with the potential to build an oil-based, Arab-Moslem democracy. There are also a great many people under-employed in Iraq and easy money might be available from the US government for an organization that said it was going to build a "Visitors Center". Not only might such a project boost the local economy, but it might be seen as a badge of Iraqi progress, either of which might prompt financial support.

The problem, of course, is that the American project in Iraq is still a long way from resolution, one way or another. On the bright side much, if not most, of Iraq seems to be relatively quiet and stable. Of course, there are savage terrorist attacks occurring daily in the unstable part of the country.

Would it make sense to build a Visitors Center in Iraq? In theory, "Yes". Is it practical? Only time will tell. The point of this section is to illustrate how energetic approaches to the problem can help to get Visitors Centers built. Indeed, in all aspects of the mission to Mars, we intend to use unusual circumstances to further larger objectives.

Outreach to the Arab/Moslem world seems necessary if one intends to build two dozen Visitor Centers around the world. Under a decision to reach out to the Arab world, access to money from the US government, combined with the relative stability of some parts of the country suggest Iraq as a possible location, especially if there exist at the time of the decision other positive attributes for an Iraqi site.

Beginning The Earth Embassy

While it may seem obvious that the earth Embassy would be an extension of one of the Visitor Centers, in fact this may not be feasible. The earth Embassy will need to be in an isolated area, most likely surrounded by several hundred square miles of unsettled or mostly-unsettled land. This contrasts with the Visitor Centers, which will mostly be near centers of population or with easy access to transportation.

Still, there are several models that might be pursued for the earth Embassy and certainly one of them will be to extend an existing earth Embassy. Figure 30 lists the most likely scenarios for the creation of the earth Embassy, in order of likelihood, where the US would serve as a host nation. While many of the contingencies remain true if Mexico, Canada, Australia, etc. were chosen as a host nation, the final possibility—that a mostly undeveloped nation becomes the mission to Mars' host nation—will provide both maximum flexibility for the mission to Mars and the maximum risk that its operations would be disrupted by external events (riots, wars, etc.).

	Biggest Advantages	Biggest Disadvantages
Ranch(es) Deeded To Mission to Mars	Relatively Easy to Obtain Ranch With 50,000 acres Large ranches will be found in very isolated areas	Ranch will need to be taken "as is", without regard to location Might be too isolated
Isolated Town or County As Sponsor	Easy to find an underdeveloped political entity in US willing to become the world's "space port" Much infrastructure already in place	Some residents will oppose the development Concerns of non-MTM residents cannot be ignored
Extension of a Visitors Center	Infrastructure already in place Presumed good working relationship with local and national governments	Most will be near urban populations, making launch sites problematic Difficult to convert 300-acre site to 50,000-acre site
American Indian Reservation	Problem of state and federal controls greatly reduced Bring industry to poor area	Tribe could change its mind "White man" taking advantage
Island	Easy to acquire rights to a site Isolated from populations	Difficult to supply? More difficult to recruit people to live on an isolated island

Figure 30: Where Might The Earth Embassy Be Located?

Demographics

This section presents a snapshot of an earth Embassy with 5,000 residents. We discuss here some details about the people living at the earth Embassy by presenting a detailed review of community occupations. First, though, let us review the demographics of the earth Embassy. While the earth Embassy is a highly unique community, united by its desire to get to Mars, it nonetheless will be a *human* community in the western world. As such, it will have a fairly standard demographic profile for ages and distribution of the sexes. Figures 31 and 32 present a general picture of this community:

"Class" of People	Number	Percentage
Workers	3198	64.0
Spouses	236	4.7
Retired	164	3.3
Children, Students	1402	28.0

Figure 31: Breakout of a 5000-Resident Earth Embassy

As a community attempting to create a self-sustaining economy, a wide variety of skills and occupations will be needed at the earth Embassy. Thus, there will be room for a great many types of people, not just engineers and scientists, but bakers and candlestick makers, too. Figure 32 provides some information about the kinds of industries in which employees will work.

Industry	Sub-Class	Number	Industry Total
Industrial			700
	Computers	50	
	Spacecraft	50	
	Civil Engineering	50	
	Fuels	50	
	Materials	50	
	Other Engineering	150	
	Other Manufacturing	200	
	All construction	100	

Industry	Sub-Class	Number	Industry Total
Commercial			455
	General Merchandising	200	
	Consulting	100	
	Clothes	20	
	Private Transportation	20	
	Banking	10	
	Insurance	5	
	Financial Services	10	
	Furniture	20	
	Electronics	20	
	Lawyers	10	
	Accounting	10	
	Bars, Pubs	30	
Infrastructure			575
	Electric	50	
	Water	50	
	Wastewater	50	
	Recycling	100	
	Telephone	50	
	Internet	50	
	Plumbers	75	
	Electricians	75	
	Bus Service	75	

Industry	Sub-Class	Number	Industry Total
Food			1050
	Farmers	400	
	Food Processors	150	
	Restaurant	400	
	Food Stores	100	
Confederation			205
	Fleet	150	
	Government	20	
	Local Government	5	
	Mars Bank	30	
Education			1050
	Students	1000	
	Administrators	10	
	Teachers, Professors	40	
Religious			38
	Christian	20	
	Jewish	4	
	Islamic	8	
	Hindu	4	
	Others	2	

Industry	Sub-Class	Number	Industry Total
Miscellaneous			527
	The Very Young	402	
	Spouses, Retired	400	
	Private Medical	125	

Figure 32: Economic Profile of the Earth Embassy

The move to Mars will require that each of these businesses, schools, and government officials plan such a move. In the case of a school, once a certain number of children are on Mars, a school can be "moved" to Mars, perhaps leaving an annex at the earth Embassy. As the number of people on Mars increases and the earth Embassy begins to shut down, the school will close its annex and operate only on Mars. Similar ideas about transplantation must be developed for all jobs and all companies.

Having an operation on earth planning its move will greatly simplify the establishment of that operation on Mars. Imagine, for example, a circumstance where one hundred people on Mars have been eating in a mess hall or in a family restaurant and that a fast food alternative—a snack bar—is desired on Mars. The BGBS solution would be to open a NASA snack bar on Mars and hope it can make money. If NASA became creative (or were forced by Congress) they might contract with an earth pizzeria to open a branch on Mars. The pizzeria program will not only be expensive, but complex. Imagine the conversation, "Mr. Signiori, you run three pizzerias in Denver. We want you to open one on Mars. How much will that cost us? Signiori thinks a moment, realizes that he can now buy that small skyscraper in downtown Denver he had always admired, and tells NASA that he thinks he can get one open for $45 million, as long as there are contingencies for cost overruns. To move a Domino's franchise from the earth Embassy to Mars, however, will not only be much simpler, there will be no expense to the Confederation. Even better, the franchise owner will be desirous of expanding to Mars and for a number of years her entire business would have been premised upon the idea that she would one day move to Mars. Neither the BGBS method, nor moving an earth Embassy business to Mars will be easy, but the person who has operated within the Confederation's rules and community will have a much easier time and require far less Confederation attention than the pizzeria that must be created out of thin, Martian air.

Figure 32 should also suggest two other matters. First, is that once a community like this is transplanted to Mars, it would make for a real colony there, not a science station. Second, is that we have described a fairly self-sufficient community. Given that resources are available on Mars to provide for raw materials and also that each of these industries will have been planning their move to Mars for many years, a self-sufficient community on Mars can be established quickly, based upon a self-sufficient earth Embassy counterpart.

So What's It Look Like?

Figure 33 shows how the earth Embassy might be laid out. The earth Embassy will be a planned community, dividing residential areas from commercial and industrial ones and, as will be discussed in more detail in later chapters, will conform to a general plan that will be used on Mars.

The main areas of the earth Embassy region include the earth Embassy proper, its spaceport, its free trade zone, and "Marsbridge, New Mexico". The earth Embassy has been discussed in general terms; in more specific terms, it will be laid out to be test the concept of a "Martian village". This area will mostly be "off limits" to non-mission to Mars personnel. Outsiders wishing to interact with the mission to Mars will usually do so in the free trade zone.

The reason that the earth Embassy itself is "off limits" is simply a matter of ease of control and protection of sensitive mission to Mars businesses. In essence, the earth Embassy is declaring itself a separate political entity from its host nation (US, for example). As foreign embassies in Washington, DC are not usually open to American citizens, so the earth Embassy will generally be closed to the public.

The spaceport is self-explanatory. The main feature of this spaceport is an intervening terrain feature (mountain, ridgeline, etc.) that will isolate the earth Embassy from most of the spaceport's noise. In addition, the spaceport will need to be located as far from other human activities as possible, especially from non-mission to Mars activities.

As indicated above, the free trade zone will be an area where there will be interactions between residents and non-residents of the earth Embassy. A classic example of an activity in this free trade zone will be the universities, where non-mission to Mars students can study side-by-side with residents of the earth Embassy. In addition, this area will host restaurants, hotels, and other mission to Mars businesses that cater to non-mission to Mars residents. There will also be some industry here, especially those that choose to hire both mission to Mars residents and non-residents.

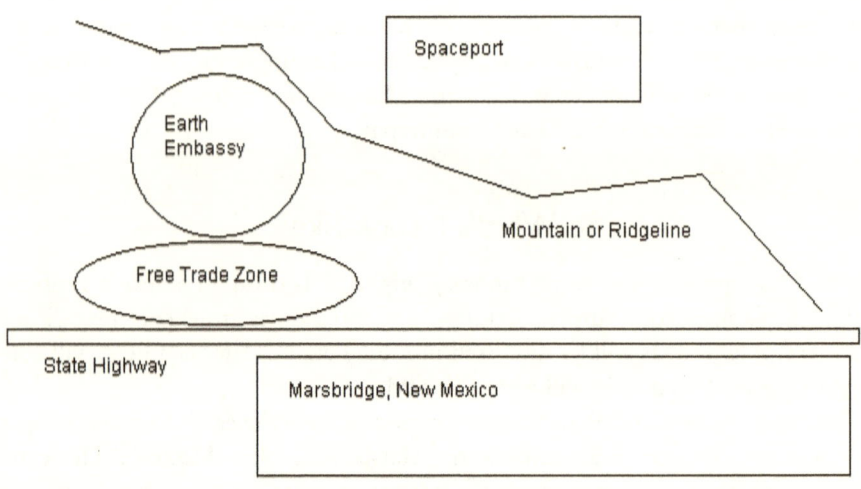

Figure 33: Layout of The Earth Embassy

Finally, Marsbridge, New Mexico is the name we have given to the new American (if the US is the host nation) town that will grow up around the earth Embassy. As there will be markets for earth Embassy businesses and its residents, a small town with a reasonable assortment of shops and services will cater to the earth Embassy. As Marsbridge might be larger than the earth Embassy, it will round out the number of economic offerings in the area. Thus, while there might be one McDonald's in the free trade zone with mostly earth Embassy employees, there might also be a Hardee's and a nice steakhouse in Marsbridge, New Mexico, whose employees will be residents of Marsbridge, not the earth Embassy. Anybody—earth Embassy resident or not—can choose to patronize any of these restaurant establishments. The one difference will be that the earth Embassy residents can pay for their meals at the McDonald's using Confederation Script.

Average Joe

The finances for an "average family" should be understood in terms of allowing everyone a chance to get to Mars. Some families will have only one breadwinner. Many people will not be married. Some people will be retired. Other families will have children and some will have none. The purpose of this section is to introduce the idea that even "normal" families can afford to participate in the mission to Mars.

Figure 34 lists characteristics of three different families at the earth Embassy. While each has two working adults and two children, we see a wide variance in the amount actually paid into the ticket program, the number of years before tickets are paid off, and that some spouses will pay off their tickets before others.

	Example 1	Example 2	Example 3
Paid In	CS $750K	CS $570K	CS $390K
"Returns" Paid on Ticket Account	CS $250K	CS $430K	CS $610K
Job Description	Software Engineer	Carpenter	Janitor
Years To pay Off Ticket	12	19	26
Return	6%	6.5%	7%
Spouse Pays Off Ticket Before Example?	No, spouse takes an additional five years	Yes, spouse paid off ticket year before	No, spouse stayed at home with kids for four years, needs four more years
Total:	CS $ 1M	CS $ 1M	CS $ 1M

Figure 34: Average Family Finances

Although we use a "progressive" rate of return in this example, paying more to lower-earning residents, there is no plan to do this necessarily. In fact, the Ticket Corporations will want to drive as hard a bargain as they can with each ticket holder. Some will get better rates of return than others. The point to take from this section is that each person—rich or poor—will be able to afford the mission to Mars' program, even if some need to be more patient than others.

Each child will need to pay for their own ticket to Mars once they are 18 years old. Those under 18 may travel to Mars with their parents under a program of paying for the weight of each child and of the extra food taken aboard the transporter vessel for the flight to Mars. Thus, if a 16-year-old wants to accompany her parents, she will need to pay CS $100 (assumed cost to get into low earth orbit) per pound that she weighs, plus CS $100 for each pound of food (two pounds per day?) to be eaten during the trip to Mars. A reasonably sized teenager might cost parents an extra CS $25–40,000, which can either be added to the ticket contract (earning large returns at the end of the ticket contract) or paid for with fat dollars earned on Mars. There will be few cases when it will be a great financial burden to take children along with parents transporting to Mars.

Confederation Script

These last sections lead directly into the next chapter, the second one devoted strictly to economics. The subjects here—"money", a "central bank" and an agency to coordinate some of the earth's embassy's economic activities—also tie directly into fundamentals of the earth Embassy. We start with Confederation Script.

The discussion a few chapters back about "Funding Star Fleet" implied the creation of "currency" for the earth Embassy. As with the Star Fleet example, the earth Embassy will use two different kinds of money. For outside purchases, the mission to Mars will use "hard currency", existing forms of western money. These purchases will be for items that the mission to Mars cannot make for itself. As a medium of exchange within the community, however, the mission to Mars will use "Confederation Script". The labor of the residents and the products of businesses at the earth Embassy will be compensated for in a free market economy that uses this Confederation Script.

"Dollars" and "cents" will be the units of this new currency. The initial value of these units will equal that of the American dollar (US $) and cent. After creation, Confederation institutions will "regulate" Confederation Script and the use of hard currency[2]. Most likely, the value of a Confederation Dollar (CS $) will fall against the US $, even if one of the stated goals of the Confederation will be to maintain a one-to-one exchange rate. People at the earth Embassy will be required to use CS $, exchanging their US $ (and other hard currencies) for CS $. In fact, it will be "illegal" to hold hard currencies at the earth Embassy, except in small amounts[3]. Any hard currency payments received at the earth Embassy will be processed through the Mars Bank (see below) and converted into CS $. Thus, the Mars Bank will have "hard currency" resources to coordinate the outside purchases discussed above.

The use of Confederation Script will go a long way to help create the earth Embassy's self-sustaining economy. If managed properly, the Confederation will have enough hard currency (US $, British pounds, etc) to pay for the needs of the earth Embassy. There will be much more about these ideas in all three volumes for those readers who doubt the ability of such a system to be effective.

The use of Confederation Script will also allow for a program to repay accounts held by those who have worked at the earth Embassy, but decide to leave. Taking

---

[2]　Discussions about these institutions will occupy many later chapters.

[3]　It will not be illegal to hold hard currency accounts "outside" of the earth Embassy.

Confederation Script outside of the earth Embassy would be meaningless. Paying hard currency lump sums to earth Embassy "refugees", however, will exacerbate the mission to Mars' financial difficulties. In fact, if it is easy for the quasi-subsidized businesses at the earth Embassy to sell out, earning a quick hard currency payday after receiving the benefits of earth Embassy financial programs, the likely result would be an intolerable drain on the earth Embassy's short supply of hard currency. This drain on the earth Embassy's ability to make vital outside purchases might even be the end of the mission to Mars.

A similar problem exists with respect to workers who "save" large amounts of money paying for their ticket to Mars. If a worker decides to leave, he will receive value for the amounts paid into his ticket accounts at the rate of one-half of all money paid into such ticket accounts. Thus, Figure 35 and the mission to Mars' "forced savings" plan.

The solution for both of these problems is to have hard currency payouts over a number of years, rather than in lump sums. This, obviously, will not be to the liking of those receiving the money, but if it is a policy at the time they entered the earth Embassy, there can be little room for complaint.

The purpose of this short detour has been to show that even those who may not be certain about the mission to Mars can decide to participate in order to test their desire to get into space. For those who decide to leave, hard currency payments will allow such individuals to walk away from the earth Embassy with a reasonable nest egg for the next chapter in their lives.

	Example 1 (CS$K)	Example 2 (CS$K)	Example 3 (CS$K)
Paid In	5	20	150
"Returns" Paid on Ticket Account	0	4	25
"Ticket Account" Total	5	24	175
Years At Earth Embassy	0.25 Years	1 Years	3 Years
US $ Paid (over Time?) Upon Departure From Earth Embassy	US $2.5	US $10	US $75
"Forced Savings" Per Year	US $10 (year equivalent)	US $10	US $25

Figure 35: The Financial Calculus of Leaving the Earth Embassy

For both business owners and workers, Confederation regulations will dictate how and when people will receive a hard currency payment after they decide to leave or sell the business. The main idea is that "exploitation" of the advantages offered at the mission to Mars will not be easy. Now, the mission to Mars hopes that dozens and hundreds of people try to "exploit" those at the earth Embassy and, in so doing, help to build up its economy. In fact, a scenario is developed in the next chapter that shows how an entrepreneur with no interest in the mission to Mars can further the overall goal of reaching Mars, just from the collateral effects of creating a business at the earth Embassy.

An example of one repayment method can be seen in the case of Vinnie (whom you will also meet in the next chapter), who wants to purchase a home in New Jersey. Vinnie and his wife will receive US $100,000 as they process out of the earth Embassy. The Mars Bank will not provide such a large sum, except over time. What the Mars Bank *can* do, however, is to become a party to the purchase of their new home in New Jersey. They might provide the US $30–40,000 that is necessary for a down payment on the home and pledge part of the monthly mortgage payment to the bank. Thus, Vinnie and his wife are able to purchase their new home and have a low monthly payment as well.

All transactions at the earth Embassy will be conducted electronically or by check. Therefore, no actual currency or coinage need be created by the Confederation. This will be slightly inconvenient for small purchases, but not impossible, and checks written for less than one CS $ might be common. Likewise, an internal "credit card" system with earth Embassy banks should not be difficult, even if CS $0.43 transactions conducted over an internal earth Embassy bank card system are occasionally made.

The most important reason that bank cards and checks will be used instead of money is to avoid problems of counterfeiting. Even assuming a perfectly honest community, the hassle of creating paper money and coins should not be lost on the reader. Given the relative ease that a system of checks and bank cards can be implemented, a truly "moneyless" society seems the best idea.

The second reason that no special money will be created is that eventually the US dollar will be adopted on Mars as Martian currency. The reasons for this are complex and will not be addressed here other than to say that it aligns with a realistic notion about what the future human economy will be like. If eventually the US $ becomes the Martian currency, there is even less reason to create a new currency at the earth Embassy.

Finally, just a word about why US $ won't be used at the earth Embassy, even though it might be convenient to do so: there remains a problem of fraud and other financing. Using CS $ allows the Mars Community to limit its risks, which would be high if US $ were the currency. For example, in the next section we introduce the topic of a "stock market" at the earth Embassy and a system of equity investments for individuals and as a means of creating capital. An account with CS $100 million cannot be stolen and taken to the Jamaica, since Confederation Script will only be legal tender within the earth Embassy. Obviously, this problem would exist for an account holding US $100 million.

A Short Note About "Celestial Investments" Accounts

The sums discussed in the last section will be supplemented by sums saved in a second account, of which we give very brief mention here. Beyond the cost of paying for a ticket, 10% of an earth Embassy resident's salary must be deposited into an account we call a "Celestial Investments" account. This money is like the 401k plans in the US, sums saved out of an individual's own payroll, which are intended to be used for retirement. These accounts will be invested in securities established at the earth Embassy to fund the growth of businesses and in bonds issued by the Confederation. This subject, as you may suspect, is rather complex and requires several in-depth discussions, which will be presented in Volume III. For now, the reader should understand that a reasonable amount of money may be accumulated both for the good of the mission to Mars community (capital formation), as well as the good of the individual, as very large sums should be expected to accumulate in these Celestial Investments accounts over time.

The total sums therefore a ticket holder might be expected to be able to receive upon departure from the earth Embassy will be larger, in some cases two or three times as large, as indicated in the last section. Such large sums will usually only appear in those cases where the member has been a long-term resident of the earth Embassy or where there has been usually successful investing of the Celestial Investments account. For the purposes of this chapter, the sums indicated in Figure 35 should be considered low estimates of what an individual might be able to accumulate under the Confederation's "forced savings" programs.

The Mars Bank

A central bank will be created at the earth Embassy for a number of reasons. This bank will be charged with ensuring that the earth Embassy's economy runs smoothly, avoiding both the fears of inflation and the enervating effects of deflation[4] The Mars Bank will be armed with a number of regulatory tools—discussed in Volume III—to affect these ends.

Most relevant to this chapter, however, will be the retention and management of the hard currency needed by businesses at the earth Embassy. Again, the details of this program will be found in Volume III, but the essence of this program will be to allocate each registered business a hard currency account—an annual budget—based upon hard currency cash flows and needs. Pooling resources in this way prevents earth Embassy businesses from buying computers and potato chips and supports plans to buy computer chips and potatoes instead, since there is no reason some earth Embassy business cannot *build* computers and another *make* potato chips. And there is every reason to support the development of such businesses, as they will be needed on Mars. The Mars Bank will support the creation of new businesses like these through the careful management of the hard currency that will flow through its coffers.

The Central Contracting Agency

The Central Contracting Agency (CCA) will serve two main purposes. First, it will provide a format for contractual relations for the entire mission to Mars. This will allow for efficient economic relations to develop between the various players at the earth Embassy by creating a non-mandatory, but useful, process to govern contracts and payments. More importantly, however, the CCA will allow for some central coordination (outside of the Fleet) for those matters being researched and developed by the Confederation. If one Fleet organization has already created a space flashlight or a certain software program, the CCA can help to ensure that such efforts are not needlessly duplicated. The goal will be to avoid the US $2,000 toilet seat problem that plagues US government spending.

While the CCA will be mostly an administrative organization, it is expected to wield great informal power at the earth Embassy. In many ways, ranking officials at the CCA will be the technical leaders of the entire mission to Mars, because they will work out the technical specifications being agreed to between, for example, the Fleet and a rocket engine manufacturer. This may sound trivial to someone who has never done contract work, but there is often a gap between what a contractor and a subcontractor mean when they use a certain term or talk about a certain problem. The CCA official will bridge this gap both legally, by

---

4   Radical libertarians will, of course, mostly oppose the idea of a central bank. Utilizing the principle of adopting proven financial and social tools to achieve the larger purposes of the mission to Mars, and given that every western nation has its own central bank, it follows that the idea of a Mars Bank makes sense.

using language they have grown accustomed to using, and technically, based upon their experience in dealing with all Fleet contracts.

The other way that the CCA is expected to wield power is to influence indirectly those projects that are being contemplated by the Fleet and others. Again using the rocket engine manufacturer as an example, if the CCA's technical administrator prefers one design or company over another, contracts can be structured not only to ensure success, but possibly to subtly sabotage the contract the CCA official did not favor. He could, for example, propose deadlines he knew were short or add technical language to the contract that made compliance more difficult, etc. These kinds of behaviors are not being advocated, except to the extent that they empower the CCA to coordinate technical matters or otherwise gently influence the direction of the mission to Mars' technical (and other) affairs.

Economics 102

The extraordinary ingenuity and industry of the Dutch in wresting land from the sea were legendary. Still the actual spectacle of all that had been contrived and built, the innumerable canals, bridges, dams, dikes, sluices, and windmills needed to cope with water, to drain land, and hold back the sea—and that all had to be kept in working order so that life could go on—made first-time visitors stand back in awe and New Englanders especially, knowing what they did of inhospitable climate and limited space. Francis Dana, when he arrived later, wrote of the Dutch living in a world made by hand. "The whole is an astonishing machinery, created, connected, constantly preserved by the labor, industry, and unremitting attention of its inhabitants at an expense beyond calculation."

—David McCullough in *John Adams.*

This is the second of three chapters in this volume devoted to economic ideas. The sub-theme here is the development of a sufficiently strong economic infrastructure to transplant the earth Embassy community to Mars. "Infrastructure" means both what will be required at the earth Embassy and what will be required on Mars to connect Martian settlements with earth. And, as with many economic concepts, infrastructure refers to the intangible as well as the tangible. The structure of this volume limits the theoretical coherence of matters economic in favor of a collection of stories, anecdotes, and discussions that paint big pictures. Chapters such as this are especially vulnerable to the charge of being untidy strings rather than whole cloth. We trust that an open minded reader will forgive whatever mental skips might be necessary as we hew out the edges of the mission to Mars' economic plans. With luck, by the end of this chapter the reader will have little difficulty sensing the economic potential of these specific ideas and their potential to help the mission to Mars transplant its earth Embassy to Mars.

Martian Real Estate

In two senses of the word, Martian real estate will be extremely valuable. The obvious, first measure is its financial value. Neither the numbers used here nor the short accompanying discussion are adequate, however, to convey the magnitude of this financial value. What, after all, is the "financial value" of the United States of America, or could this concept be captured using a single number? Same problem for Mars. The second measure is the value of making land available to the earth's poor and to those who wish to stake a claim on this new world. This is a spiritual value whose worth can only be evaluated in the mind of each individual. The first value—financial—is of such staggering portions that it is impossible to dismiss it in any conversation about Mars. The second—a value of the spirit—is of such importance politically that it must be expected to be adjudged politically incorrect and thus to be opposed by the earth's power brokers and Elitists.

Value # 1

The sale of Martian real estate has the potential to raise very large sums of money. Initially, the mission to Mars will sell Martian land to a few far-sighted people at US $10 per acre. The acreage on Mars is so vast that if it all were sold at US $10 per acre, the mission to Mars could raise US $360 billion. Terraforming will make the land even more valuable. Given that it will probably take a century or more to sell off all the land on Mars, *significant portions* of the real estate on Mars might sell for US $500 per acre (constant US dollars). If so, the Confederation might raise US $5–10 trillion dollars from the sale of this real estate.[1] While these sums may seem astronomical, the reader must remember that Mars has as much surface area as the earth and that even if 30% of this land eventually became surface waters, parks, or otherwise not for sale to private entities, the potential pool of capital to fund mission to Mars projects can reasonably be measured in the trillions of

---

[1]   The mission to Mars intends to allow each Martian settler to claim 1,000 acres on Mars as part of the ticket contract. The total amount of land the 50,000 mission to Mars' settlers will be able to "cherry pick" will represent far less than 1% of the total surface area available on the planet. The dollar amount lost will be about US $500 million. Even if much of the "best land" on Mars is taken by these settlers, three very favorable developments will result. First, taxes will be paid on this land, creating a huge annual stream of income for the mission to Mars. Second, as ownership implies an incentive for development, the development of Martian real estate will increase the available capital on Mars. The advantages to be derived from capital on Mars itself should by now be reasonably clear. Finally, the vibrant sale of Martian real estate will create markets that will benefit the local Martian economy and the accumulation of even larger pools of capital, both useful for Martian development.

American dollars. These sums will not appear overnight and the mission to Mars does not intend that they will fuel more than a few auxiliary economic engines at the earth Embassy. Rather, the discussion in this section is meant to convey an order-of-magnitude idea about the amount of money that could be raised from the sale of real estate on Mars. The reader can then speculate about what might be accomplished with such sums.

Terraforming becomes both an effect and a cause of the capital that the mission to Mars might raise. A portion of the funds from the sale of Martian real estate ($5 trillion?) will be put into a trust fund, to be administered by an independent, quasi-governmental agency, to fund the terraforming of Mars. And, as suggested above, terraforming will create a financially self-sustaining process to *increase* the value of real estate on Mars. In the strongest contrast, it should be noted that the terraforming issue will forever block NASA from making progress on Mars, because of the utter impossibility to pay the trillions of dollars it would require for BGBS terraforming programs.

The orderly transfer of property rights for asteroids claimed by Mars will likely double the numbers discussed here. It is virtually certain that asteroids contain large quantities of high-grade metal ores. While none will be suitable for habitation, there will likely be significant mining and research activity conducted on the asteroids[2]. A Confederation system to sell mining rights and to purchase asteroid real estate or even entire asteroids will not only allow for the beneficial use of these resources by humans, but will create yet another large revenue stream for the ideas in these volumes.

As the mission to Mars gains credibility, more and more speculators in Martian real estate will appear[3]. Once settlers reach Mars, settlers will have a unique opportunity to purchase Martian real estate, just as George Washington and other early Virginians supplemented their income by speculating in lands in Virginia's western counties. Mars' wealthier settlers will likely have few consumer goods to

---

[2]   Over the long term, asteroids will be hollowed out and "spun" to create a gravitied environment inside of the asteroid's hull. Such projects will be expensive, require relatively large amounts of new technology, and so remain only theoretical possibilities for a somehow, someday.

[3]   "Speculator" has a bad connotation in 21st-century English. It implies some evil person who underhandedly invests money to acquire huge profits. Indeed, to the extent an investment is "underhanded" and that inflated profits derive from improper investments, speculation is wrong and economically wasteful. The accurate definition of a speculator, however, is one who invests in a commodity not for the commodity itself, but merely in hopes of earning a profit on the re-sale of that commodity. Land speculation was an economically useful occupation in colonial America and will be useful on Mars.

buy, so money is more likely to go into the purchase of land than a new car. And, of course, as things develop on Mars, mining corporations can also be expected to purchase large tracts of land.

The virtual certainty of increasing land values on Mars will eventually attract large numbers of speculators from earth. Anyone willing to pay the Confederation hard currency for a claim to Martian land will be welcomed at the mission to Mars. The money invested will help to pay for the infrastructure the mission to Mars intends to build for its settlers. And if someone gets rich in the process (especially if it's a *Martian*), so much the better.

Value # 2

Volumes II and III tell the story of an idealistic mission to Mars that hopes to set examples of good behavior and human dignity that other humans can emulate. One such ideal will be the belief that all humans should be offered the opportunity for a western-style life, not just those fortunate enough to live in the modern West. The mission to Mars will pursue this end by offering people opportunities on Mars that may not be available in some dictator's sand-swept, socialist paradise. Appropriate to this section is the ability to obtain land cheaply for the individual, small landowner.

One would like to think that to offer land and opportunity for downtrodden masses on earth would be considered a noble idea. Indeed, the mission to Mars expects that the great majority of people on earth will recognize this to be the case. Unfortunately, not everyone will view this end to be worthy.

Putting land into the hands of people on Mars will have its detractors. Indeed, Elitists will rally to the call to *preserve* Mars—a là Antarctica—while telling the starving masses on earth to eat their cake. The preservationist call will take three major forms. The first group will consist of those who will oppose any terraforming activities on Mars, supposedly to preserve the few one-celled creatures that might—just *might*—become extinct if the planet were warmed to accommodate human beings. While the possibility of extinction is real if the planet is warmed one hundred degrees Celsius, it is equally likely that whatever life currently exists on Mars will live in environments that will be protected (deep underground) or can be preserved (equatorial bacteria reseeded at the Martian poles), etc.

The second group that will oppose Mars for "Martians" will be those scientists who argue the need to preserve Mars "for study". Their voice will be strong and will make a strong impact on many people. Indeed, *some* science *will* be made more difficult by mission to Mars' activities. If ten-million-year-old polar ice is melted or deep ravines fill up with water, opportunities for study of that ice or that dry ravine *may* be lost. The mission to Mars responds, however, with a mostly-easy-to-understand argument that scientists *actually on Mars* will increase scientific opportunities and dramatically reduce the cost to take advantage of such opportunities. Figure 36 explains some of the costs and trade-offs involved with acquiring knowledge under both a mission to Mars scenario and a NASA scenario.

	"Units" of Scientific Advances	"Cost" (US $) Per Scientist Year	Opportunities For "The Huddled Masses" on Earth
Mission to Mars	100	$ 100,000	1000
NASA Scenarios	100	$1,000,000	1

Figure 36: Hypothetical "Cost Per Unit of Science" Analysis

Figure 36 suggests that the mission to Mars will not only be ten times cheaper per "unit" of scientific advancement, but that this scientific progress will be made in concert with opportunities for a better life for humans on earth, an opportunity that will not exist under a NASA program. In essence, living and working on Mars will provide huge amounts of information that NASA could only acquire through astronomically expensive and inefficient space probes or other types of stand-off equipment.

Even if we ignore cost or opportunities to the earth's poor, we see in Figure 37 that "science" will advance much further under a mission to Mars scenario than under the NASA scenario. Figure 37 suggests that by the year 2050, the mission to Mars will likely acquire more than twice as much scientific information about Mars as would NASA, since it will have actually landed and been working on Mars for 25 years and NASA will continue to rely upon robot toys. By the year 2200, with or without a few NASA "exploration" missions to Mars, humanity will have more than 125 times as much information about a mission to Mars Mars as it would under a NASA exploration scenario.

Year	Units of Science Per Scenario	
	Mission to Mars	NASA Exploration
2050	100	10
2100	500	20
2200	5000	40

Figure 37: Total Amount of "Science" Accumulated Under MTM and NASA Scenarios

The third great opponent to settling Mars will be those political entities such as the UN and other "non-government organizations" (NGOs), which dislike anything that they cannot control. NGOs have been given massive legitimacy in the last few decades because of the growing "internationalist" movement. Some NGOs deserve respect and consideration as humans move to Mars. Others do not. The most distinctive difference between the one group and the other is the degree to which the organization has a political rather than social agenda. Thus, a Martian Red Cross and Salvation Army should not only be given respect and consideration, they should be invited to participate in the transplantation of a human society to Mars. Leftist groups—environmentalists, anti-globalists, and the like—will not be allowed to participate *as a group*, though they are more than welcome to send hard-working individuals to help build the earth Embassy society.

The reason that the UN and other NGOs will oppose the mission to Mars is that their own *non-democratic* program to influence the course of humanity will suffer. The mission to Mars will bring humans together *as free individuals* to reach into space, rather than as agents of some non-elected global authority. This is no place to wax eloquent about human freedom—we have devoted much of Volume II to that task—but it is important to point out how this issue runs square into the UN's and others' notion about the human future. *If you are seriously to discuss free, human colonies on Mars, you must account for the reality that many of today's most politically correct establishments will oppose your ideas.* The fundamental basis for one's moving forward to Mars despite Elitist criticism will be a belief in one's own freedom to act when such action creates no harm to third parties. Now, environmentalists will say we will harm Martian bacteria. The UN will say that the mission to Mars will harm its goals for "international consensus" (whatever that means). And many existing governments will fear the consequences of individual human beings actually doing what these governments cannot; and they will fear—as *status quo* groups always do—individual humans escaping their paternalistic grasp. In any case, many anti-freedom groups will rally together to oppose the mission to Mars and the threat it poses to their own versions of the human future.

Though Martian land will be extremely valuable, it will not fuel the engine driving the mission to Mars. Rather, it will be entrepreneurs, working directly on Confederation projects needed to transport its settlers to Mars or indirectly to raise hard currency, who will create the infrastructure and jobs for earth Embassy people. We now discuss several different types of entrepreneurs, their roles at the earth Embassy, their impact on the mission to Mars, and why entrepreneurs might be attracted to the earth Embassy.

The Moocher

The value of entrepreneurs to the Confederation will be illustrated by an example of the worst possible type: a person who has no interest in going to Mars and intends only to take advantage of the "free food" and "free housing" arrangements at the earth Embassy in order to get his business started. He has an entrepreneurial idea that is completely unrelated to the mission to Mars and intends to abscond as soon as the business starts making money. With apologies as needed, we call him Joey from Joy-Zee. Figure 38 lists the two main sources of financing for Joey's on-line baseball card business.

	Investments
Private Investors at the Earth Embassy	CS$ 200 K
Joey's Investment	US $15 K

Figure 38: Initial Investment in "The Moocher"

Joey received CS $200K financing from private organizations at the earth Embassy. Several investors were convinced that Joey's idea had merit and decided to back him. (The details of earth Embassy financing will be discussed in another example below.) To receive this financing, Joey needed to invest US $15,000 of his own money in the business.

Joey's business took several years to become profitable. Figure 39 shows that Joey's business lost $90,000 his first year on an income of $30,000. Thereafter, his income steadily improved and his expenses decreased until, in the fifth year, Joey made a little money[4]. At this point, Joey absconded, leaving his mostly worthless equipment, and taking his business records, software, and the choicest of the inventory with him.

Joey's plan was to resume his on-line business at another location. He believed that he would have no difficulty restarting business, as he could continue his business with just a webpage. The assumption was mostly good, as six months after leaving the earth Embassy, Joey had recovered 50% of his business. Joey's absconded business was immediately profitable, as now Joey did not hire employees.

---

[4]   The cost to create a paper catalogue, develop reliable sources of supply, a stock of quality merchandise, effective advertising, etc., all took several years. Thus, expenses were higher at first, declining as the business gained experience about how to operate successfully.

Year	Income (CS$)	Expenses (CS$)	Loan Repayments (CS$) (Included in Losses and Expenses)	Accumulated Losses (CS$)
1	30	120	12	-90
2	40	110	12	-160
3	60	100	12	-200
4	80	90	12	-210
5	100	85	12	-195[5]

Figure 39: Income/Expenses

The Confederation was dismayed about Joey's sudden departure, but it was not a devastating loss. As we shall see in greater detail in Volume III, the Confederation actually came out ahead, despite Joey running off. While there were several positives impacts, most importantly for the Confederation was that it enjoyed a reasonably large stream of hard currency while the baseball card business operated at the earth Embassy. Joey also provided jobs to one full-time and one part-time employee during this period. And because Joey's wife worked at the earth Embassy, her talents, too, helped to develop the Confederation's economy. These matters alone suggest that the Confederation benefited from Joey's time at the Confederation, but there is even more.

The Confederation had created some strict and aggressive bankruptcy laws. These laws allowed Joey's investors to confiscate the equipment, inventory, and records at Joey's business and to sell it to a new owner, who was also able to continue "Joey's" business. The fight over the domain name and web page address ended quickly, because the new business owner and Joey promptly split up the customers in a way that allowed both to stay in business and both to make some money. Moreover, carefully prepared CCA agreements, if taken into a US court, would have given the Confederation even more rights. The investors, too, were protected even though Joey's business had lost money most every year. How? Because Joey and his wife left their ticket accounts behind. The money from these accounts paid off Joey's business debts and even allowed the Confederation to forward to him a few (hard currency) dollars, albeit taking their sweet time doing it.

---

[5]   At this point, sixty thousand dollars of these accumulated losses were loan repayments.

The Angel

Now assume that it was not Joey who opened his business at the earth Embassy, but his cousin Vinnie, an honest person, a law school graduate (an oxymoron?), who wanted to make an honest go of the baseball card business. All of Vinnie's numbers remain the same as Joey's but, unlike his cousin, Vinnie wanted to get to Mars.

Vinnie and his wife had decided to open up their business at the earth Embassy because it was much less risky than opening it up anywhere else. For example, food, housing, and medical care were supplied under the ticket contracts that he and his wife signed. There was also easy access to investment money for their projects by earth Embassy investors. These investors, moreover, could not throw you out of your house if the business failed. The low likelihood that anyone on the outside would have lent Vinnie money was also an important consideration.

After five years, however, Vinnie told his earth Embassy investors that he had to sell his business. Vinnie and his wife had changed their minds; they would return to New Jersey to live the suburban life. Vinnie and his wife left the earth Embassy with US $100,000 saved from their ticket accounts, a net of US $85,000 over the money initially invested in the business. Vinnie and his wife had had a wonderful adventure, their kids would have gotten five years of excellent schooling, and they would have left with a reasonable nest egg to begin elsewhere. Such a positive outcome—we know Vinnie was terribly disappointed he never got his business truly profitable—would not likely have been possible with investors outside the earth Embassy.

"Failure" meant that they could take their US $100,000, find a nice suburban home back in New Jersey, and continue their lives, with stories to tell their grandchildren. This is a powerful scenario for any would-be entrepreneur.

More Millionaires To The Moon

This work takes the position that private enterprise alone will have a very tough go trying to make it into space. Government is necessary to structure markets in the same way that a house needs a structural framework to stand. Current space entrepreneurs lack a clear government structure necessary to allow them to survive and then thrive. The lack of structure explains the lack of entrepreneurial push into space over the past 30 years[6]. Still, these entrepreneurial ventures exist and many of them will eventually succeed *because of the mission to Mars.*

---

[6]   Much more later. This, like many complicated ideas, is explored in detail in Volume III.

The two entrepreneurs discussed in this chapter have had nothing at all to do with space. The point of this section is to suggest how supportive the mission to Mars would be for both existing and future space entrepreneurs. We examine now a not-quite-hypothetical idea to sell vacations on the Moon to earthers.

This earth-to-moon-and-back company will need many things that the Confederation can offer. For example, it will need to have spaceport facilities and access to people who can repair or provide supplies for a spacefaring operation. This describes the earth Embassy perfectly. In addition, this company may need financing. The financial organizations described in these three volumes would likely be interested in making loans to this earth-to-moon entrepreneur[7]. Another thing that many companies—young and old—look for is a way to diversify their business so that a downturn in one area does not so adversely affect the company. Thus, a company might take on Confederation contracts totaling 10–20% of a company's business, either to help a "worthy" cause or to serve as a buffer against a downturn in the company's main business.

This list could go on, but most readers should now understand the issue: those interested in entrepreneurial projects related to space should consider doing business at the earth Embassy. As a *society*, the mission to Mars would not be a competitor, but could serve as a source of financing for and as protector of the new venture. And, of course, a new "space venture" at the earth Embassy will help to diversify that community's economy. Even if some space entrepreneurs choose eventually not to go to Mars or choose not to re-locate their business to Mars, working cooperatively will present opportunities for both entrepreneurial ventures and for the mission to Mars.

Case Study: The Lester Electric Company

During the author's tour of duty in Iraq, he had many occasions to observe the "re-wiring" of Saddam Hussein's government buildings, which had been destroyed by the war or by looting, to serve as temporary quarters for US military forces. There were no screens in the windows, sometimes no windows at all, just holes in the wall. And, of course, no electricity, no sewage, no water. US military forces occupied such shells all over Iraq, fighting off the bugs and the heat and, as military vets understand, "the Joes" immediately set about to make themselves comfortable. After the author's unit moved to Fallujah, the first order of business

---

[7]   One of the requirements for such loans, of course, will be participation in the economics of the earth Embassy. A loan for CS $ 1 million means nothing if there is little or nothing this money can "buy". At its simplest, one way to participate would be to hire earth Embassy residents and pay them in Confederation Script.

was to get electricity into rooms which, at the author's end of the complex, meant using a five-kilowatt, towed generator and lots of extension cords. When someone wanted to hook up an Iraqi air conditioner, Roy Lester was called in, master electrician, on duty in Iraq with the author's Kansas National Guard unit.

Leave off now with history and take a leap into the unknown. Becoming interested in the mission to Mars, Roy talked his wife into going to the earth Embassy. Kansas was only 700 miles from the earth Embassy, but the weather was much warmer than in eastern Kansas. So, the couple took a ten-day "Christmas vacation" there in December 2008.

As was his wont, Roy ignored his wife and worked 12 hours a day during his "vacation" at the earth Embassy, eager to apply his skills in a circumstance where they would be appreciated. At that time, about 100 people lived and worked at the earth Embassy. Most of the wiring was haphazard. During his ten days, Roy was able to rewire three homes and to rework the worst of the bad wiring in some of the commercial buildings. Fortunately for him, Roy's wife became interested in some of the social agenda at the earth Embassy, so talk about moving, not divorce, moved to the top of the family agenda. As he was leaving, Roy made a promise to return to the earth Embassy after a few months to continue his work.

Roy was impulsive, however, and soon forgot about his return trip to the earth Embassy. But his wife did not. In August 2009, the Lesters returned to the earth Embassy, this time for two weeks. August in the high desert made everyone think about air conditioning, an entirely different wiring problem from what Roy had seen the past winter. Again, Roy worked 12-hour days and finished only a portion of what he had hoped to do. This time, Roy's wife worked, too, helping an attorney draft guidelines for labor regulations for matters related to children. At the end of the two weeks, Roy decided to stay at the earth Embassy to open the Lester Electric Company. His wife would return to Kansas to wrap up personal affairs for the family.

Lester Electric landed its first contract the day after his two-week visit ended. Previously, Roy had been operating under the auspices of an engineering firm. As a new resident and with a new business, Roy registered his business, signed ticket contracts for himself and his wife, hired an accountant to set up Roy's books, and found some suppliers for electrical supplies. As wearisome as the electrical work was, life at the earth Embassy made Roy feel more alive than he had felt for many years.

Roy's first big job was as the major sub-contractor for an engineering firm that was reworking all of the main electric lines at the earth Embassy. Instead of spaghetti strings of wires and extension cords going in every direction, Roy would implement plans to run appropriately sized wire throughout the community. More importantly, Roy was busy rewiring homes and businesses tying into these new power lines. It was in the matter of the new rewiring where Roy got involved with a "political" issue at the earth Embassy.

As someone who had worked with standardized electrical codes during his professional life, Roy thought that everyone should be made to rewire their homes and businesses to meet these codes. Many people at the earth Embassy, however, were either too "cash poor" to pay for rewiring or were philosophically opposed to having codes imposed upon them by the Confederation. The earth Embassy's mayor and its city council reflected the will of these people, so they, and not Roy, won out[8]. Roy felt personally disappointed that standardized electrical codes were not imposed at the earth Embassy and was justifiably concerned that there would be unnecessary electrical fires. But to no avail. Roy ensured that all his contracts met work standards he was proud to enforce and left those parts of the earth Embassy he never worked on with jumbled and poor (but cheap) wiring.

Roy was constantly looking for good workers. During the first three years, he was able to employ one man who remained with his firm for many years, but most of his workers either gave up on the earth Embassy or found it too easy to find better-paying jobs[9]. As his business continued to grow, however, Roy was able to offer sufficiently good employment opportunities to keep a core of very good employees. Many had to be trained, but in any pioneering society on-the-job training is the norm, not to mention efficient economically. Those who really wanted to work as electricians for Roy were able to do so at Lester Electric. Those who were only looking to get by did not waste everybody's time and money.

It soon became apparent to Roy that there was a great need for additional electrical generation capacity. Accordingly, the Lester Electric Company refurbished an old 300-kilowatt diesel-powered generator that Roy had scrounged in Utah, at the cost of hauling it away. It took Roy and his people nearly four months to get the generator operational, but the new capacity allowed not only for surge and back-up capacity at the earth Embassy, but got Roy interested in a whole new aspect of electricity.

---

8   The Confederation Constitution had a provision that made it difficult for the government to interfere in what were mostly economic decisions. Volume III expands greatly on the political matters such as the structure of government at the earth Embassy, building codes, regulations, etc.

9   The demand for labor at the earth Embassy and later on Mars will always be high. *Everything* will need to be built and many people who might otherwise work as employees will choose instead to exploit lucrative opportunities to start their own businesses. Accordingly, labor will command high wages at the earth Embassy and on Mars.

Lester Electric soon became the largest supplier of electrical power at the earth Embassy. He bought power from several sources of electricity[10] and bought out two smaller sources of electricity outright. Lester Electric stabilized the electric grid like it had not be stabilized before and brought electric service to the earth Embassy to a standard higher than most of the US. After Lester started his grid, there were never any black- or brown-outs and even the rare storm experience at the earth Embassy's desert location could not knock out power because of Lester's protected transmission system.

As much as Lester would have liked to have kept his electric company a "back of the truck" operation, what he had accomplished at the earth Embassy was too important, too complex, and too central to too many other businesses for it to remain simple. Even though Roy was surprised to find that he actually enjoyed the business end of Lester Electric, he realized he needed to hire some "college boys" to help him run it. A few years of effective management made Lester Electric's stock a part of virtually every portfolio at the earth Embassy. Still, Roy's staff of electrical engineers and other professionals were people Roy preferred to avoid. He preferred his work sites, making sure that the electrical work was being done according to his own high standards. And though he and his wife remained the largest shareholders, all but 10% of Lester Electric shares were owned by others. At one point, Roy and his wife were worth about CS $50 million dollars. Roy might have lost control of his company, but he was lucky. Roy's high-ranking staff was too loyal to him and never tried to orchestrate a financial coup to oust him. His staff found a way, however, to satisfy both their growing desire to "professionalize" Lester Electric and to keep Roy happy hooking up wires: they would ship him off to Mars.

Recreating the success Roy had enjoyed at the earth Embassy was a top priority on Mars. Although Roy's company lost out on some of the first contracts for electrical work on Mars, getting Roy and a small team of his best electricians to Mars made too much sense to have escaped the Fleet's notice. As a consequence, Roy and six of his employees were chosen to be on the first Caravan Transporter mission to Mars. There, Roy would have wirework enough to keep him busy for the rest of his life.

Roy's wife came to Mars on another transporter mission about a year after Roy had arrived. During that time, Lester Electric's subsidiary on Mars was again a small, but important company, generating and transmitting electric power to all the new facilities being created there. Although he was nearly seventy, Roy

---

[10]   There were two companies generating electricity by wind power, three using solar cells, and a company which generated electricity using steam from the community's waste incinerator (much more about this incinerator in Volume III).

was still working 60-hour weeks. Financial security allowed Roy to stop worrying about things he never cared much about and to concentrate on hooking up wires on Mars. His wife wanted him to slow down but, during Roy's last year, he was still working 40 hours per week. And, of course, they found his 88-year-old body at work, with a wrench in one hand, cigarette in the other, working through the lunch hour, trying to fix a conduit nobody else could get right.

You Give Me Food, I Build You a Spaceship

This section develops an idea about money and economic power that flows from the ideas of "Gilligan's Island" in the Economics 101 chapter. In that short discussion, one person grew and prepared food for the rest of the society. In return, he received shelter and ultimately a ride back to civilization. In this section, these ideas are extended, to include a brief discussion about the utility of commercial banks to build a self-sustaining economy on earth and to transplant this society to Mars.

Figure 40 uses the "One Picture Equals 1000 Words" principle to help the reader to understand how the purchase of "tickets to Mars", the necessities of life (food and shelter), and an interlocking system of contracts can create a closed economic system at the earth Embassy. If the system is organized to operate under familiar principles of free enterprise with "easy money" bank loans, there is no reason that people working together cannot build this economy and achieve their political goal of getting to Mars.

Figure 40 envisions sets of economic actors we call "Travel Corporations", "The Fleet", "Farmers", and "Civilian Contractors". Each set exchanges its service for things the others can provide. The Travel Corporations orchestrate the system and commercial banks (and Confederation/Fleet contracts) finance it. The result is a closed system where Travel Corporations sell tickets, people receive food and housing, the Fleet organizes the effort to Mars, which is planned and executed by Civilian Contractors, and Farmers provide food, through the Travel Corporations, to the people living at the earth Embassy.

	Travel Corporations	The Fleet	Farmers	Civilian Contractors
Travel Corporations Pay		The Fleet (Confederation) for Tickets to Send People to Mars	For food for ticket holders	To build homes for ticket holders
The Fleet Pays	Full Time Crewmen buy their tickets to Mars		For food for crewmen on duty	For ships and supplies for the mission to Mars
Farmers Buy	Their tickets to Mars	Contract labor (crewmen) for short periods		Various services, greenhouses, to produce more food
Civilian Contractors Buy	Their tickets to Mars	Contract labor (crewmen) for short periods	Extra food	

Figure 40: Who Pays Whom in the Confederation Economy?

The financial circle is completed by having commercial banks at the mission to Mars lend money to new operations, often accompanied by a contract from the Fleet or some other Confederation organization. These contracts will be made based upon the estimated contribution of the new business to the earth Embassy's economy, and the return the bank expects to earn for its shareholders.

Figure 41 envisions how a technical proposal might be accepted by the Fleet, which issues a contract to the (new) business, in turn allowing the business to get financing for the project. Alternatively, work that is not acceptable gets rejected in favor of other ideas or products. A commercial bank at the earth Embassy lends Confederation Script based upon its evaluation of the business proposal. If the initial contract is successful, additional contracts and larger loans can follow. If the (new) business fails, it either needs to rethink its idea or go out of business.

A CCA Story

We now present a story that was found near the end of the original work. As such, the reader does not have the background to understand all the jargon. Hopefully, this will not present an insurmountable obstacle and, in fact, will help the reader anticipate some of the later discussions in this work.

ABC Corporation (ABC) has solicited for and received a Preliminary Contract (PC) from the Fleet to test an entirely new type of launch system, to lift cargo into space for CS $50 per pound. The PC is a form document created by the CCA to assist in negotiating aspects of contracts, prior to the contract's submittal to the CCA. As directed by their commanders, Fleet comptrollers incorporate the information from the PC into their future years' budgets. The comptroller in this case had informed her commander[11] that there was not much excess hard currency for future Fleet budgets and asked about the possibility of a hard currency cost overrun in ABC's program. Since the Fleet was keen on the program, she was told the risks were minimal.

A meeting was held between the Fleet's Project Officer (PO), the two principals of ABC, ABC's attorney, and the CCA's Technical Administrator (TA). The TA informed everyone that she was up for re-election and asked everyone to vote for her. Muted laughter followed. The TA was much less certain than the Fleet about the technical merits of ABC's proposal. Nor did she foresee the availability of much hard currency during the next three years. Further complicating matters was that the TA had her own agenda for lift platforms, an agenda that did not include ABC. Though the TA administers contracts and is not responsible for them, her position gave her great clout at the earth Embassy. Not only was she able to use her contract administration position to "fix" bad technical projects and to reduce cost overruns, she had great informal power to steer technical development at the earth Embassy in one direction or another.

---

[11]   Both the Fleet comptroller and her commander were part-time members of the Fleet. Both worked about two days per month and spent several unpaid hours each week ensuring mission accomplishment. Those unwilling to spend extra time to ensure that their Fleet jobs got done would either not be promoted to important roles or would be demoted and placed into less responsible jobs.

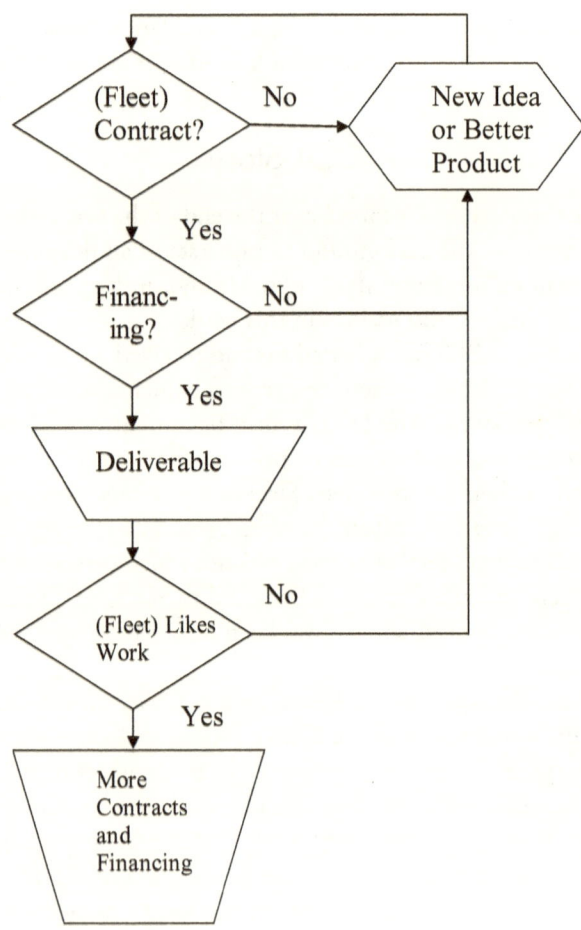

Figure 41: Proposal and Contract Life

A sidebar now to understand that Thomas Zubrin, the relative of Robert Zubrin whom we met a few chapters back, had been very interested in becoming the TA at the earth Embassy. This one person draws up virtually *every contract* made relating to the Fleet's program for space. This one person, though serving ostensibly only as an administrator, greatly influences the direction taken by the Fleet and the mission to Mars. The mission to Mars proposes that the TA be an elected position, open to anyone who decided that she could (should) influence the direction of these programs. Ultimately, Zubrin decided not to run for the job, as he was sufficiently satisfied with the responsibilities described in that preceding chapter.

One way to smooth over the problem of *too many egos* is to divide power, while simultaneously providing great responsibility to many persons. This, by the way, is the real secret to the success of democratic systems in the western world. This brief discussion about the TA provides just one example of the mission to Mars' plan to divide power and responsibility between many people and many offices.

As an elected official, the TA was vulnerable to political pressures from the community. Two things made her look bad: cost overruns and bad projects. Though the earth Embassy was still a small community, there were several people who coveted the job of TA. She had to be careful if she intended to get re-elected.

ABC and the PO were not at the meeting to discuss the technical merits of the program. Rather, they were meeting with the TA to work out the details of a 15-page, standardized CCA contract. The PO and the TA had been in many such meetings. The corporation's attorney, Jim, had only recently joined the Confederation. Although he did not have a law degree and had never practiced law, he had read most of the governing documents related to the Confederation. As he had always felt himself good with people and able to make a good argument when necessary, Jim had registered as an attorney only six weeks before. He had represented people in three other hearings and at one trial before this meeting, but this was his first meeting with the CCA. In fact, to support himself, Jim worked two other jobs, landing only five or ten hours of attorney work per week. The PO and the TA politely, but firmly, told Jim to keep quiet except when discussing specific issues of law. He was surprised by this request, but abided by their wishes.

The major point of discussion related to the specifics of the six-month and one-year deliverables. Both the PO and the TA understood how important such matters were for the contract. One of the greatest points of contention in Fleet contracts had been the lack of actual agreement between the parties about what was expected by each side. The Fleet, being a large, well-financed organization, usually won these disagreements, but the Fleet never consciously tried to over-negotiate anything. Rather, the Fleet tried to understand fully what it would get and to pay a fair price for it. Smaller corporations like ABC, especially those run by former academics[12], were often unclear about what they were promising and sometimes did not understand the significance of the deliverables. Thus, the Fleet's recent push to ensure that the specifics of the deliverables were spelled out in as much detail as was necessary to ensure both that the Fleet got what it wanted and that the corporation could deliver it to the Fleet. Jim made several points

---

[12]   The classic "theory versus practicality" problem, whereby the academic provides peripheral theories rather than core products that could fit into Fleet plans *now*.

during this discussion of deliverables, points that helped to clarify certain aspects of ABC's contractual duties with the Fleet.

At the end of the meeting, the contract was signed by the PO and ABC Corporation. It was received by the TA as an "accepted" contract, available for public scrutiny and even open to citizen law suits[13]. The contract stated that the Fleet would pay ABC CS $40,000 to deliver, in six months, detailed plans to the Fleet about its new lift technology. Six months after that, the Fleet would pay ABC another CS $110,000 for the satisfactory delivery of a proposal for a prototype and the prototype's construction schedule. This second deliverable also had to incorporate the Fleet's written comments about the design, presented by the Fleet to ABC thirty days after submission of the first deliverable. There was a contingency to release to ABC US $5,000 in hard currency (in lieu of CS $5,000), to allow ABC to purchase test equipment and supplies it might need for its second deliverable.

Two major points about the contract not discussed during the meeting, but discussed in detail between ABC's two principals and Jim, were the payment schedules and provisions for rights to the intellectual property developed. ABC's president, Pete, had been at the earth Embassy for six years. He had long been a "Mars buff" who insisted that he knew more about Mars and rockets than any ten of the fancy people working for the Fleet. Because of his personality, Pete had not been able to hold a job for very long and had bounced around the earth Embassy during most of his time there. This was his first involvement with the CCA. Pete was almost 60-years-old and he realized that if he did not set his finances in order, not only might his Mars dream die, but he might even be asked to leave the earth Embassy. Pete was constantly behind in payments on his ticket contract. Part of the CCA contract process evaluates principals of the contracting corporation, especially regarding the status of ticket contracts.

One specific purpose of the CCA is to facilitate the collection of payments to Travel Corporations. Such money was necessary to keep the lifeblood of the Confederation's economy moving, as it was counterproductive to disburse CCA money to people who were not keeping current on their ticket contracts. Thus the provision in all CCA contract's that the CCA could withhold payments to the contracted party, and direct these payments to the appropriate Travel Corporation, for a period twice the length of the contract, up to a maximum of two years[14].

---

[13]   To recover lost money, enjoin improper contracts, and to defend rights previously negotiated. Openness of process was a major factor in ensuring that the mission to Mars functioned as a community and not as a commune.

[14]   Pete's current ticket contract called for monthly payments of CS $1500. The CCA had the right to withhold up to CS $36,000 from the payment to ABC, if they determined Pete was not creditworthy.

Pete realized that *some* of the CS$40,000 for the first deliverable would not be paid to ABC, but would be withheld to continue payments under their ticket contracts. He was afraid, however, that unless he got up to date with his ticket contract payments, all the money would be withheld.

Since he did not think he could pay off what was due the Travel Corporations, Pete went to the Olympus Bank and negotiated a private loan for CS $15,000. The bank charged a high interest rate and had the right to place a lien against the payments made on ABC's contract, but Pete was able to bring his ticket payments up to date and to pay three extra months in advance. Pete's partner was not completely happy with the circumstances, but he realized that Pete could right his boat financially if their first and second deliverable were well received, something of which both partners were confident. Besides, neither partner wanted the CCA to withhold money that ABC needed to fund its operations.

ABC's Vice-President for Technical Development (there were only two employees) was "Willie" Yao Chen. Willie had only been at the earth Embassy for six months. He was a graduate of the University of Texas, with a PhD in mechanical engineering. His parents did not like the idea that he had joined the mission to Mars and they especially did not like the idea that he had started a business with Pete. But Willie has had his idea for a cheap launch capability into space since his early days as an undergraduate and he felt that Pete and ABC represented maybe his only chance to develop this idea. His PhD thesis, in fact, laid a theoretical groundwork for his intuitions. There was nowhere else in the world that Willie, only a year out of school, would be given the opportunity to test his ideas. He did not feel that he could get any grant money to continue developing his idea. His professors at UT barely tolerated his ideas as a PhD topic. Willie was cautious of Pete, but he believed Pete had a good, if impolite, heart.

Five years previously, the Confederation had negotiated a treaty—the US calls it a "congressionally-recognized contract"—whereby "all income" earned from earth Embassy sources were exempted from US and state income taxes[15]. This "treaty" was signed because public opinion in the US very much favored the Confederation's work and a clever public relations campaign organized by the Confederation helped to convince members of the US Congress that it should help, not hinder, the mission to Mars. The US, for all the abuse it sometimes gets from the Confederation, has no real desire to hinder the mission to Mars' potentially historic work. The "treaty" requires that all contracts at the earth Embassy must be reported to the American Internal Revenue Service annually (this, too, is a CCA function) as well as payments made under these contracts.

---

[15]   More about this in Volume III. For now, understand that US federal law can exempt an entity from all US taxes.

In addition, the "treaty" allows the US to examine all documents, including technical documents, related to these payments. In essence, the US has a right to review technical information that could be highly valuable intellectual property. Thus, NASA has a liaison officer permanently stationed at the earth Embassy to review such information, evaluate its potential for use at NASA or elsewhere, and to report on this information to people and agencies outside of the earth Embassy[16]. Part of the extended negotiations with the US about this "treaty" concerned the legal status of intellectual property that could be so easily reviewed by a third-party. The Confederation took the position that easy access to intellectual property at the Confederation would undermine the ability of owners of this property to claim rights to it. The matter continues to be discussed with US officials, as Confederation diplomats try to expand this and other agreements in Washington.

The matter of intellectual property and the CCA contract was discussed at length. The basic CCA contract claims all intellectual property rights for the Fleet. Jim argued that the main idea was Willie's and that ABC should be able to retain some rights to the intellectual products it would develop, but the CCA gave no ground on the issue. Although CCA contracts include the right to patent and the right to license information and other intellectual property from the Fleet, Jim explained to his clients that he thought their rights were limited. Jim has only vague ideas about the complexity of intellectual property law, but part of his fees go to a non-profit group working at the earth Embassy—a very expensive non-profit group—that discusses intellectual property issues with paying customers and works with the Confederation and attorneys to shore up the intellectual property rights of inventors, scientists, and engineers who develop them. Jim plans to spend about 40 hours in lectures and meetings with this group during the course of the one-year contract. What he learns will not help for this first contract, but subsequent contracts could be written to preserve as far as possible the intellectual property rights of his clients. Everyone realized that, as work progresses, ABC would be in a better position to re-negotiate the standard clauses in CCA contracts, else the Fleet program get side-tracked or otherwise lose momentum.

Willie and Pete were concerned about their intellectual property rights, especially if they decided to leave the earth Embassy and cash in on their work. Pete would not likely leave, but Willie was likely to leave if there were a cash bonanza to be had outside of the mission to Mars. Pete constantly rails about how the Confederation only wants to steal other people's work. Indeed, the mission to Mars was highly conscious of the value of the intellectual property it was

---

[16]   One of NASA's dilemmas was that they kept losing these liaison officers to the mission to Mars once they understood its progress and potential.

helping to create. In the end, however, Pete would always catch himself, remembering the precariousness of his own financial circumstance, and the opportunity he and Willie had. Both were confident that the Confederation would make great strides in space if they developed their project and that they, too, would be handsomely rewarded, inside or outside of the Confederation. In any case, since the earth Embassy was the only place they had the chance to develop their ideas, they mostly shrugged their shoulders and kept working.

Together, Willie and Pete put together their plans (first deliverable) package. Jim reviewed their work, not only for legal sufficiency, but as an editor at non-legal-fee wages. Unlike the author of this book, Jim is an excellent writer and given the streamlined Fleet programs, the ability to communicate in writing was important for success. With two days to spare, ABC submitted this deliverable to the PO, who certified its receipt. Within five days, ABC received payment of CS $4,000, notifying Willie and Pete that their obligations to their respective Travel Corporations have been paid one year in advance and that a check for CS $2,000 had been forwarded to Jim, as stated under the CCA contract.

On the thirtieth day after the submission of the first deliverable, the PO called Pete and asked for an additional three days to complete drafting the Fleet's comments. Pete was furious and was about to articulate his thoughts. "Just like the government never to get anything done on time. If we had been late ten minutes, they would have docked us thousands of dollars". Fortunately, Pete held his tongue and agreed to the extension, despite their need for the Fleet's comments for their work on the second deliverable. After the call was over, Pete got excited. The Fleet had to be interested in their work or it would not have asked for three more days to complete their written comments.

Willie had never seen business operate before. The Fleet continued to be interested, but it was also quite concerned about certain aspects of the design. Fleet leaders had also been, in general, displeased with the quality of the work submitted under the contract. The Fleet had been led to believe that more of the research and testing had been done when Willie had been at UT. A few members of the Fleet review committee had, in fact, argued in favor of dropping ABC's ideas altogether, cutting their losses at the second deliverable. The PO, however, believed in ABC's ideas and, despite the obvious need for more research, had faith that Pete could help translate Willie's ideas into something practical. The written comments ABC received from the Fleet indicated that significant testing needed to done, with results appended to the second deliverable and with strong indications that no additional contracts would be forthcoming without convincing evidence generated by the tests. A Contract Modification (CM) was generated by the CCA both to formalize the additional three days for comments and as regarding the need for significant testing data.

Jim was mad at himself that he had not helped ABC word the original contract better, preventing the Fleet from "redefining" much of the second deliverable. He was reminded, however, that even if he had anticipated this issue from the start, it would have made little difference. Clever wording in the original contract might have prevented the Fleet from, in essence, adding these layers to the original deliverable; but the major idea was to keep the Fleet happy, not position your client to win a lawsuit and lose future contracts. After discussing the proposed changes, Willie and Pete agreed to them.

The Project Officer

The PO was a part-time member of the Fleet. His 31 days of annual duty were spent working in this important office, usually in four- or five-day increments every few weeks. The PO also spent working many long, unpaid hours working on the two projects the Fleet had assigned to him. At the PO's "real job", he was the owner of a shop that manufactured "environmentally efficient" packaging for dozens of other businesses and for the Fleet[17]. He liked his civilian job, but not like he loved working for the Fleet. The PO as packager was a decent business-man, making a good living and an honest contribution to the earth Embassy's society. In his role at the Fleet, however, the PO felt that he was making a huge contribution to the mission to Mars, even if most of his projects were limited in scope.

His other "PO" project dealt with "bookshelves" for Mars and Deimos. The need for bookshelves may seem trivial at first, until you understand more fully what was being discussed. "Bookshelves" was the term that would apply not only to storage systems for books, but for virtually everything else carried to or created on Mars and Deimos. Most shelving systems at the earth Embassy were either bought or "refurbished" from the outside economy. One company, however, was beginning to manufacture its own plastic "bookshelves" at the earth Embassy. While the Fleet had made a tentative decision to buy several hundred "off-the-shelf" storage systems from the outside for early transport to Deimos and Mars, the PO was now working to review the movement plans to Mars for many dozens of companies that could assist with the follow-on manufacture of shelving on Mars. The idea seemed feasible for Landing Year Twelve, but many variables remained. The PO, in effect, was helping to determine the order that businesses and suppli-ers would be orchestrated for Mars, using something as simple as shelving systems for the settlers.

---

[17] Volume III discusses packaging standards and how these standards might help facili-tate recycling efforts at the earth Embassy and in space.

The PO was a member of C Company, 1st Engineer Regiment. As indicated in Figure 23 in the chapter about the Fleet, this unit had about 50 members. They were all part-time crewmen except the unit commander, who was responsible for all "project management" in the Fleet. "Project management" was a term of art referring to that litany of smaller, subsidiary, or quasi-research efforts for which the Fleet did not mount specific efforts. Thus, D Company, 1st Engineer Regiment, would organize drilling efforts at the earth Embassy and on Mars (big programs), but smaller or less clearly defined drilling projects would be managed by C Company[18].

The PO had a nominal part-time supervisor, but most actual supervision was conducted by the full-time commander of C Company[19]. He was very careful that the projects he managed offered as little opportunity for "conflicts of interest" as was possible in a small community. For example, the PO was specially chosen as the PO least likely to make a future contract with ABC. In fact, both the PO's commander and the TA—up for reelection—made the PO sign waivers about his having no current or future expectations of dealings with ABC. And, of course, those who were not to be trusted were not be made a PO in the first place. As it turned out, the PO's were remarkably good at helping the Fleet develop significant new technologies and programs.

Meanwhile, Back At The Ranch …

The PO arranged a meeting with the two entrepreneurs. His current four-day tour was over and he was not planning to be back on duty for another six weeks. The PO spent four hours explaining the CM, the Fleet's position, and the politics of the TA's office. He hoped that ABC would keep two things in mind. First, that ABC was caught in the nub of a perennial battle between the Fleet and the TA. They should be aware of such battles, but not focus on them. In addition, he explained how concerned the TA was about cost overruns, especially since the TA's

---

[18]   LT CMDR Zubrin from an earlier chapter was a C Company project officer for the purchase of simplified drill rigs for marine operations on Mars' surface.

[19]   The Fleet's chain of command handled both substantive projects and administration. In most circumstances, unit and battalion commanders would make arrangements with their subordinate leaders how to divide this workload. Many "commanders", therefore, were satisfied to ensure that the administrative duties of the Fleet were their main duties, allowing their subordinates to be direct points-of-contact for a specific project or mission. In the case of the PO, most substantive decisions were made by the full-time unit commander, while most of the administration was handled by the PO's platoon leader.

record on cost overruns would surely become part of her election campaign six months hence. Currently, the TA's cost overrun percentage was 18% for payments in Confederation Script and 6% for payments in hard currency[20].

Cost overruns were not the TA's fault; she neither decided on the terms of the contracts nor selected the parties to whom these contracts were awarded. As the earth Embassy grew larger, however, fewer and fewer people knew the TA personally and fewer and fewer understood the intricacies of the contracting system. The TA's two most likely opponents in the upcoming election would each argue that "better ideas" and "better management" were needed in the office, whatever that meant. It was the prospect for cost overruns due to the increased testing of ABC's prototype proposal that the PO warned ABC about.

Both Willie and Pete left the meeting feeling uneasy about the direction of their project. Willie understood that much of the data asked for by the Fleet would be difficult to obtain at the earth Embassy. They could do some of the testing themselves, using jerry-rigged wind tunnels[21] and computer models, but this information might be insufficient. Willie told Pete that what they really needed to do was to get some time on the supercomputer at UT, where Willie still had some friends. The problem, of course, was not only the availability of the supercomputer but, more problematic, scheduling the time quickly enough to meet their next deadline.

Pete came up with a solution, though Willie did not like it. Pete wanted to travel back to Austin to have one of Willie's friends use the supercomputer surreptitiously. That friend could use the supercomputer without raising suspicions if they could overcome two problems. First, Willie and Pete estimated they would need at least US $500 for the trip. They took US $1,000, just to be safe. This was money, of course, that would not be available to purchase other testing equipment they would need. Pete suggested making the supercomputer analysis more comprehensive to make up the difference. Willie did not like this idea at all, especially given the various time constraints, but Pete encouraged him that it would work. The other problem was that even having "a friend at UT" could not guarantee instant

---

[20]   Expenditures over an initial estimate constituted a "cost overrun". Neither the Fleet with its payments in Confederation Script nor the CCA with its payments in hard currency "paid" more than was called for in carefully crafted contracts. The CCA was expected to help manage *all* Confederation contracts to ensure maximum efficiency in the use of their limited budgets. This sometimes resulted in battles with the Fleet about the timing of new contracts or the amount of hard currency to be released to a given contractor at a given time.

[21]   There was an especially enterprising company at the earth Embassy that provided "wind tunnel" trials using a combination of room-sized wind tunnels and special mounts on fast automobiles. The results were imperfect, but usually invaluable.

access to the supercomputer. Pete made sure that he had the extra US $500 just in case it was needed to get access.

It turned out it was. The trip to Austin took two days. Another five days and US $500 to get access to the supercomputer. Willie worked feverishly on his laptop to update subroutines and to write programs that he hoped would meet the Fleet's requirements. Pete did the hustling, the driving, and the cooking. Just as the two were leaving the newly built computer facility at UT, someone questioned their credentials. They were off running and didn't slow down until they returned to the earth Embassy.

The supercomputer helped but it did not solve every problem. ABC ended up needing $2500 more in hard currency than was budgeted to complete the testing protocol. The TA was irate, but the Fleet insisted that the money be released. In the end, things went well for Willie and Pete. ABC provided its second deliverable on time and although the Fleet had many more questions in its comments, it got another contract. The second payment of CS $110,000 (minus a few withholdings) solved all of Willie and Pete's financial problems.

And, in the end, the TA lost her reelection bid, mostly for being "over budget" on too many projects.

Space Monkeys

The entrepreneurial, capitalistic system envisioned for the Confederation economy will be unlike the BGBS relationships of rotund Mr. Sam and even fatter aerospace and defense contractors. There will be no milking of the teat at the earth Embassy. This section will describe a scenario where an established aerospace engineer learns a very hard—nearly financially fatal—lesson about the Confederation's drive to get to Mars "Now!" and cheaply.

Sally Rhodes had been an aerospace contractor for twenty years before joining the mission to Mars. Sally had built her own aerospace company and was doing exceptionally well when she decided to move her business to the earth Embassy. A few loyal employees came with her, but mostly Sally had to hire earth Embassy people—she was amazed at the quality of people available—to fill positions in her transplanted company. Helping to lure Sally to the earth Embassy was a contract to build a prototype second-generation SSTO. The Confederation did not give Sally specifications to work on. Rather, her CS $10 million contract was to design and build this second generation prototype using her own ideas, with 10 intermediate deliverables to the Confederation. Some of the deliverables included: 1) basic drawings, 2) a computer simulations package, 3) a scale model, etc. Each CS

$1 million phase of the contract was contingent upon successful delivery of the preceding phase.

By the time of the Computer-Simulations-Phase (Phase II), Sally was seriously behind schedule. Her "old" business had never really paid attention to schedules and being three weeks behind did not seem like a major problem. Sally had always been very good at explaining to NASA or the Air Force why she was behind and had always been able to manipulate the federal procurement system to her advantage. Sally knew that the Confederation would be demanding, but did not expect them to be "unreasonable".

Unfortunately, Sally was in for a surprise. Sally was told that her computer simulation would be tentatively accepted, but that she had only one month to complete Phase III (and put her back on the original schedule) or the contract would be cancelled. In addition, she was told that if there were problems with her computer simulation that she would have to make these corrections to the scale model (Phase III) and still keep to the schedule. Sally was about to bust a gut at the meeting where these conditions were laid out, but ultimately she agreed. She knew, however, that it would be nearly impossible to deliver anything acceptable after just one more month of work. A few weeks later, the contract was severely modified by mutual agreement with the Confederation. Under this new contract, Sally was to serve as an agent for the Confederation to review the work of a competing company working on its own version of the second generation SSTO. For now, her own chance to get an exciting part of the SSTO project were over.

Sally's company was forced to lay-off half its employees. She felt humiliated and was thinking of leaving the earth Embassy "to save face". Ultimately, however, Sally's desire to go to Mars prevailed and she decided to use her professional embarrassment as a lesson about how to compete and survive in the ruthlessly efficient and determined world of the mission to Mars.

What was particularly galling to Sally, though, was that an engineer only two years out of MIT—one of her former earth Embassy employees—was awarded a contract substantially similar to her original contract. This engineer, named Singh, was given Sally's half finished scale model and told to make the modifications he had suggested to her, but which Sally had rejected. Thereupon, Singh essentially developed the project that Sally had started.

A Young Man Gets Ahead

Sally's business had not needed much financing. She had simply moved her business and most of its infrastructure with her when she moved to the earth Embassy.

Singh's business, however, was a different matter entirely. He needed lots of financial help. This section describes how the Confederation arranged to help him.

The biggest issue Singh had faced before joining the mission to Mars was his US $35,000 in student loans from Pennsylvania State University and MIT. Part of his arrangement with the Travel Corporations was that they would pay his student loans as part of his "salary arrangements". Singh was therefore extremely poor, even by earth Embassy standards. He lived on CS $100 per month in discretionary spending. Moreover, finance-types inside and out of the Travel Corporation were not happy about making such a large monthly payment in hard currency to a school, getting little except Singh in return. As Singh was exceptionally bright and exceptionally motivated about Mars, however, he was allowed to join the earth Embassy and to enter into contractual arrangements whereby the earth Embassy would pay off his student loans. And Sally had been happy to put him on her payroll.

Singh had no money to start a business. But he had not been happy with Sally's "business as usual" attitude toward her contract and had not understood why she had repeatedly rejected his suggestions as "too risky". It was only later in his life, when he himself was the owner of a large business, that Singh understood these risks. Though his ideas were not risky from an engineering perspective, he was only a few months out of MIT and knew very little about real world engineering. Still, Sally had been unwilling to go in any new directions and Singh was lucky that others at the earth Embassy had been willing to listen to his brash set of ideas.

Given these negatives, when he first approached a commercial bank at the earth Embassy, he did not really expect to get very far. He was surprised when they accepted his proposals. It had been another MIT graduate, working with the Confederation's CCA, who had encouraged him to sell his idea and start his own business. This same man was also able to convince a few key people at commercial banks to take a chance on Singh, too.

The earth Embassy bankers, of course, *were* taking a risk on Singh. Besides the hard currency outflow to pay his student loans, they were now going to fund his new company, even though Singh could put up none of his own money. The Confederation certainly helped by offering government guarantees on 75% of the bank's loans, but the bankers remained uncertain. When the CS $3 million contract was awarded to him by the Confederation, however, the bankers put together a financing package.

The main points of the deal were:

1) CS $1 million—loan from The First Mars Bank, released (under certain circumstances) to Singh over three years;

2) CS $1 million—investment capital from Celestial Investments, who took a 40% share of his business;

3) CS $500 K—loan from The Commerce Bank of Mars; releasable (under certain circumstances) on the first anniversary of the start of the business;

4) Hugh Jazz—attorney with start-up experience was to become General Counsel and to receive a seat on the board of the directors of the new corporation;

5) Sven Kruger—accountant with start-up experience was to become Chief Financial Officer and receive a seat on the board of directors of the new corporation; and

6) IP Freely—President of the First Mars Bank received a seat on the board of directors.

Singh was not a very good businessman at first but, given the board's guidance, his company fulfilled its contract and was awarded another. As it turned out, Singh's SSTO did not win the production contract from the Confederation. Another small company produced that design. Sally Rhoads' streamlined corporation became an invaluable part of making the new generation SSTO not only elegant, but also a far more efficient vehicle than was originally thought possible. But Singh's company continued to do well, too. He married and put his wife on the board so, in split decisions, there were three votes against two. As the working relationship in his company was excellent, however, it was usually not difficult for him to persuade one of the other three board members to vote with him. After a few years, more of Singh's company's stock was offered on the earth Embassy's stock exchange, further fueling the rapid growth of the corporation. It turned out that Singh's business was selected to go to Mars to supervise the construction of the first all-Martian space vehicles, so Singh and his key employees were seated on an early flight to Mars. He finally made it to the Red Planet 17 years after he first joined the Confederation. Singh eventually retired a billionaire on Mars, having built one of the great aerospace engineering companies of all time.

The Ark Transporters

And God said unto Noah, the end of all flesh is come before me; for the
earth is filled with violence through them; and, behold, I will destroy them
with the earth.

Make thee an ark of gopher wood; rooms shalt thou make in the ark; and
shall pitch it within and without with pitch.

And this is the fashion which thou shalt make it of: The length of the ark
shall be three hundred cubits, the breadth of it fifty cubits, and the height
of it thirty cubits.

A window shalt thou make to the ark, and in a cubit shalt thou finish it
above; and the door of the ark shalt thou set in the side thereof; with lower,
second, and third stories shalt thou make it.

And, behold, I, even I, do bring a flood of waters upon the earth, to destroy
all flesh, wherein is the breathe of life, from under heavens; and every thing
that is in the earth shall die. But with thee will I establish my covenant; and
thou shalt come into the ark, thou, and thy sons and thy wife, and thy sons'
wives with thee.

And of every living thing of all flesh, two of every sort shalt thou bring into
the ark, to keep them alive with thee; they shall be male and female.

Of fowls after their kind, and of cattle after their kind, of every creeping
thing of the earth after his kind, two of every sort shall come unto thee, to
keep them alive.

And take thou unto thee of all food that is eaten and thou shalt gather it to
thee; and it shall be for food for thee, and for them.

Thus did Noah; according to all that God had commanded him, so did he.

—*Genesis*, Chapter 6.

A signal achievement of the mission to Mars will be the design and construction of space ships called Ark Transporters. These space ships will each carry 1000 people to Mars, accommodating their transport in an orderly and comfortable fashion, with a minimum of difficulty and danger. The Ark Transporters will be the largest in the series of earth-based transport ships built by the Confederation. Now, "earth-based" refers only to the design of the vehicles and the departure point to Mars. The vehicle itself will be built entirely in low earth orbit and will itself never leave space[1]. The Confederation's SSTOs will launch workers and supplies into space to build the Arks.

The sections that follow briefly describe the preparation of the 1000-person mission, the daily and other routines in-flight, disembarkation on Deimos, and the mission's first few weeks on Mars. The section on in-flight activities will provide significant detail about this trip. The reasonability of the mission to Mars' proposal might best be illustrated by providing broad brush pictures of the entire program and exploring in detail certain aspects of the mission. Thus, a vista is presented to the reader with enough detail to evaluate the feasibility of the proposal, to highlight some of the difficult issues, and to assess whether the mission to Mars programs are well conceived.

Ark-Cyclers

The Arks will have two primary purposes. The most important will be to transport large numbers of people to Mars in an efficient manner. The secondary purpose will be to transport together a group of people to populate a new base on Mars.

The Arks will be designed for simple construction. Thus, there will be no need for huge engines or complicated hull systems. Once built in space, these ships will neither be required to reach a high "escape" velocity to free it from the earth's gravity nor will the ship need to enter an atmosphere for landing. The Ark may also have a deployable heat shield, in case aerobraking is used to save fuel[2].

The Ark is a modified "Cycler" vehicle, something that has been discussed in many proposals, both in science fiction and as a NASA project. The Cycler idea is to create a space taxicab that will be forever in space and forever in motion, orbiting on its own—in this case between the earth and Mars. The reader should understand by now that this book is not about engineering new vehicles or better

---

[1]   The Ark will "dry dock" in a hanger on Deimos. While not technically space, it is the next closest thing.

[2]   If you use Mars' atmosphere to slow the Ark down, you need not carry and burn fuel. Details like this are deferred to actual design.

technical solutions to problems. The main difference between the Cycler idea and the Ark Transporter is that the Ark is a much more conventional "space ship", doing much more conventional things, while the Cycler is an artificial celestial body that people ride from one planet to the next.

The Ark idea, rather than the Cycler, is presented here for several reasons[3]. First, is because the Ark is a much more conventional idea than that of the Cycler. The Ark will have rockets to move between the earth and Mars, into and out of orbits, and to land at Deimos. The Cycler orbits constantly and Cycler rockets would be used only occasionally to maintain its "cycling" between earth and Mars. Second, the Ark eliminates 90% of the taxi-work that would be necessary if the Cycler did not orbit the earth, but merely used the earth's gravity to slingshot its way back to Mars. While people will need to be ferried up to the Ark orbiting the earth, they will not need to "chase down" a Cycler moving at tremendous velocities and at some distance from the earth, nor will a duplicate high-velocity, high-fuel effort be required to ferry people to Mars. Third, the Ark concept builds upon a simple idea that can be used from the start: transporter vessels are built in low earth orbit, dock at Deimos, and discharge their passengers there. It is hard to imagine how the first mission to Mars could use a Cycler. Adoption mid-program of the Cycler concept would be to adopt an idea without a track record in space. Fourth, transporter vessels will allow a measure of flexibility in case of accident that will mostly be absent with Cyclers. In case of disaster, an Ark mission can be reconfigured easier than can a Cycler mission, which will be moving through space at high velocities.

The Ark in this chapter is not an engineering proposal as much as it is a solution to the problem of fulfilling "ticket contract" obligations. While there will be numerous difficulties that the mission to Mars will face, perhaps none is bigger or offers greater potential than the proposal to transport a person to Mars after the "purchase" of a ticket to Mars.

The Ark-Cycler described here may become the mission to Mars' reality or there may be some better idea developed at the earth Embassy to the same end. The larger point is that if 50,000 people contract for a trip to Mars, the Confederation must have an effective program to get them there. Widespread belief in the ticket system will come about when there exists a feasible Ark-Cycler program which, in turn, will attract people to the mission to Mars. Detailed engineering and realities at the mission to Mars will determine what kind of vehicle ultimately is built to satisfy 50,000 eager customers. The important thing for now is that ideas like the

---

[3]   As always, the technical strawmen present in these three volumes are intended to paint a portrait of one possible reality for the reader, not to pontificate upon a necessary design.

Ark Transporter are not inventions of this book, but have been in the literature for some time. As such, the idea should make sense to a great many readers.

Years After Start of Earth Embassy	First Generation Arks			Second Generation Arks				Third Generation Arks				Fourth Generation Arks			
	1	2	3	4	5	6	7	8	9	10	11	12	13	14	15
30															
31	1	2													
32			3	4											
33					5	6									
34							7								
35	8	9													
36			10	11				12	13						
37					14	15				16	17				
38							18					19	20		
39	21	22													
40			23	24				25	26					27	28
41					29	30				31	32				
42							33					34	35		
43	36	37	(38)	(39)				(40)	(41)					(42)	(43)

Figure 42: The Ark Missions

Figure 42 shows the schedule for 15 Arks. As depicted, it will require about 25 years to transport 40,000 people to Mars by Ark. Because upwards of 30,000 people will enroll at the earth Embassy only after the first landings on Mars, most people to be transported aboard the Ark might realistically expect to go to Mars soon after their ticket has been paid for.

Preparations For A New Life

Each Ark Transporter mission will coalesce a number of earth Embassy activities into a single Mars mission. People will have received generalized training during their service in the Fleet, but this training will become accelerated and specialized as Ark Transporter manifests are created. Each Ark voyage will have an important mission to improve the economy of Mars and to expand the infrastructure there. The coordination of people and resources for each Ark mission will require significant planning and pre-flight preparation. In addition, the people selected for any given Ark mission will form a group trained to establish a new base on Mars[4]. As the Fleet plans each new settlement, it will solicit businesses for transplantation to the new base. These businesses will accept the assignment based either upon a desire for early transport to Mars or as a result of some natural feature—minerals, manufacturing opportunities, potential for tourism, etc.—near the new Martian base. As corporations are selected to transplant their businesses, the Confederation will offer to transport people to support these corporate moves, supplementing the economy on Mars[5]. The Fleet will harmonize Fleet assignments with these passenger manifests. An Ark "task force" will provide for training opportunities for the people who will share the new Martian base. One year prior to the departure of the Ark Transporter, the passenger list for a given Ark mission will be finalized. As with so many other aspects of Confederation life, the Fleet will serve as a force not only to solve the technical issues of the mission, but also to overcome societal, economic, organizational, and other problems facing the settlers. For each Ark mission, the Fleet will be responsible to create as a cohesive and disciplined group as possible, with an end goal of preparing this group to achieve lasting success on Mars.

Each Ark mission will include a mix of farmers, educators, government officials, and medical personnel. Also included will be groups of business people managing the businesses going to Mars. Almost every Ark mission will include a construction company and some will have two or three. Though Martian Fleet units will have laid out the new base and built a few structures, most of the work to construct the new base will be done by those transporting on the Ark. The cre-

---

[4]   If manifests of the Arks are tentatively created three years out and finalized one year out, then the settlers can work together for up to three years, doing what is necessary not only to plan for their six-month flight to Mars, but also to prepare for their first years in the new settlement.

[5]   Several people have expressed concern over the repeated use of the word "corporations" in these three volumes, betraying a fear that the mission to Mars is a conspiracy of some evil, "Big Business". The mission to Mars envisions an entrepreneur-friendly business registration process that will serve most purposes (without the hassles) of the corporate form used in the US. Registration (for access to hard currency) implies corporate protections at the earth Embassy, thus allowing the use of the word "corporations" in the text. The term is used even if many such businesses are, in fact, small, mom-and-pop operations.

ation of new homes and the shelter's infrastructure will occupy the settlers for at least a year. Having one, two, or three construction companies with each Ark and a reasonable number of other skilled construction workers will not only provide the new settlers with jobs, but also will facilitate the rapid transformation of a spartan military base into a Martian village. A mix of people with different skills will not just speed construction of a new town, it will also help create an entirely new economic unit, with new farms, new schools, and as many other amenities as can be moved on one ship, at one time.

The *ad hoc* Fleet task force will be under the command of a senior Fleet leader, who, in his private life, may be a lawyer or a farmer or a teacher. This commander will be assigned missions before the voyage to Mars begins, during the flight itself, and for the first year[6] after landing. She will command Fleet activities on board the ship and will assist the Ark's captain with whatever tasks or problems may exist. The captain of the ship will "rank" the Fleet commander on most matters while in space[7], but the task force commander will command virtually everybody else (except the small crew) on the Ark. While civilian authority will be established at some point during the first year on Mars, the commander will play a vital role while on full-time duty during the formative months at the new base.

Once on Mars, people will not be forced to engage in mission to Mars' activities. On Mars, people will be free to live their lives as they see fit, exploring the endless possibilities that the settlers will find. While there may indeed be "Martian mining towns" like those that form the backdrop to Hollywood horror films, life on the Red Planet will never be as grim as is painted on the silver screen. What crime and social disorder were found in the old American west came from two factors that will be absent from Mars. In the Old West, there was significant friction between the European and African-American settlers and the American Indians, whose land the Europeans coveted. The scattered fighting that occurred between established American Indian tribes and new immigrants will not recur on Mars. Outlying Martian outposts are hardly likely to come under attack from anyone. The other matter that contributed to the disorder of the Old West is that it was a magnet for southern refugees escaping the physical and psychological damage of the American Civil War. Many of the gangs and much of the trouble in the west were caused by refugees who had little alternative in their mind but the life of the

---

6   The commander will revert to part-time status after the election is held for village officials.

7   The analogy is to a ship transporting a marine battalion. The captain runs the ship, even if outranked by the marine commanding officer.

outlaw. Again, this issue will not be faced on Mars. The earth Embassy experience will screen out those who might be a disruptive influence on Martian society.

It would violate an individual's human and Confederation rights to encumber unnecessarily an individual's freedom of choice and travel. The Arks are not to be prison ships, transporting convicts to a 21st-century Botany Bay. Rather, Arks will transport people who have contracted to live on Mars and in a new Martian home. Each person disembarking at a new Martian base will find a home and a job at the new base. If, however, a person or a family wants to sell the house and find employment at a more densely populated settlement or with friends at another Martian town, the Confederation will not stop them. From the beginning, there will be some cross-pollination of people at the new base. People from more established settlements will move to the new base for better employment opportunities or for adventure, while some of the Ark mission members will leave the new base soon after arrival. Though each Ark mission will be designed to create a new human settlement on Mars, only time will decide which settlements grow and which ones remain small.

Six-Months in Space

Using chemical engines, the Ark's travel time between earth and Mars will be five to eight months. During this time, 1000 people will need to eat, sleep, work, fight with their spouses and neighbors, exercise, and generally try to keep sane in cramped quarters. While the ship will not be as crowded as a World War II (WWII) submarine, it will nonetheless require large numbers of people to live together for a reasonably long period of time. The best way to pass the time will be to work. There will be two main sources of employment during the voyage. First, the Ark itself will require a myriad of tasks to be accomplished during the mission, offering numerous employment opportunities aboard the ship. In addition, with a little planning, other economic activity can also occur during the six-month voyage. A well-developed plan can allow the trip to be a productive period in people's lives, despite their isolation from the rest of humanity. Figures 43–46 show how all of this might be achieved.

Each Ark Transporter will have 1000 passengers and crew. The average manifest might reveal 600 able-bodied workers, 300 children and students, and 100 retirees and non-economy parents. If six hundred job positions can be created, then most of the adults can work during the voyage. As shown in Figures 43–46, these six hundred "positions" require less than 300 "jobs", if one structures work so that each position is filled by two employees[8]. These figures do not envision part-time work, but "virtual part-time" employment. One of the most important aspects of planning the voyage will be to ensure that most passengers follow a routine that supports this "virtual part-time" system. Thus, each passenger should spend one of the six months on "vacation", one month serving with the Fleet, and the rest of the time working a 30-hour week. The ship's captain and the Fleet's task force commander will be available, if necessary, to enforce these regulations. The remainder of this section will explain how this mini-economy might work.

Figures 43–46 have been divided into various sections of this economy. The first group of jobs will consist of those people employed directly by the Confederation or directly under Confederation contract. Listed first is the crew of the Ark Transporter. The crew has three deck officers, who will each have an eight-hour watch to supervise operations of the spacecraft. The crew also includes lower-ranking Fleet crewmen who will be assigned routine crew duties. Most likely, the captain and at least one other member of the crew will have already gone to Mars, perhaps aboard one of the smaller transporter ships. The other crewmembers will have trained in earth orbit. Moreover, since each Ark will have been constructed in space over the course of many years, each crewman will have hands-on experience operating the ship in space.

The crew will fill normal shipboard positions. The "Provosts" will serve as the chief law enforcement officers on board the ship. They will be the direct representatives of the captain and will serve as the leaders of the Fleet teams used to ensure the safety of the vessel. The "Electricians Mates" will operate and repair the communications, electrical, and electronic devices on the ship. The "Engine Room" personnel will operate the nuclear reactor and the ship's thrusters. The "Outsiders" will make any necessary repairs outside the vessel and, along with the remaining ship positions, will generally assist with ship duties. Every crewman will be a full-time member of the Fleet, having decided either to make a career of the Fleet[9] or to have signed on for an extended tour. This full-time crew will have received extensive training for their mission, even if most of them have not yet made the earth-to-Mars run. Once having made the run, of course, the crew should be highly proficient in their duties.

---

[8]   There are also about 75 "professional" positions, for which there will be only one person assigned.

[9]   By the time the Arks start their runs, there will be several hundred full-time members of the Fleet, most of whom will be posted to space and to Mars.

Professionals	# of Positions	Non-Professional	# of Jobs
Captain	1	Provost	5
First Mate	1	Electricians Mate	7
Second Mate	2	Engine Room	7
		Outsider	2
		Plumber	3
		Deck Hand	3
Staff Leader	1	Charge of Quarters	5
Staff Assistant	1		
Fire Emergency NCO	Fleet Part-timer	Shift Leader	3 Fleet Part-Timers
		Fireman	3 Fleet Part-Timers
		Policeman	3 Fleet Part-Timers
JHC Chief	1	Coordinator	9
How This Works: Nine JHC Coordinators, working 30 hours per week, equal 270 hours worked per week. If the office is open for 54 hours per week, then five people will be working at any one time. This is also the equivalent of about 7 full-time employees.			

Figure 43: "Confederation" Jobs

There will also be a five-man Fleet staff, organizing the Fleet task force on the Ark. As you recall, this task force will draw members from most of the Fleet regiments. These personnel will be organized into a provisional battalion for the flight and for operations at the new Mars base. For example, personnel from the engineer regiment will assist with construction at the new Mars base and crewmen from the military police regiment can serve as the policemen we have listed in Figure 43. There will also be crewman on board who hold fire suppression and other ratings necessary for the safe operation of the ship. The job of this full-time Fleet staff is to coordinate training, activities, and supplies of these various personnel, to support the Ark's captain, and to keep people occupied during each passenger's 31-day tour of duty during the flight.

On the Ark, the ship's provisional battalion will conduct a variety of training exercises. In addition to equipment training, exercise to combat the effects of low gravity, map reading and navigation on Mars, first aid, etc., there will also be extensive training for actual shipboard operations. The two most important shipboard operations will be the Fire-Emergency Patrol (FEP) and the Emergency Operations Drill (EOD). The FEP will be a continuous patrol of the ship, with a mission to assist the ship's provosts with fire fighting and any domestic or criminal issues that may arise. The Ark's "Fireman" position will be filled by highly trained crewmen, who will be the ship's primary fire fighters. EOD is an exercise to ensure that all persons on the Ark understand what to do during an emergency. While a fire, hull breach, nuclear reactor problem, etc. may be critical, the size of the vessel and the ship design will allow the event to be met with a low risk of catastrophic destruction of the transporter. These EOD exercises will entail mobilizing the entire ship's compliment, made rather convenient because of their organization as a Fleet provisional battalion. Two or three "all-hands EOD" drills should be run during the voyage. The Fleet and the captain will run numerous simulations and training exercises to enhance other shipboard capabilities.

The Jobs and Health Coordination (JHC) staff will assist the captain to meet two important mission objectives: to ensure that there are employment opportunities for everyone on board and to administer a formal program of physical exercise to combat any problems caused by initial exposure to low gravity conditions[10]. To accomplish the first mission, the JHC will post job openings for those wishing to find new employment[11] and will monitor these jobs to ensure that overly ambitious workers do not work more than their fair share of hours. Indirectly, the JHC will also help to monitor the one-month vacation "suggestion" that is designed to promote leisure on board the Ark. If someone has not taken this "vacation", then the JHC will act accordingly, perhaps even denying permission for a person to take a certain job.

The one-month vacation "suggestion" is meant to minimize two psychological problems facing the passengers. First, because the ship will be relatively crowded, there will be limited opportunities to vent the normal frustrations of life. An enforced vacation will allow for a much more relaxed atmosphere on board. Second, it will allow everyone to have an employment opportunity during the voyage. The vacation may also be a person's last chance to relax before beginning the multi-year task of building a new base on a strange planet. While shopping

---

[10]   Mars will be perpetual "low-grav".

[11]   Within reason, people may change jobs in this mini-economy.

opportunities may be limited (with experience, retail activity may become increasing feasible on Ark missions), there will be a library, a swimming pool, a gym, and a "hotel" area, where couples can go for real privacy and sex. Of course, the galley will always be open and likely a retail food outlet (Domino's Pizza), where people can linger as long as they like over bad coffee or a greasy pizza. The JHC will suggest strongly to parents that they make time for themselves and take advantage of the hotel and the ship's daycare facility. After all, the kids will be only a few decks away.

The exercise regime administered by the JHC will consist of swimming, "weights", and possibly racquetball or basketball courts. Swimming may be best because of the total body workout it provides. How the JHC will monitor each individual's activity or enforce standards will be left to the experience provided by previous flights. This activity will be important to ensure health aboard the Ark and will go hand in hand with the Ark's program to catch health related problems early. Getting sick at the earth Embassy will be a problem; getting sick on Mars may cause difficulty; getting sick aboard a transporter vessel could be disastrous, not only because of the possibility of contagion, but because of the ship's limited medical facilities.

Figure 44 lists some services provided on board each Ark Transporter. The most vital of these services will be the medical and dental services. A nurse and medic staff will see most patients, with a doctor and physician's assistant available for appointments. The medical clinic will probably have a day nurse and a night medic. There will also be Fleet crewman with medical training that can supplement the staff in an emergency. The health clinic will likely be a privately run operation, with each person being an employee of the clinic[12]. Most of the remaining services discussed will, in fact, be run as a private enterprise, with payments coming either from those receiving the services, from the Confederation, or from another large contractor. The new Martian base with 1,000 residents will have these 12 trained medical people, operating as a private clinic, which should be able to provide the base with a reasonable level of primary care. On the Ark, the desire is to catch trouble early, when limited resource interventions can be efficacious and to prevent, if at all possible, health problems from lingering.

---

[12] At the earth Embassy, the Fleet provides most medical care. On the Ark, the Fleet pays, but civilians provide the vast bulk of the health care services. On Mars, both insurance and providers of health care are private.

Professionals	# of Positions	Non-Professional	# of Jobs
Doctor	1	Medic	5
Physician's Assistant	1	Technician	3
Nurse	5	Medical Secretary	3
Banker	1	Teller	3
Lawyer	2	Legal Secretary	2
Catholic Priest	1		
Indian Holy Man	1		

Figure 44: Services

Other services on the Ark will allow passengers access to a lawyer, her bank, and her God. These will be private activities, branching off activities begun at the earth Embassy. The banker will allow for normal cash management for each household[13], while there should be two lawyers available on each trip. If some legal issue arises during the voyage, there will be an attorney to represent each party. As there will not likely be an overwhelming amount of legal work generated aboard-ship, these lawyers can also work on issues and cases transmitted from earth or from Mars. Most of these positions, lawyer, priest, banker, etc. represent the "professional" class of worker, which will not have the kind of structure as the "job positions". Rather, these people are expected to maintain an easy work schedule, working their vacations, Fleet tour, and 30-hour week into this easy schedule. This category of employment will not be amenable to the JHC employment rules, but JHC will nonetheless goad workaholics to relax during the voyage.

As there will be several hundred students on board the Ark Transporter, a reasonably sophisticated system of education will be necessary. At the college level, there will be 100 full-time students (18–24 years old) and 100 part-time students. The two universities (Mars and Ares) will share a pot of ten professors, each of whom

---

[13]   The financial situation of the settlers will be bright on the Ark. No more ticket contracts, a paid-for home on Mars, and free food en route. Food will no longer be free once they reach Deimos/Mars.

will teach four classes during the trip. Thus, forty college-level courses, taught during two "semesters" on the Ark Transporter, will allow most college programs to continue unabated. Probably only two or three of these college professors will hold full-time positions at their universities. The others will be adjunct professors recruited for the voyage. During the multi-year preparatory phase at the earth Embassy, these adjunct professors will be trained to ensure that the materials they present in class are university caliber. Three of the ten professors have been programmed for a seat to Mars because of their professional talents or to reunite families. We see that the musician will teach four music-related classes and was assigned a seat on the Ark to rejoin her husband already on Mars. An engineer is teaching four engineering classes (two classroom courses and two labs) and was given a seat on the Ark because he was needed at the main Martian nuclear plant. An accountant was also given a seat on the Ark to replace the Confederation's staff accountant at another Mars base, who contracted a rare blood disease and died. This gentleman will teach an accounting class each semester. In addition, as an amateur Martian areographer[14], he has trained himself (and been certified by the university) to teach Martian geography. Thus, there will be available one geography/areography class per semester. In total, there will be 35 different classes (five classes are repeats) offered en route to Mars. These classes break down as shown in Figure 46.

Professionals	# of Positions	Notes
Professor	1	
Professor	1	
Professor	1	
Professor	1	
Professor	1	
Professor	1	
Professor	1	
Professor	1	Musician
Professor	1	Engineer
Professor	1	Accountant
Principal	1	
Kindergarten	1	10 Students

---

[14]   A Martian geographer.

Professionals	# of Positions	Notes
Grade 1	1	"
Grade 2	1	"
Grade 3	1	"
Grade 4	1	"
Grade 5	1	"
Grade 6	1	"
High and Middle School	1	Specific Subject
High and Middle School	1	"
High and Middle School	1	Hockey Coach
High and Middle School	1	"
High and Middle School	1	"
High and Middle School	1	Writer
Library	2	

Figure 45: Educators On The Ark

During the preparatory phase for the voyage, students will have been advised to take classes that would not be available aboard the Ark and were advised not to take classes that would be available en route. While traditional scheduling problems would develop for many students, those problems should not be dissimilar to what most people experience at an earthbound school.

Number of Classes Offered	Types of Classes
12	Freshman- and sophomore-level classes
9	300-level courses
9	400-level courses
10	Graduate classes or seminars

Figure 46: Classes Offered Aboard The Ark

The students and the professors can help make an interesting point about final destinations for the passengers. As part of the contract with the Travel Corporations, a home on Mars will be provided for the adult voyagers, which will be located at the new Mars base. The university-settlement bound professors and a few other people, however, will have homes elsewhere. When these people arrive on Mars, they will not go to the new base, but will go to an existing village. The musician, wife of a man at another Martian base, will also depart for an existing Martian settlement. As mentioned previously, no one is required to remain at a particular destination on Mars. Such an idea is contrary to notions about a person's basic right to freedom of travel. If a person's house on Mars is at the new base, most people will choose to remain there, especially if there is a job and "life" programmed for them. The city slicker settler could decide, however, to pay for transportation to one of the more established bases and to sell or rent out this new house[15]. Freedom in action. Just as likely, too, will be incentives for people to move to the new Mars base from the older Mars villages.

Back on the Ark, the grade school will teach the younger students. As indicated in Figure 45, two of these teachers will not be full-time teachers, but will have been specially recruited to teach during the Ark voyage. The one man is a former Russian professional hockey player who will take the full-time coaching job at the University of Mars. During this trip, he will teach middle- and high-school classes in history. The other teacher is a writer who will rejoin her husband on Mars. She will teach Shakespeare to high-school students and will also attempt to teach the un-teachable: to write intelligent, comprehensible English.

Each grade, K-6, will have a separate teacher. Each teacher will have about ten children. The middle and high school grades will divide into subjects. Again, about ten children per grade will be taught, though with some clever class scheduling, class size can be larger than ten students per class.

Finally, there will be a library on the Ark (to include data links with earth and with Mars). The two librarians working on the Ark will assist with the thankless task of helping people find and use information and books. They will also help to ensure the safekeeping of private libraries, transported by Ark passengers, many of which will be made available to others during the trip, as part of the Ark library. Naturally, both the universities and the grade school will have some materials that will be used by their students. Consolidating all these items into an Ark library will create a resource similar in size to that in most American small towns, but may rival some larger libraries in terms of the quality of some materials available.

---

[15]   This house will be free of separate mortgage, although any unpaid portion of the ticket will be backed by a lien against this property.

The next major category of Ark employment is "Ark Services". We start with the recyclers who will have very interesting roles on the Ark. The easy part of their job will be to segregate waste materials into streams of various plastics, glass, metals, paper, etc. A more interesting part of the job will be to process some of these materials, like paper and plastics, into feedstocks that can be sold upon arrival on Mars. The most interesting part of the job, however, will be to accept for cleaning and re-use certain bottles and other containers that will help maintain a reasonable lifestyle aboard the Ark. For example, most passengers will enjoy the option of drinking a Coke from a bottle while watching television. These bottles will need to be recycled and refilled. Or, a man may want to relax during the evening with a bottle of beer. Same issue. Containers of fruit juices and jars of mayonnaise, ketchup, coffee, and a dozen other items would be easy to clean, refill, and reseal. The recyclers will have a major role in these operations. It would only require 5–10,000 such containers on the Ark to allow each family a small kitchen cabinet refrigerator full of Coca-Cola, beer, coffee, milk, and sandwich essentials. As will be further articulated below, a small concessionaire, and a small business to refill these containers from food stocks will help to make the Ark feel less like a barracks and more like a large apartment complex.

Professionals	# of Positions	Non-Professional	# of Jobs
Re-Cycling Manager	1	Operators	9
Laundry Manager	1	Helper	1
Galley	1	Shift Leader	6
		Breakfast Crew	23
		Lunch Crew	23
		Dinner Crew	23
Daycare Manager	1	1st Shift	5
		2nd Shift	3
		3rd Shift	2

Professionals	# of Positions	Non-Professional	# of Jobs
Gym/Pool Manager	1	Shift Leader	6
		Lifeguard	9
		Helper	9
Water Manager	1	Potable water crew	6
		Wastewater crew	6
		Plumbers	3

Figure 47: Ark Services

The laundry we discuss is self-service, though one that provides laundry services to its customers is possible. Self-service would probably be much less efficient and harder on the machines, but there would be much less fuss about lost or improperly washed clothes. The manager and her helper keep the laundry open 24/7, enforcing simple rules to keep the machines from breaking down.

Most food on the Ark will be provided buffet style at the galley. It will remain open 24/7 and will be the largest employer on the Ark. There will be some ability for families to cook for themselves and there will be a fast food restaurant and pubs on the Ark as additional choices for food and beverages.

There will be a daycare facility that will remain open 24/7. It will not be a large facility, as most parents will work only 30-hours per week and will have one month off during the voyage. Still, there will likely be 10–20 children during the day and a few "night-shift" children sleeping there.

One of the treats of the Ark will be the swimming pool. It will always be crowded, though those who have not yet realized that they are going to live a pioneering lifestyle will complain that the water looks dirty. Not only will the JHC people require exercise, but the pool should be the friendliest place on the Ark. Another 24/7 operation, the pool will actually serve double duty as a large storage tank in the water circulation system aboard the Ark. A small workout room for free weights and maybe a racquetball court or two can also be built.

The final Ark service will be the water and wastewater engineering crew. Sink and sewer waste must go somewhere and the water from the faucets must come from somewhere. In addition, water can provide some added shielding from radiation, as well as being something transported to and from Mars for various reasons. The Ark water system will provide water and compost to grow crops aboard ship (see below), the water circulating rapidly among the 1000 people and dozens of animals on the Ark.

The Ark water system will accomplish three main missions: 1) it will remove contaminants from the water, 2) it will compost these contaminants for farming operations on the Ark, and 3) it will provide as pure drinking water as is possible. The primary treatment of water removes solid objects from the water such as feces and kitchen waste. The secondary treatment of water removes the "nutrient" contaminants through biological processes. The final treatment of water polishes it to kill any remaining biological organisms and sends drinking water through an ultra-efficient filter process. Only drinking water will go through all stages. The water in the pool will probably be a combination of drinking water, water from secondary treatment, and some added chlorine. Partially treated water for crops will be piped over as "non potable".

In addition to the plumbers assigned to the Ark, the wastewater company will also have three plumber jobs.

Private enterprise will also function aboard the Ark. We have already mentioned the pubs and hotel, and suggested that a company will refill plastic bottles. Discussions about a fast food establishment and the farmers will occur in separate sections. In addition, there will also be others engaged in activities for benefit of those on board and others working on problems transmitted up from earth or Mars.

On this trip, another "plastics smythe" operation will travel to Mars. Their craft, discussed again a few chapters on, will be to fashion plastic into any shape desired. Some of their wares will be used during the flight. Other orders will be fashioned for sale on Deimos and Mars.

There will also be a seamstress/clothing manufacturer traveling on this Ark mission. A manager and three workers will offer a full line of clothes on the Ark though, once on Mars, the company plans to specialize in girls' clothing.

Professionals	# of Positions	Non-Professional	# of Jobs
Domino's Manager	1	Workers	3
Vegetable Farmer	3	Laborer	5
Chicken Farmer	2	Laborer	2
Hog Farmer	1	Laborer	1

Professionals	# of Positions	Non-Professional	# of Jobs
Pub Owner	3	Workers	15
Hotel Manager	1	Worker	1
Refill Manager	1	Worker	3
Plastics Smythe	1	Helpers	6
Seamstress	1	Assistants	3
Concessionaire	1		
Communications Svc Manager	1	TV Crew	6
		Helpers	6
		Technical Crew	6
		Internet	2
Mars On Line	1	TV	6
		Helpers	6
		Technical Crew	6
		Internet	6
Miscellaneous	12		

Figure 48: Private Enterprise on the Ark

A concessions stand will provide sundries, including some foods, to the passengers. This outlet will be run by someone who will open a drug/general store at the new base. The idea will be the same on the Ark, with the qualification that the inventory will be tiny, unless there has been an enormous amount of planning done by the owner. The key will be the ability to store toothpaste, snack chips, razor blades, soft drinks, etc. in a way convenient to the operation of a cramped ship.

There will be two communications outlets. It turns out that people from both Communications Services, Inc. and Mars On-Line are aboard this Ark mission. It was decided to let both operate on the Ark. On Mars, each will provide a variety of TV, radio, and internet services. On the Ark, only the TV and internet will operate and, given cramped quarters, the two companies will be working together. The internet is standard for all Ark missions, but there will be several TV shows produced aboard the Ark (mostly talk shows about earth and Mars news) and new movies beamed aboard from earth.

There are also about a dozen independent businesspeople on the Ark, working on projects unrelated to the Ark mission. Some receive their work by communicating with earth, some by communicating with Mars, and some by communicating with both.

Finally, a point to remember before we leave off talking about our mini-economy on the Ark. While several of these businesses will continue after arrival on Mars (the pizzeria will continue, the wastewater treatment plant will not), many businesses "traveling" on the Ark did not operate. For example, there were three small construction companies on this Ark trip. They will have an immense quantity of work before them at the new base. For the Ark mission, however, these workers hired out as plumbers or with the JHC. There was also a five-person newspaper organization. It did not operate formally during the trip although four worked for Communications Services during the voyage[16]. In all, 33 separate businesses traveled on the Ark and another 364 people have jobs with employers already on Mars. Only about 100 of the 1000 people on the Ark have not finalized work plans for Mars and, of these, 90% have careers or think they know what they will do. The mini-economy on the Ark is thus different then from the maxi-economy the new colonists will enter once they reach Mars.

All Aboard!

Five to eight months is a long time for a business to be idle and, equally important, is a long time for people to be idle and to forego the "luxuries" of life. One

---

[16]   The other worked at the daycare facility.

of the most important aspects of the large transporter vessel operations will be to prepare food to supplement the ship's galley. Here, a Domino's franchise moving to Mars will be asked to operate its business for six months aboard ship, to allow people the chance to live well as they travel.

Operating a Domino's franchise aboard a space ship to Mars may at first seem ridiculous. Consider, however, the relative ease and simplicity of this plan. Domino's might use the galley's equipment, but more likely would be allocated space to move its pizza ovens, freezers, etc. to Mars. Since this equipment is moving anyway, it may as well be used during the voyage. The food stocks for the Domino's will be extensive, but no less difficult to plan than the provisions required for a thousand people for six months. If each person is expected to eat 10% of their calories at the Domino's, then the ship's main stores are simply reduced accordingly and lots of cheese and pizza sauce brought aboard. The space needed to operate will not be much greater than that needed to store the food and equipment, but even if some extra space must be allocated, the benefits of having the Domino's operating on the ship will far outweigh the marginal cost of allocating this space. Moreover, an efficient plan might have the Domino's adjacent to the ship's galley, where Domino's customers can sit in the main cafeteria.

The advantages to this system are numerous. The Domino's business will continue to have cash flow as it travels to Mars, the business owner may be given seats aboard the vessel to transport her employees, and the morale of the ship can be improved by having a familiar—if greasy—option for food. Remember that the continuous movement of people, infrastructure, supplies, and equipment from earth to Mars is a key point about the mission to Mars' ideas. Once the ship has docked at Deimos, the Domino's started at the earth Embassy now becomes an asset on Mars. People who will have struggled for years and will have been deprived many luxuries to fulfill their Martian dream will now find that they can indulge themselves with a simple pleasure; offered up by a business owner whose thoughts will be expansion, expansion, expansion, and how much money she will be able to make on a planet hungry for pizza.

There will be about 230 missions to transport 50,000 people[17] to Mars. There will not be 230 Domino's transported. Perhaps one of the missions will have a fine French restaurant aboard. The difficulty of operating the restaurant on a cramped transporter vessel may be much greater than the difficulty facing the Domino's,

---

[17]   Most of these missions will made by the small Explorer and Shuttle transporters, which will accompany the larger transporter ships. Ninety nine percent of all people transported to Mars will travel on either a Caravan (100-person) Transporter or an Ark (1000-person) Transporter.

but *Homo capabilus*[18] will figure out how to make it work. Perhaps this restaurant will not operate in competition with the ship's galley, but in lieu thereof. Thus, every Tuesday and Thursday night the galley might be operated as a French restaurant, by the restaurant's staff, with all meals served restaurant-style in the cafeteria. The point is that as the earth Embassy creates several dozen restaurants during its first 50 years, the transportation of these restaurants will create opportunities to supplement the galleys on each large transporter mission and will allow for the quick introduction of these businesses into a vibrant Martian economy.

Bah, Bah, Black Sheep

Given four assumptions: 1) that there will be "farmers" traveling on board the Arks, 2) that terran flora and fauna will be transported to Mars, 3) that the travel times to Mars will average six months, and 4) that wastes aboard the Ark will need to be recycled, it makes sense for the Arks to provide both "farmland" and a place to transport animals. The outer layer of the Ark can provide about two acres of land, enough to produce 1–2% of the calories needed by people being transported. In addition, if the outer layer is devoted to agriculture, a collision with a micrometeor or other space debris will result in losses to plants and a few chickens rather than to humans and more precious cargo. The soil, deposited on the outermost wall directly against the exterior of the ship, should also help to absorb the energy of the micrometeor and thus reduce damage.

The actual farm work need not be ambitious. If the two acres are used to grow carrots, lettuce, potatoes, and tomatoes, ship-raised supplies might meet much of the fresh vegetable needs of the Ark. In addition, the chickens and any hogs that might transport on the Ark can assist with the ship's waste recycling. Six or seven hogs and a hundred chickens will not add much to the diet of the settlers during their voyage, but the transport of such animals to Mars will be invaluable. Once a hundred hogs and a thousand chickens reach Mars, the Red Planet will have an inexhaustible supply of pork and chicken. Neither animal requires much space on the Ark and either can easily be cared for by Ark farmers working their small fields.

The actual details about how to "farm the Ark" will be worked out through the free market, in concert with Fleet and Confederation plans to transplant the earth Embassy to Mars. The likelihood is that better and even more astonishing programs will be developed than those envisioned here. Remember that the Arks will be built in space and may be used for several years before their first voyage to

---

[18]   That sub-species of the genus (sic) that solves problems rather than complains about them.

Mars. In addition, the Arks will remain in space running back and forth between the planets for twenty or thirty years before they are retired. It may actually be possible to create a business and a homestead on the Arks themselves, with full-time farmers raising food full-time on the Arks. There is no reason why—given sufficient nuclear-generated and solar power—farmers could not use the time returning to earth "to farm the Ark"[19]. The idea of using the Arks as farmsteads represents another cost-cutting advantage of the settlement option. If people are going into space and to Mars to live, then the cost to boost food into space from earth might be lowered by 90% if the Arks are farmed. Combine this with the idea that Mars and Deimos become food exporters, too, and food costs in space will only be slightly higher than food costs in the modern west. Again, we find that the mission to Mars' idea of settling Mars and building an economy will be vastly cheaper than NASA "exploration" missions. For the mission to Mars, people build an economy; the more efficient that economy becomes, the less everything costs. For NASA, voyagers are not farmers, but scientists, chiseling rocks as they look for "life on Mars". The former add value, while the latter are economic black holes, running up enormous costs to ship food, water, and other supplies to Mars.

Arrival on Deimos

The Ark will tie down at Deimos. From Deimos, settlers will be transported down to the surface on Mars Landers. Given the limited availability of the Mars Landers, however, people will likely be housed on Deimos for a few days or weeks. There should be no lack of work for those who desire to delay their departure to Mars or those who want to earn a few extra dollars while they wait at Deimos. There will be routine maintenance activities, including maintenance on the Ark itself. Plus, there will be work in Deimos' greenhouses and work to continue the construction of the Deimos base. Both the harvest and these construction activities will have been planned to coincide with the Ark's arrival and not a little communication back and forth between Deimos and the in-bound Ark would have concerned final arrangements between workers from the Ark and "employers" on Deimos. While the Ark may have had some comforts, no doubt the passengers will have

---

[19]   After each six-month voyage from the earth to Mars, there will be a two-year layover awaiting the next launch window, then the six-month return trip. As techniques for farming in space becomes more sophisticated, the Arks themselves may be able to provide 7–10% (2.5 years with no passengers and 0.5 years with) of the nutritional requirements of the passengers. Even more food might be grown if more of the Ark than just the outer rim is given over to agriculture when there are few or no passengers aboard.

developed cabin fever during their half-year in space and will be happy to work in Deimos' pressurized domes and bays.

"Dry" docks on Deimos will be the only place the Arks will not be in space. The ability to work on these vast transporters in a pressured, gravitied environment will allow for the easy repair of the exterior of the vessels as well as facilitating more complicated internal repairs that the captain had decided not to tackle in space.

For a few, Deimos might even serve as a temporary home. Those with special skills might find a good wage working a few months on "the rock". Not only might the chance to establish a more normal routine be attractive, but the desire to allow events to unfold in the new community or to await a treasured job opportunity at an older community could convince some to sign on at the Deimos base. As Deimos is likely to be a busy spaceport as well as an industrial center, with ever-growing trade with Mars, inner asteroids, and vessels orbiting the earth, these opportunities will grow with each landing transporter vessel. And as Deimos would not likely have a permanent settlement, only permanent facilities, the demand for workers will always be great.

Arrival on Mars

Approximately 1000 people will have prepared for years, journeyed for six months in space, and endured a two- or three-week hiatus on a lonely rock a few miles above Mars, only to find their greatest challenge before them. The new settlers may land at a flower-rich town with a population of 1,100, but their ultimate destination will be a dusty plain hundreds of miles away.

The primary purpose for each Ark flight will be to land settlers to build a new community on Mars and to help facilitate a rapid and orderly settlement of the planet. Accordingly, the new settlers will be asked, after a short rest, to load buses or mount horses[20] and head off for their new homes. The system of 49 bases/towns, each town 100 miles from its neighbors and planted on a Texas-sized area of 360,000 square miles, will allow 1% of Mars quickly to become reasonably well known. The final leg of the Ark's mission will be for its settlers to go to a newly prepared base and build their new home.

Upon arrival at a new base, settlers will find a large tented area and a few buildings. Crudely divided barracks for families should also be in place and possibly a few domes and other agricultural structures, but little more. No doubt some previous settlers will have been drawn to the new community, especially as

---

[20]   Much more later.

the Fleet will not undertake most of this construction itself, but will contract with existing Martian firms for construction.

After each new base is created, the base will develop as economics dictate. Some will grow quickly, other villages, with less economic reason to exist, may struggle to keep their residents and may only become way stations on the way to somewhere else. In any case, a swath of Mars will be claimed for the Confederation, for humanity, and for the first two Martian republics. Most villages, however, should become successful. The economic possibilities at each one will be endless, as the task for a decade at least will be to build the village and its economy. Immigration to Mars will not end after the 50,000 are transported, but will increase. Most of these first 49 villages can expect to grow richer and more populous, as immigration to Mars becomes routine.

Part III

Rounding Up the Usual Suspects

Gabriel: How about cleanin' up de whole mess of 'em and sta'tin' all ag'in wid some new kind of animal?

God: An' admit I'm licked?

—Marcus Cook Connelly in *The Green Pastures*

Just The Facts, Ma'am

In the realm of Nature there is nothing purposeless, trivial, or unnecessary.

—Maimonides

Some readers will have little background on Mars. This work cannot help, therefore, but deal with some of the basic facts, issues, and technological problems that will accompany a manned mission to the Red Planet. Beyond just a primer in scientific and engineering ideas, this chapter will address some concerns any thoughtful person might have about a human mission to Mars.

Surface Area—despite being a much smaller world, Mars nonetheless has as much land area as the earth, because two-third's of the earth's surface is water. Most of the land on Mars will be available for settlement and terraforming engineers can plan to maximize this amount of useable land. It is also likely that in Mars' past there was some surface water, possibly even large lakes or a small ocean. While these surface bodies of water will probably reappear during the terraforming process Mars, ironically, may end up having more "useable" land than earth. Terraforming will decide the matter, especially if neither the vast deserts nor the large polar regions found on earth will be created by the terraformers. Again, much of this is speculative and will rely upon future political and technical decisions by those orchestrating the terraforming. Still, the reader should understand the vast potential of the Red Planet to provide immense amounts of land for settlers.

Martian Day—the Martian day is 40 minutes longer than the day on earth. This is only 3% longer than an earth day and likely poses not a problem, but a biological benefit, as humans will have an extra 40 minutes for themselves and for many, an extra 40 minutes per day—every day guaranteed—to relax or sleep.

The Martian day is commonly called a "sol", an expression that measures time in *Martian* terms.

Mars as Antarctica—though a bit of a stretch, this work uses Antarctica as a model to relate the Martian climate to the earth's. Throughout this work, this idea of Martian weather will be refined, but the reader should understand that as humans can tolerate the temperatures in Antarctica so, too, can they tolerate Martian temperatures. As there are warm days in Antarctica, so there are warm days on Mars. As temperatures and storms in Antarctica can be deadly so, too, can they be on Mars. Finally, many readers may not know how dry and desert-like much of Antarctica is. In this respect, too, Mars is like Antarctica, for the Red Planet is a very dry world. In fact, there are places in Antarctica (the Dry Valleys) that so much resemble Mars that some Mar-related research is conducted there.

Martian Atmosphere—the Martian atmosphere is very thin and without sufficient oxygen to sustain human life. In fact, the atmosphere is poisonous to humans, as it is composed almost completely of CO_2, which humans and most mammals cannot breathe, except in slivers of a percent of the total volume. Nonetheless, Mars *has* an atmosphere so, unlike smaller celestial bodies like the moon or Mercury that can never have atmospheres[1], there exists the potential for terran life to exist unprotected on Mars' surface. In much less time than most readers would suspect, this atmosphere can be transformed to allow plants, animals, and even humans to work in the open, with limited or no special equipment.

Axis of Rotation—the earth rotates around an imaginary line canted 23° off of its true north pole. This axis of rotation is nearly identical to that found on Mars. This axis of rotation helps to explain the Martian seasons, which follow a pattern like those on earth. In fact, once the terraforming process has warmed the planet and has thickened the atmosphere, one can expect that seasons on Mars can be similar to seasons in the temperate regions on earth.

The Martian Year—a year is the time it takes a planet to complete one orbit around the sun. Mars' year is much longer earth's, requiring 669 "sols" to complete this journey.[2] Because of the 23° just discussed, a Martian year has four seasons, although each season would last about six earth months. The odd part about these seasons is that they differ from the Northern to the Southern Hemispheres. In the

---

[1]   There is not enough gravity on these smaller worlds to retain an atmosphere.

[2]   Note that in terran terms, a Martian year lasts 687 earth days (an earth day being shorter than a sol).

north, temperatures are milder, while in the south they are more extreme. This relates partially to the dramatic changes in atmosphere pressure from the northern to the southern Martian summer and partially to the differences in the average elevations found in the two hemispheres (more below). On both planets, seasons also differ with latitude and because of other factors. On earth, the Atlantic Gulf Stream provides Germany, for example, with a temperate climate, even though it lies at a latitude equal to that of freezing northern Canada. And, of course, on both earth and Mars those regions near the poles have much colder climates that those lying near the equator. There is another important factor, however, on Mars. The Red Planet is not perfectly round but somewhat "pear shaped", with lower altitudes in the Northern Hemisphere and higher altitudes in the Southern Hemisphere. The differences imply exactly what they do on earth. Higher elevations are usually colder than the valleys below, thus the Northern Hemisphere's climate is generally milder than that of the higher Southern Hemisphere.

If the Martian climate were adjusted to be so warm as to allow only small amounts of ice to remain permanently at the North Pole, many parts of Mars would probably experience six months of North Carolina spring, six months of mild summer, six months of piedmont fall, and six months of a North Carolina winter. Each season would melt quite slowly one into the next. Odd? Yes. Hindering the creativity of poets swooning over the beauty of spring after a hard winter? Yes. Fully capable of supporting normal, human, social and economic life? Of course.

Such an outcome for Martian weather is not being advocated here. Indeed the terraforming efforts required to achieve these temperatures might require such large amounts of atmospheric CO_2 that the air would remain poisonous to humans for hundreds or even thousands of years. Still, the idea of a North Carolina climate on Mars or, more likely, an Ontario climate (with much less CO_2 to scrub out of the atmosphere), should help many people understand why humans should live upon and use the vast tracts of land available so close to our own world.

Olympus Mons—is the name of the largest volcano in the solar system, rising almost 20 miles above the Martian planum ("sea level"). This massive volcano is in the Northern Hemisphere and has a base approximately 400 miles on a side. It is a shield volcano, meaning that it has a mostly gentle slope and looks more like a bulge than a grand peak like Japan's Mount Fuji. The height of this feature takes effort to understand. Mount Everest on earth is tall and some climbers need oxygen to complete the climb, but Olympus Mons is far, far taller than Mount Everest. In fact, Olympus Mons reaches into space. Even after terraforming has changed Mars to allow for terran environments and humans to breathe its atmosphere, those ascending to the summit will need spacesuits to withstand the near vacuum that will be found on the top of Olympus Mons.

Valle Marineris—is Mars' "Grand Canyon". Some speculate that this system of massive chasms, hundreds of miles long and 30,000 feet deep[3], was formed after an asteroid collision caused the catastrophic draining of a large Martian lake or an underground reservoir. Whatever the cause, the subsequent release of dam-burst forces carved out several systems of valleys so large they have no terran counterpart. The floor of the Valle Marineris might make an excellent place for an initial settlement, as the atmospheric pressure at the bottom of the valley is higher than any other part of Mars. As a thicker atmosphere generally means means a warmer climate, the Valle Marineris might also be the warmest spot on the planet. The fact that these valleys lie so close to the Martian equator will only add to the potential for warm temperatures.

There is another important benefit that the Valle Marineris can offer the mission to Mars: it also offers great protection from harmful cosmic radiation. Not only will the thick atmosphere help to screen out much more radiation than other parts of the planet, but the sheer cliffs to be found all over the Valle Marineris will also limit the amount of cosmic radiation that the first settlers would face at a cleverly-placed settlement on Mars.

Northern Hemisphere—the Northern Hemisphere is the darling of Mars. All the early US landings were in the Northern Hemisphere and most books about Mars talk about the north as if they were written by real estate agents selling plots. The Northern Hemisphere certainly has some advantage over the Southern Hemisphere, not the least of which is a generally warmer climate. Much of the excess Martian permafrost expected to be melted over the centuries will find its way to the Northern Hemisphere, which could become very swampy or flooded. Thus, large settlements in the Northern Hemisphere will either need to be very carefully cited or named Atlantis.

Southern Hemisphere—most discussions of Mars give the Southern Hemisphere short shrift. The rationale is that the south is a sort of "lunar landscape", supposedly unfit for initial settlement. The Southern Hemisphere has a higher average altitude and has more seasonal variation than the Northern Hemisphere, which has many vast plains. Admitting that the north has its attractions, one cannot help but wonder, however, whether the cratered south might not ultimately be better for early settlement, as the list of possible settlement sites will be far greater. Why?

---

[3]   This is six times the average depth of the Grand Canyon.

If one criterion for a large settlement is the existence of a draw, valley, or crater that can be enclosed with a plastic roof, surely the south is a far better candidate than the north. The reader will hopefully grasp without additional explanation the enormous advantages to be gained in covering a half-mile-wide crater and filling it with oxygen. At a single stroke, people are living as on earth, in the open and in a small village. Granted, this village will have fewer cows than in Austria and it will have a nuclear generator and the latest "high-tech" equipment. But such a crater or valley could be found every hundred miles or so in the south. Similar land formations are much less common in the Northern Hemisphere and thus settlements using this technique in the north would be much further apart.

Dust Storms—are the bane of almost every book or movie about Mars. Indeed, dust storms are interesting weather events on Mars, sometimes covering the entire planet. The problem for the Nay Sayers, however, is that neither the energy involved in a dust storm nor its effects are likely to be worth noting any more than is a spring thunderstorm on earth. The reason for this lack of drama is the low atmospheric density on Mars. There is simply very little "stuff" in the atmosphere to push around. Thus, a hundred mile per hour gust on Mars would feel like a ten mile per hour gust on earth. While the effects are not to be ignored completely— there will still be films of dust on surfaces after a storm and it is quite likely that dust fines will find their way everywhere during a storm—visions of Saharan-like walls of dirt coming at you need to be shelved. Even in the heaviest of dust storms, visibility is likely to be reasonable and probably greater than visibility in most ter- ran fog. Dust storms will be an issue for engineers as they design equipment for Mars, not life-threatening events to cause settlers to cower in fear.

Space is Not Healthy For Flowers and Other Living Things

This section will help the reader understand the kind and purpose of shelters needed on Mars and Deimos. Even assuming that it begins very early, terraforming Mars will take a long time. And even if the planet warms up in a few decades, it will nonetheless be without a breathable atmosphere for humans for at least a hundred years and probably several hundred years for most terran mammals. To understand the discussion about sheltering on Mars, the reader should understand a little about the conditions the early settlers will face. Here, we begin with the nature of space, the current environment on Mars, and how these environments will affect the human body.

The two main dangers in space for humans are the absence of atmospheric pressure and the extreme cold. Human blood carries oxygen (and other gases) to every cell in the body. These cells operate in a way that creates "pressure" inside the cells. Unless there is a sufficient atmospheric pressure on the human body, gases will boil out of the blood and cells will burst because of the pressure gradient between the cell and its surroundings. In either case, even a short exposure to space's vacuum can be deadly, because it lacks any atmospheric pressure at all.

The extreme cold found in space quickly freezes exposed parts of the body. Frostbite is the freezing of surface layers of the skin. Usually, the hands, feet, and exposed parts of the face are involved. It is very rare for the skin on the arm or the back to get frostbitten because these parts of the body rarely receive the kinds of heat-exhausting cold necessary to cause frostbite. Humans facing a freezing environment almost always have some kind of warm clothing to help retain heat. The fingers, toes, or nose may lose the battle against the cold, even if not directly exposed, if such clothing does not retain sufficient heat. As the finger, toes, nose, etc. have far less quantities of warming blood than the body cavity, the brain, etc., these areas are at greater risk of frostbite. It takes the body several weeks to recover partially from frostbite and we are told that human bodies never completely recover from it.

The cold of space is not only much colder than anything found on earth, more insidious is that exposure to space would almost always come when humans were without warm clothes, and would probably expose parts of the body, like the arm or back, not normally affected by frostbite. Invariably, we are talking about exposure to space after some catastrophic event. In space, exposed body parts would freeze almost immediately and unless immediate relief were provided, the frostbite would spread rapidly throughout the body. Thus, if a person's faceplate popped open in space but was quickly resealed, the frostbite on the face and nose might be reversible. Most of the time, however, exposure to space—a catastrophe aboard a spaceship—would affect the entire body and the frostbite would be immediate, extreme, and massive. Death from the cold of space and especially from the lack of atmospheric pressure, would be expected within a few seconds.

At present, Mars is "midway" between the conditions of space and conditions found on earth. Mars is a sort of Antarctica with very low atmospheric pressure. The atmospheric pressure on Mars is currently 1/100th that of the earth. During the first few years on Mars, humans would die very quickly without pressure suits. Cell trauma and boiling off of gases would occur almost as quickly as in space. The essence of the terraforming process, however, is to thicken Mars' atmosphere with CO_2, to allow for a much enhanced greenhouse effect. This process will increase the ambient atmospheric pressure and temperature, slowing the kinds of trauma just discussed. After the atmosphere is thickened, when the weather is warm one afternoon, humans on Mars may feel a desire to run outside without their oxygen masks, either from cabin fever or simply to express the joy of their Martian experience. The doctors on Mars can study the effects of this new "extreme sport", but as terraforming proceeds, the pernicious effects from exposure to the Martian environment will diminish dramatically.

The fact that the CO_2 in the Martian atmosphere will be *poisonous* to human beings for many years means that humans will need to protect themselves from direct exposure to this atmosphere. In more technical terms, an atmosphere with more than about 1% CO_2, will begin to interfere with human blood's uptake of oxygen[4]. Thus, ideas about creating a super-saturated CO_2 atmosphere to warm Mars quickly must be balanced against the hope that eventually the CO_2 will be reduced to levels that humans and other animals can tolerate. As has been mentioned, the Martian atmosphere (mostly CO_2) can be dramatically thickened by a runaway greenhouse effect on Mars. Unfortunately, warming the planet in this way may leave the atmosphere poisonous for a very, very long time. If humans hope to walk on the planet's surface with oxygen assist in just a few decades and to walk in the open after a few hundred years, the level of CO_2 in the atmosphere must be taken into account. This complex issue will be dealt with again in Volume III, during this work's chapter on terraforming.

So what kinds of personal protections will humans need on Mars? In Antarctica, people might run from their warm shelters to warm vehicles without needing layers of clothes. Frostbite will not occur so quickly that a person cannot endure five or ten seconds outside in shirtsleeves. At first, that will not do on Mars; though, as terraforming begins and the atmospheric pressure and temperature on Mars rise, people may be tempted to expose themselves to the Martian environment, to run from shelter to shelter, as they might in Antarctica. Initially on Mars, people will need to wear the same suits as they do in space. As the pressure and temperatures rise, however, much lighter suits can be worn, mostly to protect against low pressures on Mars. Eventually, on some afternoon when the air temperature reaches 25° Celsius, someone will take the short jaunt across Mars' surface, holding his breath so that the CO_2 causes no harm. Within a few decades after landing, it may be possible to wear normal clothing on Mars, along with some "oxygen assist" mechanism that also keeps out the CO_2. It should be noted that there is nowhere else in the solar system where the environment will be so forgiving as it is on Mars. Everywhere else, humans will be in moon suits virtually forever.

---

[4]   In 2005, the US standard for CO_2 is 5,000 ppm (parts per million). This exposure limit is for a healthy adult for long-term exposures. Further consideration must then be given for children, people over 65, and people with specific health conditions. Here, we assume that short-term exposures to twice this limit for healthy adults would likely be acceptable on Mars and that short-term toxic effects arise only when there are exposures beyond this 1% limit. Suffocation occurs quickly when CO_2 levels exceed 15%.

Less mysteriously, perhaps, is that the human habitations on Mars—the shelters—will need to protect against Mars' low atmospheric pressure and the dangerous levels of CO_2 likely to be present for at least a century or two. These shelters, of course, will need to retain O_2 for humans and their companion life forms as much as they need to keep the Martian environment out.

Terraforming

The subject of terraforming will be covered in Volume III, in one of the longest chapters of the entire project. In the original version of this book, the topic of terraforming was scattered throughout many different chapters, as well having one of its own. The original organization had merit. A discussion introduced over several chapters allowed the terraforming material to be reviewed from several perspectives and also helped to prevent the kind of 30-page monstrosity that awaits the diligent reader in Volume III. Even with this one chapter devoted to the task, however, minor discussions about terraforming cannot be avoided in other volumes. The need for perspective about this broad subject must be gained or associated discussions are impoverished.

The purpose now is not to discuss terraforming with any measure of technical precision. Rather, the task at hand is to ask the reader to accept the idea that terraforming Mars is not only possible, it is also practical. As such, we discuss now the *process* of terraforming, rather than specifics of it.

The first terraforming job is to drive CO_2 and release other gases into the atmosphere, both to increase Mars' atmospheric pressure and to increase the planet's average temperature. There are physical, biological, and chemical processes that can be used toward this end. Physical processes include the proverbial "space mirror" technology to melt frozen CO_2 at the poles or to drive out CO_2 adsorbed onto Martian rocks. Biologically, bacteria and lower order plants can be seeded onto the planet's surface to release oxygen or other gases into the atmosphere. Factories can also be built to release boutique greenhouse gases that can increase Mars' heat retention capacity without damaging the planet's ozone layer.

The second step is to introduce higher order plants onto the planet's surface. As earth arctic and sub-arctic climates are created on Mars, terran plants will add increasing amounts of oxygen to the atmosphere. Concurrent with seeding Mars with terran plants, attention will be paid to reactivating the planet's hydrosphere[5]. Plants, especially CO_2—breathing trees, need water. In addition, since water vapor is itself a powerful greenhouse gas, melting the Martian permafrost or otherwise bringing water to Mars' surface will be a priority.

---

[5]   The circulation of water on a planetary scale.

Finally, animals will be introduced onto Mars' surface. Most mammals will have a hard time living in Mars' young CO_2-rich atmosphere, but that will not be true of many insects, fish, and amphibians, especially species amenable to genetic engineering or selective breeding. Doubtless there will be many failures, but terran-style environments will slowly take over Mars' surface. Likely, the adaptability of life itself will surprise terraforming scientists and engineers, as species will themselves adapt to the ever warming and the increasingly oxygen-rich atmosphere on Mars.

Life on Mars

The most common theme of books about Mars—and the greatest obstacle to its settlement—involves the question of life on Mars. Zubrin writes about the search for life as the great magnetic pole drawing humans toward the planet. In this case, however, Mr. Zubrin is gravely mistaken. A misguided desire to search for extra-terrestrial life has actually *blocked* the way into space, allowing humanity to think well of itself if it sends one unmanned spacecraft to the planet every four years. Humans should be ashamed by such paltry efforts. The chimera of life on Mars retreads questions as scientific, not a societal, matters. If you are obsessed about "life on Mars", you lose sight of the real, *societal* possibilities and the sensible objective of settling a planet so much like our own. We cannot send rugged explorers to search for life on Mars—men and women with limited scientific training—because the search is not for woolly mammoths. Rather, the hunt is for bacteria or the most minute evidence of life found in fossilized rocks. Thus, the "scientific" search for life on Mars becomes part of the great BGBS tar pit of grants, papers, scientific jealousy, and bureaucracy and it excludes the hardy types who would enjoy the challenges posed by a new world. The forces that have been most involved with the search for life on Mars are the same forces that have prevented humans from living on Mars in the first decade of the 21st-century.

The mission to Mars takes a vastly different approach. While it does not intend to hinder a search for life on the Red Planet, its only interest in the subject will be to further its goal of settling the planet. Thus, the search for life on Mars, if properly harnessed, can assist the mission to Mars, even if the mission itself takes a cynical view of this search. If money and scientific assistance for its ends can be provided to the mission to Mars, it will tolerate those who seek scientific bric-a-brac for their office bookshelves.

The science of life on Mars must wait its turn. The mission to Mars stakes the claim that it will find life many, many years before BGBS would, because it will get there so many years ahead of BGBS and, at some point, will simply stumble across some Martian critter.

The search for life on Mars seems comical not only because of the fury of its proponents, but because straightforward scientific understandings allow the confident assertion: *Of course there is life on Mars.* One must be pretty stubborn or thick headed not to believe this. The cross contamination of the earth and Mars by microscopic organisms has been theorized for one hundred years and has been established as a scientific certainly for fifty. Actual ejecta from Mars with evidence of life[6] have been examined. Once the mission to Mars gets "home", no doubt we will find debris from earth on Mars. We know that bacterial spores can live for long periods of time, perhaps millions of years, in space. The chance that the exchange of material following major asteroid and comet collisions would not contain the kernels of life is absurdly low. In addition, since Mars probably has many sites where conditions are similar to terran sites rich with life, assertions that life failed to develop *somewhere* on Mars—independently or as a result of cross-fertilization with the earth—must be viewed skeptically.

Preaching aside, it is not scientists who are the doubting Thomases here, but those whose religious or philosophical views would require major repair if earth was just one of several—or several billion—planets onto which God had breathed life. The whole Martian-NASA farce is acted and produced by people who know that the probability of life on Mars is exceptionally high, but who will only be satisfied when they receive science's "Olympic Gold" for finding it first. As Puck has explained, we mortals be strange creatures. And so the search for life on Mars continues. The child-scientist's desired end product will be some fossil, or better, some rich soup of Martian bacteria, which will end all speculations about the question. Fortunately for the mission to Mars, the "search for the soup" can be harnessed to speed our success, as will be discussed below.

---

6   Evidence is not the same as proof.

Cultural Issues

A culture is in its finest flower before it begins to analyze itself.

—Alfred North Whitehead

A few highly important cultural questions regarding Mars and space-faring humanity will need to be answered early and, in most cases, without complete information about the future or the implications of the decision. The answers to these questions may create hard-to-break precedents for future societies on Mars, especially in a Martian culture that may not tolerate much direction from political or other authorities. The questions addressed in this chapter include basic issues of language, units of measure, the genetics of future humans, and measures of time. Some of these matters may appear to be so mundane as to be trivial. Once a decision is taken one way or another, however, that decision will have the greatest and gravest of impacts.

As these issues are overwhelmingly cultural, free human beings will ultimately decide for themselves how they will live and what they will do. If the Confederation's decision to adopt one system or the other is not popular, then people may force the Confederation to rescind its decision. The history of attempts to introduce the metric system into the US can serve as a prime example about how difficult it is to introduce cultural changes into free societies.

Standard English

The most basik of thes questshuns be, "what languij speak we?" The obvious anser be "english", tho we shoud probably spin this anser to expres this languij's utility to mankind. Moreover, the obvius anser—it be the languij of siens, finans, and of several of the most powerful erth-bound nashuns—will invok som resentment aganst nativ english speakers and aganst the chois of a languij that, after all, be the languij of only a tiny frakshun of the world's populashun, that kam from a smal iland off the koast of the smalest kontinent.

Standard English has been a staple of science fiction for many years. In many ways, this vision of a language of space presents a case with considerable merit. Standard English is defined here to be modern spoken English, with a simplified written language. Over time, certain oddities of English grammar might also be rooted out and replaced by simpler rules. For example, the use of the present tense "I am coming" and "I do come" might be eliminated as archaic usage, leaving "I come" as the only correct present tense form of the verb "to come". Simplifications such as these would make English much easier to learn. "Do I come down the road?" "Is Paul coming down the road?" "Judy is coming today", etc. would all be replaced by sentence forms such as "Judy come today" and "Come Paul today?" The end result being that today's *lingua franca* could be adopted for use as a kind of universal language in space. The simplification of spelling and grammar would make it all the more palatable for non-native English speakers, while the simplification rules would not be difficult for speakers of American or British English to learn. All humans in space could communicate directly with each other, strengthening bonds between them, perhaps to replace something spiteful and resentful and prone to violence. The creation and "administration" of Standard English could be done by the University of Mars alone or in association with English-speaking universities on earth.

This language would not replace German or Russian for use at home, in schools, or perhaps even for local government. Standard English would, however, become a *lingua franca* for Mars as well as staking a claim as the official language of all human space travel. In addition, by changing English slightly, the resentment can be reduced toward a language having originated at one small spot on earth and spoken as a native language by just a segment of the earth's population. In fact, the more the Americans and the British protest the Standard English changes, the greater the likelihood that other people would embrace Standard English as a language entirely their own. The creation of an administrative body for Standard English would help to ensure that the written language was kept standard throughout Mars and in all space. In addition, it would ease the standardization of pronunciation, so that people living on Mars and someday on Mercury, etc. would speak a mutually intelligible language.

Though a decision creating "Standard English" may seem controversial, there have been many "official" changes to language throughout history. The history of Martin Luther is, in many ways, a story of a man who incorporated spoken languages into a new church, languages which the old church would not recognize.

During the Reformation, Martin Luther began to use German and other local languages in his "Lutheran" churches and printed Bibles in the vernacular. The High German of today is, in fact, a compromise between many forms of German, spawning a national language that all Germans can accept, even if it varies widely from the German spoken in German homes. American Indians and others have had alphabets created for languages that had not conceived the idea before Europeans arrived. The alphabet and subsequent documentation of grammars and vocabularies creates an "official language" that is not always the same as all the people speak. A closer real-world analogy to the adoption of Standard English might be, however, the adoption of a written language in Norway that is similar, but not congruent, to the language spoken by the Norwegian people. In essence, in Norway, there are two languages, Landsmal and Bokmal, the latter being the literate language; the former being the one that will eventually be replaced by Bokmal. Finally, the revival of Hebrew as a spoken language should be considered in this context. Not only was a "dead" language revived, with new words reconstructed from the old root words to deal with a modern society, Hebrew also replaced Yiddish, a language dear to most of its several million speakers. So successful has the revival of Hebrew been that it is now fully accepted as a modern language, while Yiddish is being lost as a spoken language.

"Acceptance" is always a difficult thing to measure, though is easy to argue that the Norwegians have accepted their dual-language system and Germans their High German. Thus, a decision to adopt Standard English by the Confederation can be based upon historical examples of the feasibility of the idea.

The timing of the adoption may be difficult. To suggest the change at the outset of the mission to Mars would be difficult and would likely reduce interest in the program. Americans and other English-speakers may think twice if they had to live in a place that used a different form of their language. To delay the implementation of Standard English, however, would be to diminish its importance versus other mission to Mars' projects. If Standard English were important—the argument might go—then it should have been adopted in "Year 5", not "Year 55", when it becomes merely a social experiment by a government already, perhaps, starting to ossify.

Units of Measure

Units of measure might also be contentious, though, once decided upon, not overly divisive. The mission to Mars proposes to adopt the old English terms, foot, pound, mile, rather than the metric system—meters, kilometer, etc.—as its linear measures. The latter are all tied into direct measurements of the planet earth. There is no reason that Mars should embrace these measurements out of some misguided desire for standardization. Mars, after all, is *not* the earth. As we will see later in the chapter, adopting wholesale standard measures found on the earth is not possible and may, in fact, lead to ridiculous results in space. Earth measures are not necessarily bad; but the ones to be adopted must be selected with care. The mission to Mars proposes that Martians will early on desire to distance themselves from earth, not only in simple things such as how they measure distance, but in much more fundamental ways, too. Mars is not to become Vermont. Mars will become a new and *different* plantation for the human species. As such, the willy-nilly adoption of things terran will not only be rejected, it will confuse the ultimate purpose of being on Mars. While the English system is obviously also terran, it suggests a separation from earth's "internationally-" though not "interplanetarily-accepted" metric system.

If some version of English were to become the language of choice, then the English terms foot, pound, etc. might follow naturally. In addition, as these terms are already familiar to billions of people—especially to Americans who are likely to make up a large proportion of the early settlers—these English terms would likely fit easily into the Martian vernacular. Although eventually Martians may choose to create their own "Martian" measurements, doing so at the outset would seem overly artificial and contrived.

In contrast to the measurement of distance and weight, temperature should be measured using the centigrade thermometer. Temperature, after all, on Mars will be the same as temperature on earth and water will melt/freeze at zero degrees Celsius there, too[1] The mission to Mars proposes to follow the lead of the scientific community and most of the earth to adopt this aspect of the metric system. American recruits to the mission to Mars will have to get used to 32° being a warm day.

The timing of a decision about units of measure can be made early in the life of the mission to Mars. As in America, people can choose to use whatever system they like. Likely, the work of science and engineering will favor the metric system. Depending upon the size of the American contingent, the use of the old English terms will be favored by the people.

Mexican-Swedish-Americans

The segregation of the human gene pool into Germans and Italians and Jews and Chinese is a relatively recent development in the history of the *Homo sapiens*. Indeed, it is something of an historical accident, when several thousand years ago

---

[1]   The melting and boiling points of water are also dependent upon the ambient atmospheric pressure, which will be different on Mars, but the point stands.

humans segregated themselves during periods formative of language and culture. Especially in space, these gene pools will intermarry and the cultures they represent will be overtaken by new, better cultures, perhaps even an overarching space or Martian culture. Though the point is disputed by those writers of science fiction who foresee a New Italy or a planet populated only by ethnic Chinese speaking the old tongue, the mission to Mars does not see its mélange in terms of fuzzy-navel-sociology but rather as an extension of the current trend of internationalization visible almost everywhere on earth.

In fact, the likelihood of a new culture replacing the old animosities and hatreds has real significance for the settlement of Mars and, more generally, for the future of all humans. The transfer of earthbound problems from the earth into space would be a tragedy of the first order. One would hope to see space and Mars free of ancient prejudices and hatreds and that the brave, new humans in space would identify themselves first as *humans,* second as Martians, and leaving only as a footnote their ancestors' origins from a particular corner of the globe.

On the other hand, the creation of 70 independent republics on Mars may allow the transplantation of the hatreds and rivalries found on earth. If this transfer occurs, the chance for war and strife on Mars will increase, especially if certain groups or nations are able to bend the settlement of Mars to parochial ends. Nonetheless, other factors, likely to be much more powerful than sinister earthly imports will erode most Martian analogues of problems on earth. Certainly, the riches awaiting the Martian settlers will corrode many earthbound resentments and hatreds; a simplistic way of understanding this idea is to say that Martian society will be too rich to harbor traditional (self-destructive) hatreds. And, of course, there will be the positive effect of shared experiences living in a pioneering society.

This section proposes three additional reasons why humans indeed might check their hatreds at the SSTO, so that humans on *Mars* and in space will be just that: *humans* on Mars. Certainly, the precedent set by the American experience as burial ground for Old World hatreds and animosities augurs well for the future. There will always be those who say that human nature itself mandates cultural and political hatreds. The mission to Mars, however, posits most of the blame for such calamities with three factors we will examine in a moment. To introduce these factors and to illuminate the subject, the story of "M" is presented.

M grew up in a small, rural community in the Third World that was the antithesis of a Norman Rockwell portrait. Thoughts of society did not relate to funny friends and chummy neighbors, but to clans animosities and tribal hatreds. Thoughts of government did not reveal notions of chubby, talkative mayors and headstrong assemblymen, "government" was when uniformed men periodically trucked in from the city to take things from M's family, offering nothing in return.

Sometimes these "soldiers" do more than just take things. Sometimes they beat or even kill people in his village, especially members of the minority tribe who live there. As there is no real government, almost every conflict—such as all humans are fated to engage in—are resolved not in courts or with laws, but in the streets and with weapons.

M had a cousin shot dead in a blood feud with a rival clan and his father tells proud stories of the gun battles he has engaged in for honor. Unfortunately, M's father has little pride in most other aspects of his life, as he works only occasionally and then for little money. What little food can be counted on is grown at the family home.

The violence does not stay outside of the house. M's father routinely beats his mother and when he is not beating her, he lords over her with the cruelty of a beaten dog. M, too, used to be beaten when he was younger, but now that he is older and bigger, his father simply orders him around. The family does not talk about love, but rather about one's role in and duty to the family.

What really holds things together is a belief system based on a fundamentalist interpretation of a world religion. This belief system grants power to the father and demands obedience from the remainder of the family. The village's particular religious interpretation tolerates or even condones the outbreak of violence to uphold honor or when seeking to avenge an insult.

M will soon be old enough to marry. The community no longer practices arranged marriages, but M must choose wisely if he and his bride are to obtain his father's blessing. M knows about western culture, which he hates for being "decadent", but secretly longs to go to the US or Europe. It is a dream, however, that he knows will never be realized. What is certain for M is that after he is married, he will beat his wife and children and will use his rifle to resolve disputes with his neighbors.

The first and least controversial factor that should alleviate traditional hatreds on Mars will be the presence of effective democracies, to allow all Martians to vent their political and other passions in constructive, rather than destructive, ways. For M, the lack of democratic safety-valves results in feuds, gun play, and the death of people he has grown up with. On Mars, all people will live in democratic societies. While not by itself a solution to all human troubles, the ability to resolve disputes non-violently, as well as the ability to shape the society through political action, will combine to curb the worst kinds of outrages that are seen in non-democratic societies.

The second factor is much more problematic: the demands of a given religion, both on earth or on Mars. The difference will be that the stranglehold that radicals have on some religions will be mediated by the wealth and the democratic forces just discussed, resulting in far more moderate religious views than those held by fundamentalists. Just as Islam in the US appears to have a larger share of moderates than in Pakistan or in Saudi Arabia, so religious fervor on Mars will be reduced as adherents live in a First World society on Mars. More moderate religious views will result in much less heated religious debate and such debate as occurs will be directed toward constructive rather than destructive formats.

Finally, since the course of historical time has nurtured many Old World hatreds, where the older generation raises new generations to hate just as they raise long established crops or domesticated animals, the short history of Mars will prove infertile grounds for hatreds. M learned to hate from his father, just as his father learned to hate from his father and so on back through time. On Mars, there will be no history of such hatred. Happily, there may not even be nearby objects of hatred, as Jews may not live in proximity to Arabs on Mars, nor might an otherwise hateful Hindu find a Moslem neighbor. With neither a history nor a nearby object of hatred, such feelings should die quickly and disappear in a generation or two.

The mission to Mars will not offer a social panacea for all human ills. It should offer an environment, however, that can reduce to manageable levels the ugliest acne of the human complexion. Squabbles about land will not be about *sacred* land. Troubles between individuals will not become troubles between groups, if traditional groups are not found on Mars. The emphasis on individual responsibilities and individual rather than group experiences will reduce the old hatred to distant memories …

… or the author could be wrong and hate could reach its ugly hand to Mars or anyplace else that humans call home.

In any case, it will be difficult for Mexican-Swedish-American-Martians, products of a new genetic mixing bowl, to organize their hate into group action that threatens Martian tranquility.

"Can you wait here a second?"
"Did you mean an earth second or a Mars second?"

The measurement of time is intriguing to the point of justifying some detailed analysis. By way of hope and rationalization, the next few sections' novel ideas might also help the reader free his mind from constraints that might prevent a rational mind from choosing the mission to Mars over NASA.

Fortunately, two closely tied measures of time can follow the earth tradition. They will be dealt with first to clear away some of the intellectual underbrush in this subject. The first, "days of the week", can keep their traditional names. Units of time are intimately bound up with human economies. Having a seven-day week, where people might be expected to work five days and rest on two will serve as an important anchor in the transfer of human culture to Mars and into space. Thus, the "week" can remain a period of seven sols, a measure of time readily transplantable to Mars and useful to pad the jolt of life on an unfamiliar world.

Second is the joyous news that a Martian day and an earth day are identical, at least to within 3% variability. Humans use days for two main reasons. First, it marks a complete rotation of the planet. A rhythm of sunrises, sunsets, and darkness has given rise to biological activities with four billion years of history and genetic programming. Humans will have a near insurmountable obstacle if they hope to overcome these rhythms. Fortunately, on Mars such a struggle will not be necessary. Though each day will be 39.6 minutes longer than a day on earth, biologically, humans will likely not notice the difference. The second reason humans use days is tied in closely with the first. As the rhythm of mankind's biological life is tied into a day, so is our economic and cultural life tied together. A day is fundamental to an employer-employee relationship and other economic relationships. You work five *days*, you rest for two *days*, and then you come back to work another five *days*. We rent a movie at a video store for three days. The cultural importance of "the day" is unquestioned: your parents allow you to see your boyfriend three days a week. If you see him every day, then there will be trouble. We eat three times a day, even as mealtimes are a cultural, rather than a biological, necessity. The fact that an earth day and a Martian day are so similar will allow for the easy transfer of these patterns of human life from the old planet to the new.

Hours are used to divide the day. We generally work for eight hours, use eight hours for our private lives, and sleep for eight hours. Having a day with only twelve hours might be possible, but it would likely be inconvenient and culturally disorienting. As the Martian day is about 24 hours long, however, dividing the day into 24 hours should be simple. We begin here, however, to get close to the pulse of a difficult question. Should these hours (and minutes and seconds) be the same length as their earthbound counterparts or should they be adjusted to fit the difference in the two planets' rotation?

Certainly, Martian seconds (this is what we will call this issue) will not be disorienting. No one but sprinters will notice that a Martian second is 2.75% longer than an earthbound second[2]. The same applies for minutes and hours, though recent immigrants to Mars, sitting in a boring physics class may somehow sense that their hour-long class is actually 1.5 earth-minutes longer than their

---

[2]   If there are 24 hours in a day, each with 3600 seconds, and Mars takes an additional 39.6 minutes to complete its rotation, then the Martian second must be longer than an earth to compensate for this additional rotation time. Later, an idea of a Null Hour will be discussed to account for this additional time.

boring physics class on earth. People on Mars will have no problem addressing one another with the phrase, "Can you wait one second?" and a plane from one Martian city to another (yes, planes will work in a thickened Martian atmosphere) can have flight times of 3 hours and 12 minutes, without even considering that these are Martian, not earth, measures.

There are, however, some very big problems with changing the length of hours and minutes and seconds. The first problem is the precedent it sets for changing, somewhat arbitrarily, the basic temporal measures humans use. Besides Mars, the mission to Mars proposes to establish bases on three or four inner asteroids, on Deimos, Phobos, and possibly on the earth's moon. On these rocks, which measures should be used? The mission to Mars might favor Martian units, but since the moon is part of the celestial earth, perhaps earth seconds would be more appropriate there. And what happens when people go to some world with a 20-hour rotation? Is a third or forth "second" to be created for that circumstance? Even worse, what about the trip from one such space rock to another? When does the space ship change over from "Martian seconds" to "earth seconds"? At some point all of this becomes unworkable.

Even if these practical matters might be resolved, a larger issue is that science and engineering use the earth second as a fixed measurement. Granted, scientists and engineers are used to converting measures, but one wonders if there is a straw and a camel somewhere about. Since virtually everything in science and engineering is, at some point, tied to a fixed second, significant interference with routine work may exist if minute differences in seconds are hidden everywhere. One would like to avoid the explosion caused by the apprentice engineer who did not realize that his calculations were supposed to use earth seconds and not Martian seconds[3].

All in all, it seems better to leave the second fixed, rather than to create all sorts of new, utterly confusing measures.

Of course, using the earth second on Mars creates problems for our intrepid settlers, for if Martians adopt earth seconds, their day will have to account for the extra 39.6 minutes discussed above. One solution would be to create an additional "null hour" on Mars, a short 25th hour each day that will match time with the rotation of the planet. Certainly, digital clocks will have little problem accounting for this extra hour. Culturally, however, people will need to account for the time, no matter how it is allocated. At first, the Mars bases, all being located in one small part of Mars, might be able to have null hour fall deep in the night, allowing it to be virtually unnoticed during the early years of settlement and having little

---

[3]   Recall the Lockheed Martin engineer whose mistake converting measures lead to the crash of the Mars Polar Orbiter.

impact on the settlers' activities. Later, as settlements spread across the planet, the null hour might cause other problems, though one that probably will be of interest rather than a bother. Will an office worker really care that the third hour after lunch is only 39.6 minutes long? No. She will only care that when the clock reaches 5 PM (or 5:20), she can go home to watch her "space operas". You know, the TV show where that strong, handsome Martian falls in love with the purple mutant from Jupiter who, though no one suspects, is really the Martian's cousin, conceived during her uncle's torrid love affair that occurred during the Titan wars between Neptune and Saturn.

Null Hour and Siesta

One of the great "stressors" in life in the modern west is the incessant demand to be somewhere at a given time. Were Americans and others not so driven to conform to a little tick-tock, life might be much fuller and more enjoyable. Frenchmen everywhere agree.

Two circumstances offer the mission to Mars a wonderful opportunity to design a better[4] society to avoid this modern pitfall, without a resulting lack of economic productivity. The first circumstance was discussed in the last section: the null hour. The second circumstance is the high-likelihood that the earth Embassy will be built in a desert somewhere. Both circumstances suggest, after a little thought, an extended "lunch", whereby the Martian Null Hour might be absorbed and whereby the heat of the day can be addressed by an afternoon nap.

Remember, dear readers, the last thing this book is about is suggesting that Mars can only be conquered through the creation of weird new social orders. Some might argue that a long-lunch-siesta falls pretty much center-of-mass of that idea. This book offers nothing if not a picture of suit-and-tie businessmen, accountants, and lawyers resolving issues and putting together deals in a compli-cated society, yelling at their kids to finish their studies in higher mathematics before they can watch reruns of "F Troop" on television. It therefore is offered as an idea, rather than as a mandate necessary to get humans to Mars. Accountants and businessmen will get us to Mars. Siestas will not. Still, a siesta *might* be nice. Here is why.

Imagine a hot summer's day in a desert. The workforce has reported at 9 AM and has worked a good four hours. It's now 1 PM and the outside air temperature is 42° C[5]. Just then, a whistle blows and everybody knows that it's lunch-siesta-time. Everyone is off until 4 PM. People at the earth Embassy now have three

---

4   Without tinkering with basic ideas underpinning that society.

5   Americans: that's really, really hot.

hours to shop, to meet friends, to have sex, or to sleep. They can relax out of the way of the sun until 4 PM, when the workday will resume. Most will have eaten a good meal and, as the kids will get home from school by 2:30, will have been able to kiss their young-uns and come back to work in a good mood. The workday continues until 8 PM, when the sun has finally started to relent. A quick evening meal, more time with family or friends, and off to bed for another day.

Now take this to Mars. The first base there has decided to put its Null Hour in the middle of the day. Thus, there is a 40-minute hour that occurs right after 1 PM. Same start to the workday. Same break at 1 PM. Now, however, the three-hour break runs from 1 PM until 3:30, which is two-and-a-half hours plus a 40-minute Null Hour. Same schedule as on earth, only now the Null Hour issue created by the physics of Mars' rotation is swallowed by a culturally accepted siesta.

This entire issue of siesta is meant to convey flexibility, rather than compliance with an arbitrary schedule. Factory workers will still work eight-hour shifts, if it costs too much money to shut off equipment in the middle of a shift. And remember, "shopping" was going to be possible. No one can shop if the shops are all closed for siesta. Rather, shops and restaurants will operate a flexible schedule for their employees, both capturing the siesta business and allowing siesta for some employees. Other people will simply want to work eight hours and go home. These people can man the phones, etc. during the time everyone else is off sleeping. Finally, in an increasingly global world, having a business that is connected to the rest of the world for eleven hours rather than eight, will enable earth Embassy enterprises to catch three more time zones when they are open for business.

The siesta will allow for an increased emphasis on family and community that many people might find attractive. If one parent took siesta and one worked an 8-4 schedule, a family might never have kids at home without a parent, even if both work. Having time *each day* to spend with friends or family, rather than lunch being a hectic break between meetings, might be an attractive tool for recruiting people to the earth Embassy, where life would be *good,* as well as exciting. Even if siesta is an idea without sufficient merit to implement, it will be ideas like *siesta* that will make life at the earth Embassy a pleasurable and realistic option for people who might otherwise have only a limited interest in Mars.

Both the idea of an undetected null hour and the longer Martian second have merit. The null hour idea probably has a slight advantage over the longer Martian minute, but ultimately this book takes no position, as this decision should be decided by democratic and cultural process and not by some author. In any case, the length of the day will be decided by the planet's rotation and not by debate. The next topics, the Martian month and year, can be approached without reference to a decision about the length of a Martian second.

A Martian Year is Longer Than a Month of Sundays

The concept of the "month" is used in many contexts in western society, beyond the obvious purpose of measuring time. Culturally and historically, a system of months was created to help farmers plant their crops and to worship the gods at the proper time each year. The extensive celestial observations made by the Babylonians and Egyptians and the complicated calendars they developed were specifically designed to time the planting of crops for optimal harvests. Thus, months, which are sub-parts of the year, are used to predict "seasons", especially seasons important for agriculture and religion. Accordingly, January weather at any point on the earth is similar, year after year, decade after decade.

As societies developed, months became ingrained into human economic life. Apartments are rented by the month. The fact that February is 10% shorter than July is viewed as a minor inconvenience in this respect, though one suspects that if the lengths of months differed by 25–50%, this "inconvenience" would have significant economic effect. The local "Rent-a-Center", for example, might be forced constantly to change all its prices to cover the economic differences between monthly rent for a 20-day February and for a 40-day July. Monthly meetings become much more problematic if months were of dramatically different lengths. The monthly appointment with your parole officer, for example, would be even more interesting if some months were significantly longer than others.

This book will contrast three different systems to name and measure the months. It advocates the second version, its own home-brew but, like the decision concerning the length of a "second", defers this decision to future political and cultural processes. Figure 49 below will help the reader to understand these three systems.

Zubrin has proposed a system of months that we call here "Alternative I". His system is interesting for two reasons. First, it uses the signs of the zodiac as the names of the months, allowing westerners a ready understanding of the names of the month. Second, this alternative proposes to divide each month into equivalent arcs of the orbit of Mars around the sun to promote the "seasonal predictor" aspect of each month. Since seasons are dictated by the planet's position in orbit, dividing this orbit into twelve equivalent months allows for the months to correlate with seasonal changes occurring on Mars. This means that each season on Mars would be three months long, just as they are on earth. As indicated in Figure 49, a significant problem with this system, however, is that some months are much longer than others.

This book's proposal, "Alternative II", creates a system of months where the variation between months is minimized, while accounting for the physics of Mars' elliptical orbit. In addition, it avoids the possibility of legitimizing astrology with its use of terms common to astrology. Equalizing the lengths of months is fairly mechanical. This system has sliced days off Zubrin's longest months and has added days to his shortest months. Much more importantly, this system adopts a highly logical, culturally diverse, and sonorous system of using the Chinese names for months as the names for Martian months. This will alert almost every human (possibly excepting the Chinese) to the fact that Martian months are being used. In addition, the system is highly logical, being simply the use of Chinese cardinal numbers followed by the Chinese word for month (one-month, two-month, etc.) Finally, if the reader practices saying these words a few times, one will be surprised how easily the words flow. These are not German words, concatenated endlessly, or Russian words with harsh tones. They are the eastern equivalent of soft French tones, adopted to pay homage to 25% of humankind. Yes, one man's diversity is another man's ethnocentrism but, so far, the mission to Mars has adopted little from the great Chinese culture. This is a good start.

Month	Alternative I		Alternative II		Alternative III	a	b
	Zubrin's System	# of Sols	Mission to Mars' System	# of Sols	Earth (English) System	# of Sols	# of Sols
1	Gemini	61	I Yue	57	January	61	57
2	Cancer	65	Er Yue	58	February	65	58
3	Leo	66	San Yue	58	March	66	58
4	Virgo	65	Si Yue	57	April	65	57
5	Libra	60	Wu Yue	56	May	60	56
6	Scorpio	54	Liu Yue	55	June	54	55
7	Sagittarius	50	Qi Yue	55	July	50	55
8	Capricorn	47	Ba Yue	54	August	47	54
9	Aquarius	46	Jiu Yue	54	September	46	54
10	Pisces	48	Shi Yue	54	October	48	54
11	Aries	51	Shi-I Yue	55	November	51	55
12	Taurus	56	Shir-Er Yue	56	December	56	56

Figure 49: The Mars Solar Year — Names and Alternatives

"Alternative III" is simply to adopt the English language system (or the names in whatever language is spoken on Mars) to the longer Martian year. We present it here as two sub-alternatives. Alternative IIIa uses the Zubrin method of short and long months. Alternative IIIb adopts the mission to Mars' system to make the months more equal in length. Either system has the advantage of using months that are already full of meaning to speakers of the language (January, February, etc.) Seasonal meanings can thus be understood slightly more easily than using either of the other two systems. The potential for confusion, however, especially in interplanetary concerns, is enormous. If it has escaped the reader's notice, the months on Mars will not correspond in any way to the months on earth[6]. If the spacecraft is to dock on Mars on July 10, everyone concerned must clearly understand that this means Martian July 10 and not earth's July 10, which will not occur on the same day. Theoretically, this might be simple. In practical terms, the problems would be legion. If the months had different names and the schedule says that the spacecraft will dock on Leo 10 or San Yue 15, no confusion is possible.

"Alternatives III" are mostly used as strawmen for purposes of comparison and are not the system of choice here. Note that Alternative II's use of Chinese names will confuse Chinese speakers, much like Alternative III's names will confuse English-speakers. The hope here is that documents written in Chinese can easily distinguish between the Martian months and the earth months by an addition of a mark or character in these documents, something that could more easily be forgotten in English (M-July for Mars' July might easily lose the "M" somewhere). Moreover, as the Chinese economies integrate into the world's economy, they will increasingly use English month names more in international transactions and their own months' names less and less.

The Martian Calendar

The current method to keep track of years should probably not change as humans move to Mars and into space. As the dating of modern years is tied to the traditional birth of Jesus[7] and not some celestial event, there is no reason why humans

---

[6]   Even if the months were set arbitrarily to coincide with earth months, because the orbits and rotation of the two planets are different, the calendar on Mars would soon be completely unrelated to the changing seasons. If "January" on Mars was always to coincide with "January" on earth, some Martian Januarys would occur in the Martian summer and some Martian Januarys would occur during the Martian spring, etc.

[7]   The current practice, reasonable even in a diverse and shrinking globe, is to divide human history into two parts. If we consider the birth of Jesus as the separation

in space might not also use the same "year reference" used on earth. Thus, the date on Mars, using Alternative II and the Chinese names for months, might be Wu Yue (fifth month) 16, 2075 and the next day might then be Wu Yue 17, 2076, if it corresponds with the New Year on earth. The confusion on Mars or elsewhere in space would be minimal and the tie back to earth would likely serve a strong emotional (and perhaps political) purpose. Humans, after all, do not stray badly if they think of themselves as one body corporate and retain some common references. The use of a universal numbering system for "years" seems to be a prudent measure for all humans and no one on Mars should put up too much of a fuss if the end of the "year" does not always come at the end of the last month of the Martian calendar. On Mars, the months will measure the seasons. The year will be useful to measure longer events and to tie Martian civilization together with the earth's. A person born in 2067 and dying in 2174 will not much care that he only had 51 Martian birthdays. As a Martian, he can tell earth-born humans that he lived over 100 years. He can also understand someone from earth who says that she has lived 37 years. Again, the realities of celestial physics would make site-specific years unmanageable. Imagine the confusion if someone in the far future, living on Uranus and having told everyone that he was nine years old, appears to be an old man. That kind of cultural cross-talk would hinder, rather than help, the cause of building bridges for all human beings.

Having an earth year (2112), a Martian year (38), a year for people living on asteroids (6), etc. makes no sense. While ceremonial "years" such as the Jewish enumeration and the Islamic years since the Hegira can be retained, such systems are not practical for everyday usage. Moreover, these years would be less confusing than years of different lengths, based upon different orbital periods around the sun.

As with the other cultural issues, these matters can be decided upon in time by the people themselves. Unless some serious objection can be found, however, it would seem that the simplest solution is to adopt in whole the earthbound method of counting the earth years since the traditional year of Jesus' birth.[8]

---

between the two eras, we have dates annotated as BCE (Before the Common Era) and CE (Common Era). Thus, 100 BCE and 100 CE. This replaces both BC (Before Christ) and AD (Anno Domini or In the Year of Our Lord), both of which make religious statements unpalatable in our times.

[8] Many scholars believe that Jesus was actually born in the year 4 BCE.

Part IV

On Mars

Ombey was the newest of Kulu's eight principality star systems. A Royal Kulu Navy scoutship discovered the one terracompatible planet in 2457 ... After an ecological certification team cleared its biosphere as non-harmful, it was declared to be a Kulu protectorate and opened for immigration by King Lukas in 2470 ... Unlike other frontier worlds ... which formed development companies and struggled to raise investment, Ombey was funded entirely by the Kulu Royal Treasury and the Crown-owned Kulu Corporation. Even at the beginning, it couldn't be said to have gone through a purely agrarian phase. A stony iron asteroid ... was manoeuvred into orbit before the first settler arrived, and navy engineers immediately set about converting it into a base. Kulu's larger astroengineering companies brought industry stations to the system to gain a slice of the military contracts involved, and to take advantage of the huge start-up tax incentives on offer ... By the time the first farmers arrived, the already substantial government presence produced a large ready-made consumer base for their crops. Healthcare, communications, law enforcement, and didactic education courses, although not quite up to the level of the Kingdom's more developed planets, were provided from day one. Forty hectares of land were given to each family, along with a generous low-interest loan for housing and agricultural machinery, with the promise of more land for their children. Basic planetary industrialization was given a high priority, and entire factories were imported to provide essentials for the engineering and construction business. Again, government infrastructure contracts and civil workers arriving during the second ten-year period was equal to the number of farmers ...

—Peter Hamilton
The Reality Dysfunction (Part 2)

The First Eight Years

There may come a day when the courage of Men fails ... but today is not that day.

—J.R.R. Tolkein in *The Return of the King*

The transplantation of a society on earth to Mars can be better understood in the context of specific plans, during specific time periods[1] The following sections outline the first two four-year periods on Mars, when there will be "military" missions, under Fleet commanders. During each year, two missions will launch for Mars[2] some missions with two or more spacecraft convoying together. The first four years will see movement to Mars by two Explorer Transporter vessels per year, each capable of carrying four people and cargo. The second four years will see more arrivals, with two Explorer Transporters and two Shuttle Transporters—

---

[1] To provide for simplicity for us earthers, unless otherwise noted, henceforth in this volume the term "year" refers to earth years.

[2] The average of two ships per year will not necessarily mean that a ship will arrive every six months. A discussion in Volume III about "launch windows" will explain that launches for Mars *might* only occur every two or three years. What is more likely, however, is that technological and administrative advances by the mission to Mars will dramatically open these launch windows. For example, ships may start to use nuclear engines, achieve far greater thrusts, and become capable of launching at nearly any time. Or, "solar sail" vessels might be used to augment transporter flights, where ships—and possibly crews—accept three-or four-year flights between the planets. Finally, there may be options to refuel before insertion into a Mars orbit. If the fuel needed to reduce velocity sufficiently to be captured by Mars can come from Mars or Deimos and not from earth, travel times between the planets can be reduced immensely.

vessels capable of transporting eight people and significant amounts of cargo—arriving on Mars every year.

Deimos will serve as the main entrepôt for Mars. Deimos is both very close to Mars and has just a scintilla of gravity. A spacecraft that goes into Mars' orbit close to Deimos literally will be able to be roped down to the moon, docking like a large yacht at a lakeside pier. In addition, because Deimos seems to be covered with a layer of fine dust that might cushion a landing, much of the moon might be immediately suitable for spaceport operations, without a great deal of infrastructure.

Year One

A four-man crew will land on Deimos and begin creating a Fleet base on the moon. Deimos' gravity is so low that even the first landings there will resemble a space docking operation rather than the old "moon landings". No difficult re-entry operations and no complicated re-entry equipment[3] for the transporter ships means much lower costs than for high-octane BGBS vehicles. Cargo can take the place of the heavy and bulky re-entry equipment, cargo to begin those operations essential to transplanting the earth Embassy to Mars.

Matters of infrastructure will be a first priority after landing on Deimos. The greenhouse there will grow crops that can flourish in a low-gravity, high-radiation environment, to establish a food supply and sources of oxygen for the settlers. The greenhouse will also use wastewaters from the base's wastewater system to eliminate the base's wastes. The first nuclear power plant for Deimos[4] might be one of the power plants from the Explorer Transporter, detached and modified for its new, sedentary role. The energy needs of the new base will be small; the ability to refit that first Explorer Transporter vessel for a return to earth will be accomplished leisurely, as the crew awaits a favorable launch window for the return voyage.

The first shelters on Deimos will accommodate ten people easily and 100 for short periods. They will be built with an eye to becoming the center of a larger base. The shelters will be very simple, as will be the base's design, providing little more than a few small offices, shower and bathroom facilities, and a few bays

---

[3]   As discussed in Volume III, an aerobraking procedure may be used to save fuel. This procedure will eliminate the complication of carrying extra fuel and firing rockets to reduce velocity, but will require a (deployable?) heat shield to protect the spacecraft from high temperatures.

[4]   One purpose for a mission to an inner asteroid will be to obtain nuclear fuels. These fuels might be processed and employed in space, away from any jurisdictional issues that the earth Embassy's host nation might raise.

for cots. The shelter material can be constructed from the standard Kevlar-type plastic that will serve as an outer cover for all ships and Martian shelters[5]. Perhaps someone will have invented an inflation system allowing the shelter to set itself up or a mechanical system to the same end. The furniture, cots, and other items can mostly be made of lightweight plastic. Piping can be PVC. There may, in fact, be only a limited need for metals at first, metal wires and some specialized equipment for the base. If implemented, a program to establish tiny bases on two or three inner asteroids will provide years of experience to create the Deimos base efficiently and comfortably.

	Explorer 1	
	Deimos	**Mars**
Tasks	Create Greenhouse	
	Set Up Nuclear Power	
	Create Shelter For 10/100	
	Create Deimos Infrastructure	
People	4	0
	Explorer 2	
Tasks	Assemble Mars Lander	Drop Off Supplies
	Descend to Mars	Drop Off 1 Person
	Return to Deimos	Set Up Nuclear Power
	Spaceport Construction	Create Greenhouse
People	"7.5"	"0.5"

Figure 50: Year One

Some components of the Mars Lander will be loaded onto this first ship to Mars. These components can be prepared, partly assembled, or otherwise serviced, though most of the components for the Mars Lander will await the arrival of the second ship. This second Explorer Transporter ship will dock on Deimos six months after the first Explorer Transporter lands.

The second Explorer Transporter will carry the balance of the Mars Lander parts to Deimos. The main task for the eight people now on Deimos will be to

---

[5]   Properly cited to reduce radiation exposure to manageable levels.

complete construction of the Mars Lander and to prepare for the descent to Mars. Peripheral tasks like the preparation of more permanent spaceport facilities and "dry dock bays" for Explorer Transporter maintenance can also proceed. Fuel for the Mars Lander might be fabricated from the electrolysis of water into hydrogen and oxygen or from other raw materials found on Deimos.

Food, water, and oxygen will be in short supply before the descent to Mars. The plants grown on the Explorer Transporter and in Deimos' greenhouses will provide only a fraction of the food and oxygen needed by the first settlers[6]. There will be something of a race against time to get to Mars, where oxygen and water can be extracted from the regolith and where large-scale food production can begin.

The Mars Lander will make its descent carrying a two-man crew and lots of cargo. The exact location of the landing area on Mars will have been further refined during the 11 months of work on Deimos. Final selection, of course, will be made only after the crew has flown low over the proposed landing site[7]. The two people on this first Mars Lander will work together for a week or two on the surface, mimicking initial activity on Deimos. They will set up a small nuclear power plant, solar panels, or "windmills" to generate power from the Martian winds. In addition, the crew will set up equipment to extract from the Martian regolith carbon, oxygen, and hydrogen-rich substances that can be processed into fuel. The Mars Lander will not have carried sufficient fuel to return to Deimos, so this fuel processing system will be of extraordinary importance. And one of the by-products of the fuel processing will be oxygen for the new shelter[8]. After a few weeks, and after a sufficient amount of fuel is processed on Mars and loaded into the Lander, the Lander's pilot will return fully loaded (fuel, oxygen, water, etc.) to Deimos, while one crewman continues to work at the base on Mars. This crewman will be alone on Mars for a few weeks, until the next Mars Lander flight, scheduled to launch just after the third Explorer Transporter arrives on Deimos.

---

[6]   The greenhouses on Deimos will never grow large unless new kinds of cheap shielding can be developed to solve the problems created by solar radiation. While this radiation will be several times less than radiation striking the moon (where it will be difficult or impossible to raise plants), carefully bred plant varieties and clever shielding strategies will nevertheless be required to grow plants there at all.

[7]   Can the Mars Lander be built to allow it to hover or otherwise to attain non-ballistic flight? Probably not. Still, the confluence of human ingenuity, low Martian gravity, and *some* atmosphere may conspire toward this end.

[8]   Processed carbon dioxide from the Martian atmosphere will provide some of this oxygen, as will other materials processed from the Martian regolith.

Stranded

Assume now that something goes wrong and that the two people on the Mars Lander are stranded. Perhaps they cannot find the resources to make fuel on Mars or perhaps the Mars Lander gets damaged and cannot return to Deimos. While this "disaster" may interrupt plans, it will not, in fact, be disastrous. If fuel cannot be manufactured, fuel can be sent down from Deimos, in a "cargo-parachute" system developed to deliver cargo "one-way" to Mars. If the Mars Lander needs to have a part replaced, the cargo-parachute system can deliver parts instead, possibly manufactured by the six people who had remained on Deimos. If a part needs to be repaired, equipment sent down with the Lander might provide for it. Remember, the cargo loaded onto the Explorer Transporter will include economically "valuable" equipment, not science junk. Thus, there may be a few lathes, toolboxes, and other equipment that pioneers need, rather than equipment to please a visiting scientist. There should be plenty of food and, anyway, a major priority will be to get greenhouses up and running on Mars.

The two "stranded" pioneers on Mars' surface will be Fleet crewmen, on a military mission. They will have received significant survival training during their tours with the Fleet. Part of their training will have been to survive under circumstances just described. As crewmen, they will be members of a military hierarchy stretching back to the earth Embassy. Once informed of the problems on Mars' surface, missions scheduled every six months will be reconfigured to account for these new developments. There will be no hurried "rescue mission" from earth, as the military settlement plan will already have accounted for such periodic "disasters". If the descent is timed right, the two who descended to Mars will need wait only one or two weeks for the arrival of a new Fleet transporter vessel with four more people and lots more cargo. If that new ship cannot help solve the problems on Mars, then the next vessel, launching the week after the Mars landing, will. And if neither can help, the brass will reconfigure the transporter due to launch six months after that.

In a sense, the two on Mars will not *be* stranded. In fact, they will be at the very spot in the solar system they and thousands of other people at the earth Embassy will have been working so many years to reach. There is every possibility that one or both of the pilots to Mars could die, just as such a possibility will exist for everyone contracting for her ticket to Mars (or that someone you know could be killed during a rainy commute this week). More likely, however, will be that these two intrepid explorers will be rescued, for later harpooning by their buddies back at the pilot's club on Deimos, where they will need to explain again and again how it was that their Mars Lander was damaged as they skimmed the crest of the mountain, forever afterwards named for the errant pilot.

The Old-Fashioned Swimming Hole

During the author's last few months of active duty in Iraq, he was stationed in Fallujah, a very dusty, "anti-American" town. There was little fresh food, the average afternoon temperature was 110° F (46° C), and hard duty was made harder by the uncertainty of ever-changing home redeployment dates. In the last few weeks at Fallujah, the command filled an old swimming pool, probably with water straight out of the Euphrates River, dumped a few gallons of bleach into the mix to create a deep-green soup, and told the soldiers they could swim. While it was not idyllic, the author could not help but enjoy this small luxury and to imagine palm tree lined, domed, swimming areas on Deimos and Mars, offering newly-minted Martians a wonderful taste of the resort life, 100 million miles from earth.

Thus, one of the things that should be planned for the first or second year on Deimos is a swimming pool, lined with some kind of flowery vegetation, to allow those first settlers an opportunity not only to store large amounts of water, but to give them a place to relax each day, rough-housing in the water and maybe looking a second or third time at the comely young Ensign assigned to the second transporter ship.

Ridiculous? Not at all …

The components for a Deimos' "Hilton Resort" are simple: a shelter, water, building materials, and flowers or flowery vegetables. Building shelters will be a primary mission of those early months on Deimos. There is no reason why a swimming shelter could not be brought on one of the first two transporter vessels or, much better, constructed from native materials found on Deimos. Water can almost certainly be harvested from resources on Deimos. Having storage "vessels" for water is not only necessary, it will be critical as more and more people arrive on Deimos. A swimming pool might easily be integrated into this system. "Clean" or mostly clean water could be pumped into the pool and the water could be pumped into a local filter[9] or back into a secondary treatment reactor in Deimos' main wastewater treatment system. As vegetation will be a necessity on Deimos—to remove CO_2 from the air and to release O_2—there is no reason the pool could not be lined with *something*. Palm trees on the first few transporter flights might be a bit extravagant, but *something* attractive can surely be planted. The reader should keep in mind that what will happen on Deimos will happen on Mars, enhanced

---

[9]   This poolside filter could be nothing more than some Deimos rock-and sand-sized particles in a plastic container, where a pump could lift the water a few feet to spill into this container. Colonies of earth-bacteria (seeded or hitchhiking on people and equipment) would soon establish themselves, feeding off the dripping water and removing the nutrients that make the water "dirty". Much cleaner, filtered water could then be returned to the swimming pool.

by a factor of two or three. As we shall see, trees will be a necessity during the early years on Mars, even if trees will be mostly absent from Deimos. So some type of vegetation, likely something both practical and beautiful, can be planted around the pool. Finally, a sign should be posted: "No High Diving", as those attempting such dives in Deimos' weak gravity may hurt themselves by hitting the top of the shelter.

The reader is assigned the task of listing all the advantages wrought by creating this pool-shelter, but consider these additional ideas. Raising the shelter's temperature to 35° C and keeping it very humid would add considerably to its tropical ambiance. In addition, having a swimming pool will reduce dramatically the need for individual shower facilities. Instead of building large numbers of showers throughout the Deimos base, one or two showers might be built next to the pool to service all residents and visitors in transit to Mars. The pool would remove most of the dirt and grime; the shower used to get especially clean. Yucky? Remember, we are trying to build a *pioneering* society here, not a vacation home for NASA astronauts. Finally, consider the impact on newly arriving immigrants from earth. After six months on a space ship, they will have at last arrived at Deimos to be invited, after getting settled into the reception and sleeping bays, to "have a swim". Creaky bones will no doubt linger for many hours at the pool, looking up through the transparent ceiling filled with the red vistas of Mars, soaking in the warmth and humidity while happy children splish and splash. There will work a-plenty in front of these intrepid settlers but, for those few days spent lingering at the Deimos beach, life would be very, very good.

Year Two

The arrival of the third Explorer Transporter will provide the equipment necessary to begin in earnest the creation of the first Mars base. In addition, this ship will bring more things for Deimos to continue the process of making Deimos a fully operational base. This ship might also bring the first civilian to Mars: a "farmer" who will create a real farm on Mars and will serve under contract to supply food for Confederation galleys. Components for a second Mars Lander might also be aboard.

As indicated by Figure 51, this third Explorer Transporter will bring equipment needed to detect, prospect for, mine, and process metals on Deimos. Plastic components for a large hanger on Deimos will arrive to allow for its continued construction. This hanger will be needed when larger transporters begin to arrive in a few years. With a second Mars Lander, missions from Deimos to Mars and back will become frequent and routine. Everyone will go down to Mars, except a four-person duty staff for Deimos operations, who will shuttle back and forth for 30-day tours. Mars will begin to produce supplies of food, water, rocket fuel, and oxygen for stockpiling both on Mars and on Deimos.

During Year Two, shelters to house 200 settlers will be sited and construction begun. To reduce the amount of construction, a hillside or even a sharp cut in the terrain might be selected. The more Martian terrain that is used, the lower the settlers' radiation exposure. While it will take several years for there to be 200 settlers on Mars, a larger shelter will let people feel more at home right away. The shelter will take many years to complete and initially it may be used more for crops than for habitation. Nonetheless, a domed area one to two hundred yards long will make for a comfortable existence. A shelter of this size will be needed in a few years, for once the Caravan Transporters begin to arrive, over 100 settlers will arrive with every mission.

| | | Explorer 3 | |
	Deimos	**Mars**	**Civilian Arrivals**
Tasks	Mine For Iron/Metals	Commercial Farming	Farmer
	Set Up Ore Smelter	Lay Out Shelters To House 200 Settlers	
	Bring Up Water, Fuel From Mars	Create Mars Infrastructure	
	Start Hanger For Transporters	Fuel processing Facility	
People	4	8	
		Explorer 4	
Tasks	Finish Transporter Hanger	Create Large Fish Ponds	Electrical Engineer
	Prep For Return Voyage (Ex 1)	Create Electrical Utility Company	
	Water Processing Facility		
	Mining Facility		
People	4	12	

Figure 51: Year Two

The size of the Deimos greenhouse will dictate how much and which supplies must be transported to the moon. In a sense, a tiny "trade" will develop between Deimos and Mars. Say, for example, that "sufficiently engineered" beans and lettuce can be grown in Deimos' tiny gravity and high radiation. The rotating staff from Mars will care for these beans and lettuce—perhaps a 1000-person-year supply can be grown each year, between times spent servicing other equipment on Deimos. The beans and lettuce will be processed on Deimos, some of it sent down to Mars in exchange for oxygen and other food items on the Martian menu, and some of it stored for use by future settlers. Deimos will remain dependant on shipments up from Mars for most of its needs, but it will also become a warehouse to supply the transporter vessels that will skip between Mars and earth in ever-increasing numbers. This 1000-person per annum supply of beans and lettuce will feed all the people on Mars, on Deimos, those in transit between the earth and Mars, and those working in orbit around earth. As there will be no cost to "lift" these beans and lettuce into space, the cost to feed people in space will be accordingly reduced. Naturally, if beans and lettuce can be grown on Deimos, maybe chickens can be raised there as well as other properly engineered crops. In a few years, Mars and Deimos may be able to supply most of the needs of those traveling in space. Much of the balance will be grown on the transporter vessels themselves, re-circulating human wastes and providing some of the oxygen the human passengers will need.

The real beauty of the "greenhouse on Deimos" idea is the ease with which the operation could be sold to some crusty, old libertarian anticipating his own move to Mars. If the Fleet sells off its greenhouse operations on Deimos to a trustworthy person in exchange for CS $1 million worth of food over five years, the mission to Mars reduces its costs, it "transplants" an essential business off-world, and provides not only a reason to settle Mars, but an incentive—an operating farm for sale—to get one's hairy libertarian butt away from the bureaucrats on earth.

Of course, just to anger that libertarian with the sweet deal on Deimos, a similar Fleet greenhouse on Mars will be sold off to compete with him. Repeat this operation 50 times in 25 different businesses, move a few hundred other businesses that got their start at the earth Embassy, and soon Mars becomes a real world with a real economy, peopled by freedom-loving souls living in a truly *free* society.

The other civilian enterprise that will begin in Year Two will be the electric company. Western civilization runs on electricity. Mars will be no different. Getting a commercial operation started from the first will not only help to reduce costs, but it will serve as a source of employment, as the demand for electricity will not likely be met for many decades.

Year Two will be a milestone for Deimos. Not only will the first "hanger" become operational, but routine operations will begin for the handling of space traffic. There, the first returning Explorer Transporter will be prepped for its return voyage. With its departure, the first two people (pilots) will return to earth to tell their tales of Mars, bringing back Martian artifacts and test materials for closer review by scientists, engineers, and businessmen at the earth Embassy. Scientific and commercial opportunities, as well as ever more refined plans to settle Mars can be explored as these artifacts are studied and the crew is debriefed on their adventures. Their return will not only create a firestorm of excitement on earth, but will ignite dreams about their own trips, by settlers still working at the earth Embassy.

Much more mundane will be a water-processing facility that will extract water from Deimos. This water will be used on all space missions, alleviating a huge "gravity" cost: the cost to lift water from earth. As water weight will be a large share of essential supplies needed for transportation between the earth and Mars, the dramatic reduction in the cost of water used (and reused) on the various transporter mission vessels will result in much lower overall mission costs. As with the other nascent businesses suggested by Figure 51, the water plant on Deimos will provide for civilian employment, serving as a strong incentive to get people to Mars and Deimos as quickly as possible.

Year Three

The fifth transporter mission will bring two more civilians to Mars. The Fleet's mining operations on Deimos will have met some of the demand for metal for construction. Commercial mining operations will begin in earnest, however, when the third civilian arrives, a person to supervise the metals operation on Deimos and to start operations on Mars. The military will detail some labor (30–50 hours per week) to assist this operation, but most of the technical work will be done by this new civilian arrival. Hopefully, fifty to one hundred pounds (earth weight) of useable metals will be created each week, in three or four basic shapes and sizes. Much of this metal will be stockpiled for later use. The other civilian will be a water engineer who will help process sizeable amounts of water on Mars[10]. Her job will be to establish both the engineering programs that a private water company will need, as well as to transfer from the earth Embassy the administrative programs necessary to run a profitable water company. As needed, both of these people can shuttle back and forth between Mars and Deimos.

Fleet operations on Mars will concentrate on two matters. First, will be the creation of a communications infrastructure. This infrastructure will lay the ground-

---

10   The Fleet will have already been processing small amounts of water, using "field" equipment. This water might be obtained by drilling, processing the permafrost in the regolith, etc.

work for Martian person-to-person communications (telephone), computer networking, internet access, entertainment, and Mars-to-earth data and voice communications. The second Fleet project will be to upgrade the chemical manufacturing operations from one of simple fuel creation, to multi-chemical processes to create some of the suite of chemicals needed by modern societies.

With the arrival of the sixth Explorer Transporter, two more civilians will begin their lives on Mars. One will be a construction worker and one will be a metals fabricator (machinist). The metals fabricator can use the stockpiled metal to create metal parts and structural metals needed by the growing community. The construction worker will use this Martian metal to build much more elaborate structures. Finally, the parts for a marsmover (earthmover) will arrive in time for a Martian Christmas. Batteries not included!

		Explorer 5	
	Deimos	**Mars**	**Civilian Arrivals**
Tasks			
	Explorer 1 Returns To Earth	Large Communications Apparatus	Mining Engineer
	Continue Operations at Base	Chemical Plant	Water Engineer
	Commercial Mining	Water Utility	
	Prep For Return Voyage (Ex 2)		
People	4	14	
		Explorer 6	
Tasks		Commercial Construction Begins	Construction Engineer
	Prep For Return Voyage (Ex 3)	Lay Out Martian Spaceport	Metals Fabricator (Machinist)
	Explorer 2 Returns To Earth	Assemble Marsmover	
People	4	16	

Figure 52: Year Three

Chickens will also arrive on Mars during Year Three. They will likely be brought as embryos though, if necessary, a dozen or more live chickens can be brought on the transporter. The chickens can be used as a source of eggs and fresh meat. If available, embryo technology will allow for the immediate sale of chicken products. Otherwise, "old fashioned" methods will be used to increase the chicken population to one or two hundred, before eggs and meat appear for sale.

The actual economics of raising chickens on Mars will be explored in more detail in Volume III. As will be discussed there, the most likely case will be that a contract for the production of chicken will be created between a Chicken Farmer and the Fleet before the Farmer transports to Mars. One aspect of this contract, of course, will be the purchase price for the Chicken Farmer's produce. Another aspect of this and similar contracts will be the expectation to *out produce* consumption. Such provisions will pay cash to the Chicken Farmer and others to operate their businesses on Mars—even during the first few years when demand is low—and to create stockpiles of products for future use.

A well-designed contract system can provide price stability on Mars early on and allow the Fleet to plan its imports from earth. If one farmer cannot negotiate a satisfactory contract with the Fleet, then another farmer, someone who *can* negotiate such a contract, will get that seat on the Explorer Transporter. These contracts will all have termination dates, so that after the second and the third chicken farmers arrive on Mars, free market principles, rather than Confederation contracts, will set prices.

Year Four

Year Four will see two more commercial developments on Mars. First, a commercial plastics smythe[11] will arrive to complement the work of the metals fabricator. Though some plastics feedstock will still be brought from earth[12] there will now exist a significant ability to fabricate needed items on Mars. Both the plastics smythe and the metals fabricator will be under contract to the Fleet for their products and they will both live at a military base, run by a military commander, but the beginnings of a real Martian economy will be starting.

---

[11]   A plastics smythe is a person who creates objects and parts out of plastic, much as a blacksmith in the past created objects from iron. The advantages of having a manufacturing capacity where most any object can be fashioned are obvious.

[12]   The chemical plant discussed in Year Three will create feedstock plastics for the plastics smythe.

The second commercial operation will be a new place to eat, as two "restaurant workers" arrive. They will purchase their food from the farmer and from the Fleet commissary and sell meals to the settlers now on Mars. A scaled-back Fleet galley will continue to operate, to offer a second option for food.

Prior to the arrival of these two restaurant workers, members of the Fleet will have spent a great deal of time working in the mess hall. Although there will still be lots of prepared food shipped from earth in Year Four, a growing amount of food will be harvested from Deimos and Mars. In the beginning, most food will have the advantage of being fresh and the disadvantage of needing to be processed before it reaches the kitchen. Thus, corn will have to be shucked, tomatoes sliced, fish gutted, and potatoes peeled. And when warm-blooded food reaches Mars, even more expertise will be needed to slaughter and butcher it before it reaches the mess hall. All of these time-consuming chores will be accomplished by crewmen detailed from the Fleet. Naturally, as farmers and food processing businesses arrive, the Fleet can be excused from such duties. But the ability *simply to eat* on Mars will be a major subordinate goal of the mission commander. Indeed, it will also likely fall to the Fleet commander to ensure that the excess foodstuffs discussed above be canned or otherwise preserved, again requiring Fleet labor.

On a different tack, the first babies may be born on Mars during this time. Efforts will have been made to transport one or two young married couples on prior flights. Albeit a delicate subject, the mission will want an opportunity to care for a newborn baby. The ability to care for children and infants will be important for the new base, especially as, starting in Year Five, sixteen new settlers will arrive annually and children will be borne every year or two. Moreover, in Year Eight a small group of children will be specially transported to Mars (The Stork Mission), to populate the planet with youngsters.

		Explorer 7	
	Deimos & Phobos	**Mars**	**Civilian Arrivals**
Tasks	Continue Base Operations	Commercial plastics operation	Plastics smythe
People	4	18	

		Explorer 8	
	Deimos & Phobos	**Mars**	**Civilian Arrivals**
Tasks	Continue Base Operations	Open restaurant to compete with galley	Restaurant Owner
	Initial Landing on Phobos		Restaurant Worker
People	4	20	

Figure 53: Year Four

Several other "infrastructure" milestones will also be reached during Year Four. First, that communications systems started in Year Three will now provide a significant transmission capability between Mars and earth. Not only will news broadcasts become available on a daily basis, but recently released movies will be transmitted to Mars. Email will always have been available, but now it will become virtually unlimited and a significantly improved internet capability will become operational at some point during the year[13]. The shelters started in Year Two will be completed, giving Mars the ability to shelter 200 people. This new shelter will be especially important as the population on Mars begins its exponential rise in Year Five.

The other big development during Year Four will be the landing of Fleet personnel on Phobos. As stated earlier, Phobos will become a "Fleet" moon while Deimos will develop as a "commercial" moon. Both rocks, of course, will be thoroughly surveyed for their resources and plans amended accordingly. The Fleet commander on Mars will shortly be replaced. The expedition to Phobos will be the last achievement of his four-year command on Mars. The first missions to Phobos will not result in a permanent base there, but plans

---

[13]   Remember that there will always be a few minutes lag between Mars and the earth. The author's idea for an internet connected with earth is to have a central Martian server store each page viewed. After the page is delivered, other users on Mars can also call up that page immediately. If a non-viewed page is requested, the Martian internet will give the user an approximate time (minutes) when the page will become available. As viewed pages become dated, they are scrubbed out of the Martian server. While this system is crude and will be frustrating until there are reasonably large numbers of users, a community of 10,000 might have almost immediate internet access for many pages, even if they must get used to waiting for seldom used pages from earth.

for that base will arise from this "exploration" mission undertaken by the Confederation.

Year Five

The arrival of the first two Shuttle Transporters will double Mars' population from the beginning of the year to its end. Besides the use of these larger transporters, Year Five will have several other significant developments. The arrival of a new Fleet commander on Mars means that the base will operate from a different person's perspective[14] Not only will the planned relief of Fleet commanders on Mars manifest the Confederation's policy to share opportunities in space with as many people as possible, but it will also help prevent the development of cults of personality in this area or that. The old Fleet commander will have done things one way. The new commander will do things differently. The new Fleet commander will do the mission to Mars another good turn as well. The old Fleet commander can go back to the Military Academy at the earth Embassy[15] to teach or otherwise share his wealth of knowledge about Mars. And the new Fleet commander can bring in more advanced ideas—from the earth Embassy—about how the development of Mars should proceed.

	Explorer 9 & Shuttle 1		
	Deimos & Phobos	**Mars**	**Civilian Arrivals**[16]
Tasks	Continue Base Operations	Commercial Plastics Operations	Plastics Engineer
		New Commander for Mars Base	Plastics Worker
People	4	30	

---

[14] An example of the mission to Mars' efforts to divide power as much as possible. If the first Mars commander was an intolerable martinet who made everyone's life miserable with his perfectionism, the second Mars commander may be a mild-mannered "techie" who is more interested in *building* than *controlling* things.

[15] The first commander will also be a pilot who will pilot one of the Explorer Transporters returning to earth.

[16] From Year Five forward, these are only selected examples of civilian arrivals. Each semi-annual mission will bring in eight more settlers, some on active duty with the Fleet and some civilians.

	Explorer 10 & Shuttle 2		
	Deimos & Phobos	**Mars**	**Civilian Arrivals**
Tasks	Continue Base Operations	Open Wal-Mart	Wal-Mart Worker
		Start commercial goods manufacturing	Commercial Goods Manufacturer
People	4	40	

Figure 54: Year Five

The plastics business will continue and accelerate. No longer just about simple processes, a more formal plastics industry will begin, building upon plans long laid at the earth Embassy. A new "plastics factory" will begin to mass-produce some things. It may have a contract to make 100,000 non-disposable plastic forks or 25,000 glasses, selling some through the Fleet warehouse, but stockpiling most for sale to retailers about to arrive. This business might also make sheets of plastic for use in building new shelters and other construction projects.

Perhaps associated with this plastics business will be Mars' first commercial goods manufacturer (CGM). This person will represent a vast array of enterprises that are needed to create those goods demanded by the consumers of any western society. For example, this person might manufacture furniture from the plastic that the new plastics operation will create. While transporting one person to make furniture on Mars is a small step, the existence of such an operation will mean that no more furniture will need to be imported to Mars. Only a small victory in Year Five, in Year Eight when there are 200 people on Mars and numerous new businesses setting up shop, there will be a significant demand for furniture. This one person—perhaps in a business with five employees at the earth Embassy—represents two ideas. First, it represents how a self-sustaining human colony can slowly transplant itself to Mars. Idea two is that this furniture business will eventually need all five employees on Mars to meet the demand for furniture. The transport system discussed in an earlier chapter explains how this business can be transferred to Mars in a reasonably orderly fashion.

The first retailer will arrive in Year Five. Called "Wal-Mart Worker" in Figure 54, this retailer will create an outlet for goods produced on Mars and a place where the settlers can go to purchase what they need. Some of the products created by the CGM and the plastics factory, for example, will be sold to stock the Wal-Mart. The first five years of this enterprise will be both frantic and frustrating. Eventually (and especially after a competitor arrives in a year), the buying needs of the settlers will begin to be understood. This retail-generated knowledge of Martian consumer buying habits will help the earth Embassy program consumer related businesses for priority transfer to Mars.

The arrival of these new commercial ventures presents another opportunity to illustrate the power of the settlement idea and the paucity of the exploration idea. At the end of Year Five, Mars will have a farmer, a metal manufacturer, a metal fabricator, a plastics smythe, a chemicals manufacturer, a plastics "business", a commercial goods manufacturer, one restaurant and the Fleet's galley, a construction business, and a retail outlet. The cost to get to Mars will have been borne by these businesses themselves, not by some line item in a NASA budget. The only cost to the Confederation will be the cost of contracts to ensure that these businesses stay busy on Mars (stockpiling). The cost differential between the mission to Mars' program and a similarly constituted "economy" on Mars might reasonably be US $5–20 billion or more. Costs will decrease, too, as more and more businesses come in from the "business" pipeline at the earth Embassy. In fact, while NASA's ability to create a diverse economy on Mars must be questioned soup-to-nuts, an attentive reader here can see that the Confederation economy on Mars will be spring-loaded, awaiting the arrival of that next ship for each business owner's chance expand her business and to help lower the cost of creating Martian colonies.

Year Six

Life on Mars during Year Six will start to become routine, as additional CGMs arrive. The goal, of course, is to make Mars self-sufficient as quickly as possible. The two new CGMs will sell some of their products to the Wal-Mart and the others will go a Fleet "warehouse". One might manufacture clothing and the other soaps and shampoos. The availability of resources on Mars and the needs of the settlers will determine exactly what kinds of commercial manufacturers should be sent. Brick, glass, and other construction manufacturers might also be considered for these slots. Some raw materials will continue to be imported from earth, from Deimos, and possibly from the inner asteroids, but the desirability to reduce the importation of simple goods, in favor of complex items, will become more and more apparent. A "pricing system" for goods bought at the Fleet warehouse will encourage domestic (Martian) production[17]. The plan is not to manufacture the easiest or the most needed product on Mars, but *to encourage the manufacture on Mars of those items it makes the most economic sense to manufacture there.* In other words, the Confederation will bring to Mars those businesses that will most help to reduce costs for the Fleet and for the nascent Martian society. As the years go by, more new businesses and new manufacturing processes will arrive on Mars to fill economic gaps. Which manufacturing capabilities are added when on Mars will be decided by a combination of free market principles and common sense management by the Fleet. The end goal will be to create economic self-sufficiency as quickly as possible, realizing that Martians will need to import light, highly complex things like advanced medicines or computer chips for many years to come. It will be bulky, heavy, or mass producible items that will be manufactured on Mars.

	Deimos	Explorer 11 & Shuttle 3 **Mars**	**Phobos**	**Civilian Arrivals**
Tasks	Continue Base Operations		Return to Phobos and begin base operations	Construction Worker
		More commercial manufacturing begins		CMG II (Commercial Goods Manufacturer)
People	4	46	4	

---

[17]   Volume III contains a detailed account of how the Fleet's warehouse on Mars will help to establish a pricing system to this end.

		Explorer 12 & Shuttle 4		
	Deimos	**Mars**	**Phobos**	**Civilian Arrivals**
Tasks	Continue Base Operations	Open Wal-Mart	Continue Base Operations	K-Mart Worker
		More commercial manufacturing begins		CMG III
		Begin Terraforming Operations		
People	4	56	4	

Figure 55: Year Six

The reader may be confounded by this last paragraph; perhaps even a bit put off by it. Economic self-sufficiency can be interpreted in many ways. *How can you possibly decide what "makes the most economic sense" to manufacture on Mars?* This is a complex question that will be discussed at length in Volume III. The Fleet's warehouse will be part of the answer, as will the earth Embassy, whose growing financial strength will be attractive to investors. The mission to Mars asserts that Mars will import a great deal, export some food and lots of fuel to space, while nonetheless returning a fine profit for all concerned[18].

The first base on Phobos will be established during Year Six. Phobos will develop differently from Deimos. After Phobos is surveyed for metals and other substances, a final determination about the use of the moons will be made by

---

[18]   Consider that if the mission to Mars has growing colonies on the Red Planet and a viable and partly self-sufficient earth Embassy, many corporations on earth will contemplate investing in the mission to Mars. A strong mission to Mars balance sheet will pay for imports. Combine this with the wealth being generated on Mars and the result will be large investments by those who hope to cash in on a world that will grow for generations to come. Much more about this specific issue (imports, exports, etc.) in Volume III.

democratic process. As on Deimos[19], military crews will rotate back and forth from Mars to Phobos, on 30-day tours of duty.

To assist with continuing construction projects, a representative from a second construction company will be sent to Mars. Although some competition between the two construction companies will be useful, negotiating contracts at the earth Embassy will be equally useful to this end. Low-bidder gets sent to Mars. A continuously growing capacity for construction will exist on Mars, but the nature of these first few years will require mostly centralized planning and centralized pools of labor to ensure progress. Naturally, if the farmer needs some new irrigation piping or the Wal-Mart wants to build an unplanned addition for its store, such private contracts can be bid between the two construction companies on Mars. Even in these circumstances, however, the work would likely need to be coordinated with the Fleet commander, to procure supplies from the Fleet warehouse and, as discussed, to allow the construction supplies to be anticipated, purchased, and shipped to Mars.

The arrival of a K-Mart worker will mean that two retail alternatives to the Fleet's warehouse will exist. More and more commercial and personal products will be obtained through the Wal-Mart and the K-Mart and fewer and fewer from the Fleet's warehouse. The arrival of a Wal-Mart competitor will also advance the emergence of a free market pricing system on Mars.

Not all of the retail work will be "behind the cash register". Both retailers will be on Mars to help transplant their businesses from the earth Embassy. A major aspect of their work will be to establish reliable supply chains (such suppliers to be given priority on arriving transporter missions) and to estimate the kinds and quantities of goods that will be carried by the stores. Considering the vast, on-going construction projects on Mars, construction goods and basic supplies would seem a good baseline inventory.

And finally, for the science starved, holding their collective noses about all this "economics stuff", we finally arrive at something straight out of science fiction. The chemicals plant (run by the Fleet) that was established during Year Three will have produced a supply of chemicals needed to create powerful greenhouse gases. In Year Six, greenhouse gas emitters will arrive on Mars to begin the process of terraforming Mars. Volume III will go into much more detail about these gases and processes, but the hope is to create a (relatively) thick atmosphere on the Red Planet just a few decades after arrival. The earlier the terraforming operations begin, of course, the earlier humans on Mars can enjoy the benefits. Releasing large quantities of gases that are 10,000 or more

---

[19]   There will be one semi-permanent civilian miner and maybe a semi-permanent farmer on Deimos, who will share duties on Deimos with rotating crewmen.

times more effective greenhouses gases than CO_2 means that, over the course of ten or twenty years, Mars' ambient temperature might be increased a few degrees, helping to set off the runaway greenhouse effect that many expect is possible on Mars. The key will be to design light, simple reactors, train Fleet crewmen to operate their simple controls, and most importantly, to locate and process native resources to serve as feedstocks for these reactors. The fluorides likely to be critical to this operation are thought to exist on Mars. The Fleet can send marines to find the fluorides, have the mining engineer create a program to mine them, and the chemical business to process the ores. Thus begins the warming of Mars.

Year Seven

A different approach will be used to cover life in Year Seven. Here we take an eagle's eye perspective of what has been happening on Mars, both to evaluate the potential success of plans laid and the potentialities for life as human society continues to unfold there.

By year's end, a total of fourteen Explorer Transporters and six Shuttle Transporters will have been sent to Mars. Figure 56 shows how the supplies carried by these ships can be broken down into four main components:

Category	Examples
Consumables	Dehydrated Food
Parts and Supplies	Nuts, bolts, machine parts
Machines and Components For Fleet Operations	Reactor vessels to make rocket fuels
Machines and Components for Commercial Manufacturing	Small smelters to make pig iron, aluminum

Figure 56: Categories of Supplies Brought Aboard Transporter Ships

The Fleet will have had a part in handling almost every supply to have arrived on Mars. Unless something is shipped directly to a civilian operation, it will pass through the Fleet warehouse. For example, the Fleet warehouse will have received almost all foodstuffs shipped to Mars, "selling" it to the galley or to the owners of the private restaurants, as they need it. With the arrival of these restaurants and the two retailers, less and less food will be handled by the Fleet until, by the end of Year Seven, it may only handle about 10% of the food sent to Mars. Other parts and supplies will have been similarly processed by the Fleet and similarly transferred to civilian control. Remember, by the end of Year Seven, there will be those two retailers and four CGMs on Mars, as well as numerous other civilian activities on Mars. These entities, not the Fleet, will handle most imported goods.

Machines will be handled in two ways. Some of the early machines will go directly to civilian operations. Many of the machines, however, will be assembled and operated by the Fleet, using crewmen labor. For example, the "marsmover", brought to Mars in Year Three to help with construction projects will be operated by a Fleet crew; the terraforming plants referenced in the last section will be assembled by the Fleet and operated by crewmen. Thus, the Fleet will facilitate the creation of assets and capabilities, all with a mind to transferring these assets and capabilities to a private operation. The Fleet will have about 60 personnel available to it on any given day during Year Seven[20]. As much chief executive officer as base commander, Mars' Fleet leader will be responsible to run a dozen different operations in a dozen different fields.

Figure 57 lists some of the operations that the Fleet will run during Year Seven as well as suggesting how the Fleet will interact with the growing commercial community. The twin ideas of "living off the land" on Mars and the intelligent creation of commercial operations there will combine to affect an almost immediate self-sufficiency on Mars. And to belabor the point, the transport of additional people, resources, and supplies will allow for increasing self-sufficiency, will reduce overall mission costs, and will make the human society on Mars increasingly robust.

Finally, note that the Fleet warehouse will continue to maintain a supply of "rental" equipment for occasional use. If the farmer needs a saw or something to dig a hole, he can rent it from the Fleet warehouse. This circumstance will continue until a commercial "rent-all" establishment can be opened on Mars.

---

[20]   The dozen of so civilian settlers will not have lost their obligations to serve 31 days per year. The base commander may, as she determines, simply conscript 1/12 of the products and services of their daily lives as the simplest expedient to account for such service.

Operations	Responsibilities	Notes
Farm Operations	Construct and operate numerous agricultural sites and operations	On Mars, Deimos, and Phobos
Fish Farms	Harvest fish from ponds operating as part of wastewater system	Entirely Fleet operated
Water Production	Assist water company to operate	Occasional conscription of Fleet labor
Construction	Joint construction operations with commercial entities	
Electricity Production	Assist electric company to operate	Occasional conscription of Fleet labor
Oxygen Supplies	Entirely Fleet operated	To hand off to "oxygen" company
Rocket Fuels	Entirely Fleet operated	
Chemicals	Entirely Fleet operated	To hand off to chemicals company
Mess Hall		Continue galley operations, i.e. occasional meals and food for remote activities
Medical	Medics, dental technicians, doctors	Entirely Fleet operated
Consumer Goods Distribution	Fleet warehouse	Fleet working to get out of commercial warehouse business
Terraforming Operations	Assemble "factories" to release greenhouse gases	Entirely Fleet operated
Surveys For Specific Resources	Fleet operations	Working with earth Embassy engineers and scientists

Figure 57: Major Fleet Operations on Mars During Year Seven

		Explorer 13 & Shuttle 5		
	Deimos	**Mars**	**Phobos**	**Civilian Arrivals**
Tasks	Continue Base Operations	3rd Store Opens	Continue Base Operations	3rd Retailer
	Explorer 9 and Shuttle 1 Convoy Back to Earth	Hotel Preps to Open		Bed & Breakfast Operators
People	4	64	4	
		Explorer 14 & Shuttle 6		
	Deimos	**Mars**	**Phobos**	**Civilian Arrivals**
Tasks	Continue Base Operations	Open Burger King	Continue Base Operations	Burger King Manager
	Explorer 10 and Shuttle 2 Convoy Back to Earth	More commercial manufacturing begins		Burger King Worker
				CGM IV
People	4	72	4	

Figure 58: Year Seven

Year Eight

Year Eight will be graduation year for the mission to Mars. This will be the last year that the first Martian settlement will be under Fleet control. At the start of Year Nine, the first Caravan Transporter mission (and 108 new settlers) will arrive, doubling in a moment the number of people on Mars. Just after they arrive, an election will be held for mayor and council, which will govern Mars' first village. The Fleet will redouble its efforts at other settlements as Mars' first civilian society emerges at that village.

Beyond just a graduation, the real highlight of the year will be the arrival of six or seven children on a special Shuttle Transporter (The Stork Mission). These will be the children of settlers already on Mars. Family reunions will be important for the mission to Mars, but the historical importance of the children's arrival will be the new requirement for schools, facilities, and other kinds of infrastructure for children. Children will have been born during the seven previous years, so these will not be the first children on Mars, but the arrival of six or seven more children will mandate schools on Mars, employing our newly arrived school teacher. With a dozen or so children, the associated joys and dilemmas of raising children will begin, as will the unique issues of raising these children on a new planet. Naturally, these children will go through some hard times, but these hard times will be shared with a close-knit community. What they might lack in material goods and options to "hang out" at a mall will be made up for in the opportunity to live on Mars and in the social interactions of the Martian village that will help to raise them. Of course, these children will have every right to return to the earth when they are 18, as part of the mostly empty vessels that will continue to fly back to earth to pick up more settlers. And if they decide to return to earth, they will arrive as the most widely traveled "earthlings" in history.

		Explorer 15 & Shuttle 7		
	Deimos	**Mars**	**Phobos**	**Civilian Arrivals**
Tasks	Continue Base Operations	Company Arrives to Assemble "Golf Carts"	Continue Base Operations	Vehicle Manufacturer
		Lay Out Mars University		
		Prep Base to Turn Over to Civilian Authorities		
People	4	80	4	

		Explorer 16 & Shuttle 8		
	Deimos	**Mars**	**Phobos**	**Civilian Arrivals**
Tasks	Continue Base Operations	Open School	Continue Base Operations	School teacher
		More commercial manufacturing begins		Six or Seven Children Arrive
		Lay Out Second "Village"		CGM V
		Turn First Mars Base Over to Civilian Authorities		
People	4	88	4	

Figure 59: Year Eight

Two noteworthy aspects of Martian society will be the arrival of people to start assembling "golf carts" on Mars and the creation of facilities for one of the Martian universities. American readers especially will love the idea that Martians will begin to have the option of having "wheels" and no doubt the demand for golf carts will be great. The earth Embassy will have at least two universities in operation and the need to plant the seeds of scholarship and higher education on Mars will be important. At this point, we leave off with describing the remaining activities depicted in Figure 59. Having 96 people moving between three sites on Mars and the two Martian moons, and especially with school now in session, the reader should clearly understand that a vibrant new society will have begun on the Red Planet. This new society will be ready to accept the first of the large transporter vessels, carrying not just new settlers, but new businesses and new families to an accommodating, new world.

The First Caravan Transporter

At the end of eight years, there will be almost one hundred people on Mars, Deimos, and Phobos. The character of the first base on Mars will have been transformed from that of a military expedition to a town ready for self-government. Most of the difficulties associated with survival on Mars will have been solved and the infrastructure for a real society will be apparent not only in the customs and habits being formed in daily life, but in the physical hardware—streets, sewer lines, electricity, etc.—visible everywhere at the base. An economy on Mars will not only have taken root, it will be ready to expand and prosper. Flowers will be seen everywhere and trees will have transformed themselves from saplings into maturing specimens. The 90 people on Mars will have a common area resembling a suburban shopping mall[21], with tidy residential areas just a hundred meters away. Everywhere there will be bright lights; snug, aesthetically pleasing buildings; and tidy by-ways. Soon after the arrival of the first Caravan Transporter convey in Year Nine, carrying 108 new settlers[22] for the new base, the military commander will officially end military administration at the village, in favor of its first duly elected mayor.

The arrival of these 108 new settlers will bring more than just people to the Red Planet. There will likely be a dozen new business organizations arriving to establish their footholds on Mars. Overnight, there will be a new diner, two new fast food establishments, two seamstresses, a Jewish Rabbi, three schoolteachers, a hairdresser, etc. Farming operations will be augmented both by new farmers and new food processing businesses. Several new engineers and technicians will arrive to bolster the infrastructure of the growing community, as well as miners to begin in earnest the harvesting of metals from the Red Planet or from Deimos.

The medical treatment of the settlers during the first eight years on Mars will have been spotty. This pioneering society will have to rely upon the good sense and ingenuity of only one or two medical professionals to deal with the myriad of injuries, diseases, cavities, and problems that will present themselves during that time. This problem will be greatly ameliorated by the arrival of several more medical professionals aboard the first Caravan Transporter and—hurrah!—a real dentist. Henceforth, the mission to Mars will strive to force feed medical professionals into the Mars pipeline. Teams of new medical professionals will not only allow for the adequate treatment of most medical problems, but also for the research effort to solve the unknowns about humans in space and on Mars. No doubt many different effects will be noticed as people (and their pets!) live in low gravity conditions. None are likely to seriously threaten the mission to Mars' main efforts, but there will be an increasing demand to understand how

---

[21]   Maybe those teenagers will have a mall to hang out at after all.

[22]   Arriving as follows: two on an Explorer Transporter, 6 on a Shuttle Transporter, and 100 on the Caravan Transporter.

life—human and otherwise—will adapt and differ on a planet with only one-third earth's gravity.

The arrival of the Caravan Transporter will not result in any loss of missions for the Fleet. Rather, the Fleet will redouble their efforts as they begin to lay out a new Martian base and the university. The 108 new Martians are, after all, not the end of the settlers, but just the beginning of them. Six months after their arrival, another 108 people will arrive. And in a few years, the Ark Transporter missions will begin, with 1108 settlers[23] arriving every six months. The military may continue to have only a few dozen full-time members on Mars, but Fleet service will continue until the first Martian republic is granted its independence[24]. Thus, people on Mars will continue to owe the Confederation thirty-one day's service every six Martian months (once every earth year). The Fleet's normal strength on Mars will rise from a few dozen to hundreds. Using these personnel, the Fleet will continue to man the base on Deimos, to expand Fleet operations on Phobos and, most importantly, to continue to build new bases, laying the foundations of the Martian society.

---

[23]   The convoy of one Explorer, one Shuttle, and one Caravan Transporter will carry 108 settlers and the Ark itself will carry another 1000 settlers.

[24]   When the new republic will decide what military service, if any, is required of its citizens.

Creating a Viable Society

> Impossible is nothing.
>
> —advertising slogan for Adidas footwear

This chapter has a twin a few pages on. Both provide glimpses of human aspects of the settlers' life on Mars. We take first a broad perspective and discuss attributes of a viable society on a new planet. Later, we talk about life and experiences for individual Martians. As with so much of this first volume, the canvas is large and the brushstrokes are broad. We offer, of course, a view that excludes Hollywood's perspective of Martians working for "oxygen rations" or in bleak mines. Instead, we present what the reader should be coming to view as much more persuasive: the mission to Mars' vista of a growing, pleasant, western world, familiar to most readers; as exciting an opportunity as can be imagined, and open to ordinary people.

A viable western[1] society has several recognizable characteristics. First, it can furnish most of those basic goods and services necessary to support the society. Second, it can sustain and grow its population. Third, it can create new services and products for sale within the community or for trade with populations beyond its borders, especially in order to import foreign products and services[2]. Fourth,

---

[1] The modern west includes South Korea, Japan, etc. The mission to Mars does not contemplate a strictly European-style culture. Rather, Mars will be a modern western society in the sense that people with genetic material from across the earth will live together based upon principles most of us in the US, the UK, etc. would recognize.

[2] This not a repeat of the first point. Being able to provide basic necessities differs from being able to create modern goods and services. This difference is between a modern western society and Algeria, where people live their lives without much fear of starvation, but without many modern amenities.

it should have some redundancy to allow for regeneration, even if a catastrophe should occur. Fifth, it can defend itself against probable threats. Finally, a viable society should have a recognizable culture, not necessarily unique, but should see itself as some kind of collective whole, not a Balkanized set of humans occupying, quibbling, and possibly shooting each other over the same space.

The American state of California, by itself, might be called a viable society. Running down the list just created, we see that California has the markets and government infrastructure to provide most of the goods and services its citizens demand. It can easily sustain its own population with its base of 30 million citizens. California has a viable currency, the US dollar, and economic connections worldwide. California trades for those products or services not created within its border, such as diamonds, chromium, or most automobiles. California defends itself from internal and external threats through the American federal government (army, Federal Bureau of Investigation, etc.) and through its own extensive system of law enforcement. Despite the tragedies that occur when there is an earthquake, California quickly rebuilds better than ever. Finally, California has its own unique culture as well as participating as a subset of the larger American culture. And unlike the group of itinerant scientists found at science stations in Antarctica, California's culture and society self-identifies as *Californian*.

Necessities

The creation of a viable society on Mars will be an iterative process. From the very beginning, the settlers will strive to provide all of their own necessities. "Necessity" here is a matter of definition. Like California, Mars will likely never produce every good or service its people desire, but will always trade for something, whether it is the latest cancer technique from earth or Elvis Presley memorabilia. Within just a few decades, however, Mars should be in a position to grow and thrive without depending upon imports from earth for survival. Consider four necessities measuring such a level of economic development: food, basic materials, water, and oxygen.

American astronauts are not thought of as farmers, but agricultural talents may be the most important skills for the first Martian settlers and, during the initial phases of most Martian bases, Fleet commanders will likely *require* their crewmen to tend their own small gardens. After all, the world's most famous Martian scientist will end up a world famous *dead* scientist if no one can produce food for her. And as tasty as most fruits and vegetables can be, it may not be worth living on Mars if the settler cannot occasionally enjoy a greasy slab of bacon or a half-raw cheeseburger. Thus, raising fish, chicken, hogs, and cattle will be a priority.

Agriculture can create its own circle of success. The more corn that can be grown on Mars, the more feed that might be provided for animals for slaughter. Fish can grow in ponds that are part of the wastewater treatment system. Successful farming practices for Mars will not simply appear, but will require research and practice at the earth Embassy. With effort and determination, however, Mars will be able to feed its own population just a few years after humans arrive.

The creation of most basic materials will also be possible on Mars. Small quantities of metal will be readily available even during the early years. Raw materials for plastics and other chemicals can also be created early on. Wood will become available in small quantities after a decade on Mars. At first this will only mean limited quantities of cellulose from branches trimmed from growing trees. Later, the apple or maple trees planted to enliven an early Martian street can be harvested after a hundred more trees are planted to replace each one cut for lumber. Martian homes can be created adobe style from specially treated Martian "dirt". Mars has resources. The mission to Mars will create the programs to allow Martian society from the start to create most of the basic materials it will need.

The most important resources that humans will need on Mars, of course, will be water and oxygen. Fortunately—and unlike earth's moon, the moons of the gas giants, Mercury, or anywhere else in the solar system—these resources will be readily available. Water is thought to be available in large quantities almost everywhere on the planet, mostly trapped in a kind of Martian permafrost. Oxygen can be extracted from the CO_2 in the Martian atmosphere and will be a by-product of the chemical and rocket fuel industries created on the first day on Mars[3]. And, of course, O_2 will be created by growing crops.

Will water and oxygen be cheap on Mars? Probably not, at least initially. Indeed, there is a section in Volume III entitled "$10 A Gallon Water and $3 Per Day For Air" that discusses how Mars can be both a rich society and still demand hefty prices for things we take for granted on earth. The point of that later discussion is that just as Americans are not poor because they have to pay for food, so Martians will not be poor because they have to pay for food, *water*, and *oxygen*. Indeed, water and oxygen producers will be creating high-paying jobs for decades into the future to meet the ever-growing demand for their products.

Finally, a word about basic services. There should be no doubt that a society that can propel itself to Mars can also provide itself with basic services: sewers, electricity, police protection, etc. Again, electricity might initially be more expensive than in the US, but this will not preclude the creation of a modern or wealthy society.

---

[3]   Small chemical reactors will be carried on the first ship to Mars. The mission to Mars will not look for rocks, it will look to make rocket fuel and chemicals.

Given the kind of detailed planning necessary to move a *society* to Mars, no one should doubt the Martian's ability to provide themselves with the necessities of life once they get there.

Martian Population Growth

The creation of a sufficiently large genetic base to sustain an independent Martian population will likely be accomplished during the first ten years on Mars. A very diverse genetic base will be created after only a few decades. Once the population reaches 50,000, there should be few questions about the capability of such a genetic base to sustain the human species, especially as additional genetic material arrives with each new ship of immigrants.

Population growth is a slightly different matter, but even factoring for the occasional disaster, it will increase between 5 and 25% per year as the settler pipeline feeds people to Mars[4].

Decade Starting	Total Mars Population (M)	Native Martian Births	Immigration From Earth
2050	0.05	0.01	0.02
2060	0.08	0.02	0.04
2070	0.14	0.03	0.07
2080	0.25	0.05	0.12
2090	0.42	0.09	0.20
2100	0.71	0.16	0.34
2110	1.20	0.26	0.58
2120	2.04	0.45	0.98
2130	3.47	0.76	1.67
2140	5.90	1.29	2.84
2150	10.03	2.20	4.82
2160	17.05	3.73	8.19
2170	28.97	6.34	13.91
2180	49.22	10.78	23.64
2190	83.64	18.32	40.16
2200	142.11	***	***

Figure 60: Martian Population Growth

---

[4]   Note that even the fastest growing populations on earth grow at only 2–3 % per year.

Figure 60 provides an estimate of Martian population. It assumes a 6% growth rate per year after the mission to Mars' "50,000" have arrived. This adds a high birth rate (2% over death rate) to a 4% (from earth) immigration rate. Both rates are very dynamic, almost to the point of being unsustainable. The mission to Mars posits these numbers, however, because it believes that once a real opportunity to get to Mars is presented, humans will flock there.

As will be discussed below, the dramatic increase in Martian population does not necessarily imply impossibly turbulent settlements and cities, since most of the new arrivals will be located in new communities. What is important for this section, however, is to understand that the mission to Mars' program allows Martian population to reach the point of viability just a few years after landing.

Martian Villages

The mission to Mars' solution to the problem of stable communities in light of extremely rapid population growth is to grow the population through the creation of new villages, rather than simply adding new "streets" or "developments" as is common in the US. This idea of village development will be addressed briefly now and in greater detail later.

Figure 61 shows how Mars' first "town" might be designed and grow. From the earliest day on the planet, to a settlement with 5,000 people[5], growth will be managed by adding new village developments, according to a long-term development plan. The creation of this size town might take fifty years. Again referring to Figure 61, the first development might have been one of the agricultural villages. When additional growth was indicated, a town center might have been added. As more growth occurred, the industrial village might have been added, followed by a second agricultural village. Finally, a university village might have been created to continue the orderly and efficient development of this Martian town. Not every village need be fully developed before another is begun and, in the case of the agricultural villages, some villages might grow along with the size of the larger towns: first 100 people, then 200, then 400, and then 1,000 when a modern food processor establishes itself in the village. At some point, we have a five-thousand-person town, one thousand living in the town center, one thousand living in each of two agricultural (and food processing) villages, one thousand living "on campus" at a university, and one thousand living in the industrial village.

---

5   The Confederation's plan for the village "circles" addressed in this section will be used across those parts of Mars controlled by the Confederation. As each republic becomes independent, development will proceed as that particular society sees fit.

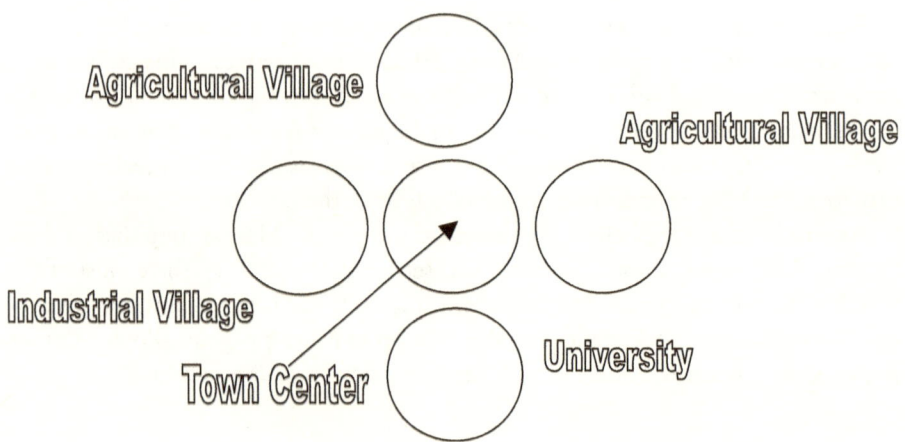

Figure 61: The First Martian Town

Each of the five sections of the town is a socially and economically self-contained area. Hence the word "village". Each will contain shops, restaurants, services, and will be partly self-contained for employment purposes, even as each village is designed to meet a different economic need. Thus, the agricultural villages might have several larger food processing operations to help round out their economic lives, while the city center will have many more shops and restaurants than the other villages.

Each village will be about one-half mile in diameter, equating to about 160 acres. Every home will be within walking distance of the village center, with its shops, restaurants, industries, and services. Thus, each village will be small enough not to need intra-village bus services (neighbors will actually meet each other during the course of the day) and close enough to each other that inter-village bus services will connect the five villages quickly, allowing residents to take advantage of the larger economic diversity of the entire town. In a word, human beings on Mars will live in areas designed to create neighborhoods in a larger urban/suburban environment. The agricultural village might have a social atmosphere different from the "New York City" tilt of the town centers, but there will be a real sense of social belonging, first to one's particular village but also to the larger interconnected "town".

Martian towns will grow not as unplanned urban blobs, chewing up pristine Mars, but as interlinked neighborhoods, longed planned[6] to account for growth.

---

6   Even purists as to free market development will concede that Mars will need to have some measure of planning because for several centuries almost everyone will need to live in a shelter.

Suitably sized shelters will keep the terraforming environments outside and will allow the use of modern human engineering to create a high quality of life, as in the best suburbs of the modern West. The five-village town depicted in Figure 61 might grow to nine thousand residents with the addition of four more villages, or even larger as economic developments dictate.

The "village" principle will be balanced by another; one designed to preserve population on Mars in the event of a major catastrophe. That principle is called the 4 To 1 Principle. As this principle is explained the reader will understand that not all shelters need be one-half mile in diameter and that some settlements will grow much slower than others.

The 4 To 1 Principle For New Settlements

The 4 To 1 Principal anticipates that for every 4 persons in Settlement A, there should be one person in Settlement B. Accordingly, for every 32 persons in one settlement, there should be eight persons in some secondary settlement, and two people in a tertiary location. More practically, using groups of ten, a scheme of development becomes clear. After the first 40 are settled into one base/village, a new base/village with ten new settlers should be established at a secondary location. To spread the population in this way is to ensure the continuation of human life on Mars, even after a disaster. Less macabre and much more practically, the 4 To 1 principle strongly supports increased commercial development across the planet, as it constantly opens up new areas to mining, travel, and other commercial activities.

The "patch" idea (new base/villages established in a counterclockwise rotation out from the first landing site) for settling Mars has already been discussed. While the main purpose for creating patches will be to settle an area of sufficient size that something of Mars might become immediately known[7] and understood, it can also now be seen to contribute to the stability of Martian population, especially as regards catastrophe.

The most easily understood catastrophe would be the sudden and massive collapse of a Martian shelter. While the mission to Mars takes the position that such a catastrophe can be mostly ruled out if easily understood engineering protocols are implemented, for the purposes of these next two sections, we posit the catastrophic collapse of a shelter, and its resultant loss of life, as the prime bogey man the 4 To 1 Principle avoids. The original English colony, Roanoke, Virginia,

---

7   Each patch will be the size of Pennsylvania; large, but not so large as to be overwhelming. Remember, the entire settlement area for the mission to Mars will be on 1% of Mars, somewhat larger than the size of Texas.

disappeared completely, leaving historians puzzling over the reasons. The 4 To 1 Principle should allow the mission to Mars to avoid the Roanoke problem, even in the event of some catastrophe.

We begin by discussing the growth of Mars in terms of specific settlements and by making the following assumptions:

1) The majority of new settlers in these scenarios will be transported to Mars by Caravan Transporter vessels, accompanied by a few (one to three) smaller transporter ships;

2) Corporate and other economic organizations will move with each Caravan Transporter mission, allowing new settlements to become economically vibrant within a few months of their arrival on Mars;

3) The "patch" idea will be adhered to generally, especially when it makes sense economically and enhances the exploitation of resources on Mars;

4) The Confederation continues its plan to "develop" Phobos and Deimos, by creating small military outposts, searching for resources on each, and creating a small economy on Deimos using temporary workers and newly arrived settlers from earth; and

5) The settlements contemplated will not be structurally interconnected, so that a catastrophe in one settlement will not carry over into the next. The settlements discussed in these two sections may be a mile from one another, or they may be one hundred miles apart.

Growth From 500 to 1000 People

In this scenario, the growth of the Martian plantation proceeds along three fronts. First, the main settlement/village continues to grow rapidly, adding many technical and economic capabilities to Mars. Next, Village #3—perhaps intended to be agriculture-based—gets built to provide a more stable supply of food, trees, and other plant materials. Many of the first "residents" of Village #3 will be temporary residents—construction workers who stay for six months and then move on to other projects or back to families at the main settlement. Full-time residents arrive as construction proceeds and as the village's economy (agricultural or otherwise) sprouts. The university (Settlement # 2) increases its population slightly as workers who had been building the university begin to leave and are replaced by additional students and professors.

# of Settlers—Before	Location	# of Settlers—After
2	Nuclear Power Plant	2
8	Phobos	8
32	Deimos	32
25	Village 3	333
100	University—Settlement 2	125
333	Main Settlement	500
500	**Total:**	**1000**

Figure 62: Location of Population As It Grows From 500 To 1000

Growth From 1250 to 2500 People

Figure 63 shows how our notional population of 1250, distributed between six Martian and two off-world (Deimos and Phobos) settlements, grows to 2500. The main settlement will have become a fully populated village of 1000. The next largest settlement will be at the third village, where construction crews, military personnel, and others will comprise a population of 550. Even though much of the population in this settlement would be transient, having 550 people living temporarily away from other settlements serves the same societal purpose: to lower risk of disaster and to increase development of the planet's surface. The third largest settlement is Mars University with a population of 450 students and staff, perhaps a bus trip away from the main settlement. There is also a new settlement, still mostly Fleet and construction personnel, preparing to receive the first Ark Transporter mission. There will be two settlements on Mars' moons. Deimos will grow its temporary worker population to 200 to handle the huge number of immigrants once the Ark Transporters begin to arrive and Phobos' small military base will have a staff of 32. Finally, a small way station with ambitions in the tourist industry will be created between two of the settlements. This way station might be a husband and wife team who have opened a restaurant and lodge for travelers or tourists. Even with these relatively compact settlements and two existing off-world, a surface are of 5–10,000 square miles on Mars and much of the surface of the two Martian moons will have become well-known and opened up for development.

While the changes in population due to commercial activity, births, and deaths—not to mention the freedom of choice any settler will have to live where she wants—will not allow for exact population management, Confederation authorities should be able to manage the populations of the various settlements so that the 4 to 1 ratio is maintained.

This scenario suggests that the first of the Ark Transporters vessels will arrive soon after the Martian population reaches 2500. The majority of the 1000 new settlers will be placed *en masse* in Village 3 or 4. Some of the construction workers and engineers living temporarily at that village will leave for new jobs elsewhere when the 1000 new settlers arrive and some of those on the Ark Transporter will decide to take jobs in different villages.

# of Settlers—Before	Location	# of Settlers—After
0	Way Station—Resort	2
8	Nuclear Power Plant	8
32	Phobos	32
100	Deimos	200
10	Village 4	258
350	Village 3	550
150	University—Settlement 2	450
600	Main Settlement	1000
1250	**Total:**	**2500**

Figure 63: Location of Population As It Grows From 1250 To 2500

Western-Style Goods and Services or "Ski Mars!"

As noted above, creating goods and services for trade is different from creating basic necessities. An easy way to grasp this trademark distinction of a western society is to ask whether that society could produce something that somebody else in the world wants. Even little New Zealand with its minute population and relative dearth of resources is able to export agricultural products and world-class goods produced from sheep and wool, as well as world-class expatriate workers. Mars will only be a viable society when it has the ability (even if the trade is miniscule) to produce goods and services that other western societies desire.

The mission to Mars' goal is to create a self-sufficient society on Mars. High transportation costs and low initial quality imply that, at first, few things manufactured on Mars will be exported to earth. What pioneering societies also invariably produce, however, are hundreds of new products and ideas. Eventually these products and ideas will be sold not only all over Mars, but to purchasers on earth as well. Transmission of ideas will be relatively simple, since Mars will be tied closely back into earth's internet (albeit with a multiple-minute time delay) and communications networks[8]. Thus, ideas, movies, and books created on Mars can be transmitted back to earth as simply as they travel today from New York to Los Angeles.

The transportation of products will not be a significant engineering problem. Transport ships for settlers will be empty as they travel back to earth for more settlers and will be able to accommodate cargo. Once Fleets of ships are shuttling between earth and Mars, there will be regular cargo runs. The use of Deimos as a warehousing and manufacturing center with its next to zero-g environment will allow for Mars to be the premier supplier of goods manufactured for space. "No-boost" means cheaper goods. No one on earth will be able to compete with Deimos. Nor will analogue manufacturers on the earth's moon have anywhere near the same kind of raw materials available as will those available to Martians, their moons, and their asteroids. Thus, Deimos will serve as a wonderful way station for all Martian products going into space.

While of necessity Martian manufacturing will focus upon "domestic" markets, at some point there will be trade back and forth between the planets. This trade will enrich both societies. One interesting difference between international and interplanetary trade will be that Martians are not likely to become inundated with consumer goods having little real importance in people's lives. On Mars, a consumer screwdriver sets might consist of four good drivers, not a set with twenty-two sizes, most of which get lost in the basement. Much more on the (mostly positive) impact of the limited availability of consumer goods on Mars in later chapters. In either case, Mars will quite soon in its history become a viable western society in the sense that many of the products available to westerners on earth will also become available on Mars and that the people of earth will want, if nothing else, the chance to become a Martian tourist.

---

[8]   The exact time will depend on the ever-changing distances between Mars and earth, as the two planets revolve around the sun.

The Fleet

The Fleet has already been discussed many times in this volume. Little more is necessary to make the logical leap of Fleet as general contractor to Fleet as defender of Mars. From the beginning, the Fleet will have a military organization, with ranks and units, and it will offer a quasi-military experience. Once Fleet pilots are traveling routinely between earth and Mars, little imagination is needed to remold a few ships into a force that can protect the settlers from whatever external threat may exist.

The redrafting of this work into three volumes has diluted in Volume I the degree to which the mission to Mars will be a *political* movement. There will be many who will line up against a group audacious enough to claim Mars as their own world. There may even be a few 21st-century nations that would choose to threaten Mars militarily. In any event, a western society looks first to itself for defense. The mission to Mars' Fleet will offer the firmest foundations to ensure that those who settle Mars need not fear that someone on earth will decide to take from them all that they have built. Mars, of course, will not be a major military power for many years. To return to the idea of New Zealand, that island might not be able to fend off an attack from a powerful adversary, but New Zealanders would certainly make it a costly venture for anyone foolish enough to try to conquer it.

The mission to Mars' political tightrope will be to placate the Elitists on earth who will see Mars as their own playground while sponsoring huzzahs from the vast majority of humans on earth to see its efforts as noble steps into the future. A few nasty sergeants with space guns will seal the deal for Mars.

We Are The Martians

We now present the idea that some of those reading this volume will become "Martians". Indeed, it will be *Martians*, not earthlings, who will settle Mars. Both will be, of course, *Homo sapiens* humans, but the difference in attitudes, desires, and hopes for Mars will be marked and of real significance. The mission to Mars foresees no difficulty creating a new Martian culture, but the dynamics articulated in the following except from Kim Stanley Robinson's book, *Green Mars*, offers a different perspective about those who might choose the Red Planet:

> *People had left … for many reasons, and that was important to remember. All of them had thrown everything away, and risked their lives, but they had done so intent on very different goals. Some hoped to establish radically new cultures, as in Zygote, or Dorsa Brevia, or in the Bogdanovist sanctuaries.*

Others, like the Sufis, wanted to hold on to ancient cultures they felt were under assault in the terran global order ... There was no obvious reason why they should all want to become one single thing. Many of them had been trying specifically to get away from dominant powers ... all the total-izing systems of power. A central system was just what they had gone to great lengths to get away from ... what Boone finally concluded was that it wasn't possible to invent a Martian culture from scratch. He said it should be a mix of the best of everyone that came here ... "I don't think we can invent [a Martian culture] from scratch, and I don't think there will be a mix. At least not for a very long time. In the meantime, it will be a matter of a lot of different cultures coexisting, I think ..."

The mission to Mars disagrees about the difficulty of creating a unique Martian culture. Mandatory service in the Fleet, a common experience at the earth Embassy, and the highly unique experiences of traveling to and living on Mars will form tangible cultural bonds, regardless of the strength of a person's culture of origin. Robinson, too, seems to reverse his original point during the final installment of his series, *Blue Mars*, where he tells us:

So that there were new native "conservatives", and old settler-family native "radicals", one might say. And this split only occasionally correlated with ethnicity or nationality, when these still mattered to them at all. One night Art was talking with a couple of them, one a global government advo-cate, the other an anarchist backing all local autonomy proposals, and he asked them about their origins. The globalist's father was half-Japanese, a quarter Irish, and a quarter Tanzanian; her mother had a Greek mother and a father with parents Columbian and Australian. The anarchist had a Nigerian father and a mother who was from Hawaii, and thus had a mixed ancestry of Filipino, Japanese, Polynesian and Portuguese. Art stared at them: if one were to think in terms of ethnic voting blocks, how would one categorize these people? One couldn't. They were Martian natives. Nisei, sansei, yonsei[9]—whatever generation, they had been formed in large part by the Martian experience—aeroformed[10], just as Hiroko had always foretold. Many had married within their own national or ethnic back-ground, but many more had not. And no matter what their ancestry, their

---

[9] Japanese words meaning second, third, and fourth generations, usually referring to "genetic" Japanese living outside of Japan.

[10] Aeroforming is the counterpart to terraforming. Aeroforming is the process of Mars changing human culture, values, and society.

political opinions tended to reflect not that background (just what would the Greco-Columbian-Australian position be? Art wondered), but their own experience. This itself had been quite varied: some had grown up in the underground, others had been born in the UN-controlled big cities, and only come to an awareness of the underground later in life, or even at the moment of the revolution itself. These differences tended to affect them much more than where their Terran ancestors had happened to live.

The mission to Mars believes that its program—and Mars itself—will "aeroform" humans to create a recognizable Martian culture. The mission to Mars foresees this new *Martian* culture as one that does not intend to terraform Mars into a "New Ontario" or have the kinds of dispirited, disconnected human communities described in Robinson's books. Rather, the mission to Mars envisions that the love of Mars will serve as an overwhelmingly powerful and unifying force for its settlers. The economic incentives to terraform the Red Planet will be met by a very strong ethic to conserve the wild, magnificent Martian wilderness. And the human struggle to survive on Mars will be such that people will band together, not conduct divisive social experiments in search of some always-on-the-horizon utopia. While Robinson's stories about superrich Martian settlers fighting for their version of Mars makes for a good novel, it lacks any connection to real people; like the friend you met this morning for coffee who, in a few years, will actually have a choice about whether or not to go to Mars. Economic systems that can be harnessed to settle Mars will not be so flexible as to create large classes of political idealists and puttering scientists, able to crisscross Mars at will on their interesting but incredible journeys of personal and social discovery. Instead, humans will have a real world option to live a life of adventure and enterprise, as neighbors in a culture where it will be necessary to co-exist. And together people of all races and creeds will build a new, better society on Mars that will advance all of human civilization.

The Confederation's earth Embassy will be specifically designed to help create a common, self-reliant, *Martian* culture. The broth of culture will simmer, too, in the pioneer setting of life in Martian towns and settlements and because of the point just made, that *Mars itself* will re-forge the mélange of human material into a new, Martian metal. At a minimum, the mission to Mars predicts that humans on the new planet will quickly lose whatever residual desire there may have been at the earth Embassy to create a new Canada or a new Italy. Most likely, human settlers will fall in love with their new world and will remold it gently to fit their new, arriving culture. Holdouts for the idea that the physical parameters of the Martian climate and culture should mimic the earth will likely die out after a generation. People will surely long for the day when terran-derived ecosystems will thrive on the planet's surface, but as those ecosystems will become *Martian* ecosystems, so humans born on the *earth* will become Martians.

The idea that *we* are the Martians will be explored in much more detail in Volume II. No sci-fi story, no Martian invasion film, no *Outer Limits* television program has imagined the irony that Martians will not be an indigenous life form; but will rather be the diploma bestowed upon those *humans* who choose to earn the accolade.

Thus, the final aspect of a viable society on Mars. Mars will become the home of Martians, a unique kind of human culture. Though born at the earth Embassy, once transplanted to Mars, it will hold station as a bold, new subset of human-kind. The mission to Mars' settlement program implies that a viable Martian society can be created after just a few years. Indeed, the mission to Mars' entire program is premised upon the idea that a viable society can be created there. This chapter has been written to help the reader understand that such a step is not 250 years into a NASA future, but just a few, dedicated decades away.

The Search For Specific Resources

These are the times that try men's souls. The summer soldier, the sunshine
patriot will, in this time of crisis, shrink from his duty. But he who stands it
now deserves the love and thanks of man and woman.

—Thomas Paine in *The American Crisis*

This chapter goes in two main directions. First, it offers some general thoughts
about how to find vital resources on Mars and how the settlement strategy for
the mission to Mars can adapt to facilitate the prospecting for and mining of
these resources. We choose eight "vital" resources, representing some of the things
Martian settlers will want or need and see how settling Mars can help to provide
a supply of these materials.

The second part of this chapter describes the search for four specific resources on
Mars. The four resources are chromium, uranium, nitrates, and a silicon-bearing
ore for the glass industry. In three of the cases, the Confederation sponsors the
search for the resource. In the case of the silicon ore for glass manufacturers, three
businesses have gotten together to sponsor the search for a specific type of silicon
to enhance the capabilities of the Martian glass industry.

Vital Resources

This section discusses three aspects of the plan to obtain "vital" resources. First,
we show how the availability of resources in one area or another will influence
the order in which settlement sites are selected. Second, we mention how having
many settlements reduces the needs to focus NASA-like on some "perfect" landing
site. Third, the obvious connection between vital resources and the start of
manufacturing is made. Finally, the reader should note that this is an introductory

discussion about vital resources. The subject will receive a fuller treatment in Volume III. We introduce it here to show that the mission to Mars has not ignored the absolutely crucial role that Martian resources will play in the move to the Red Planet, even as we discuss how economic incentives can help obtain more mundane resources. Before we begin, we should define what we mean by a vital resource. Vital resources are those that will most assist the transplantation of the earth Embassy to Mars. Non-vital resources include those that will eventually be needed on Mars, but are not of the highest priority. One main difference between the two types of resources is that non-vital resources might be able to be imported for long periods of time. Vital resources, on the other hand, are those which will cause Martian bases significant difficulty if they must be imported for more than just a short time (see discussion below about gold). Water is the obvious example of a vital resource.

Figure 64 begins the discussion by showing the availability of eight vital resources near three settlement sites on Mars[1]. Settlement A has already been established; the mission to Mars is considering the advantages and disadvantages of Settlements B and C, to include the ability to obtain vital resources at these sites. Even though the next settlement will still not satisfy the need for all vital resources, once both settlements B and C are added, sources for all vital resources except for gold should have been found on Mars.

Note the efficacy of a settlement approach to Mars. If one plans to create several pioneering settlements quickly and dozens over the course of a few decades, the importance of finding every vital resource at a given "landing site" shrinks into the background. Eventually, a reasonable source of supply of everything that Mars offers will be close to a settlement. There is no need to study endlessly the problem of a perfect landing site, a site that will have flaws however much care is taken to avoid them. Having a planned series of settlements allows the advantages of each site to be exploited and allows other settlements to compensate for flaws at existing sites. While NASA is free to spend their money studying the "landing site" problem, the mission to Mars will take its best guess at an initial landing site and then inexpensively move to additional settlement sites to find the resources it needs.

Part of the criteria for selecting each new settlement on Mars will be that settlement's ability to help establish new businesses. If there is a need to improve the

---

[1]   While gold will not be viewed as a vital resource to most engineers, this work discusses gold in the context of a project in the real world. "Gold" will not be necessary to survive on Mars. Once gold is discovered there, however, not only will whatever financial difficulties remain for the mission to Mars evaporate, but a 21st-century gold rush will occur. Any seats left to be filled will be taken by those intoxicated by the yellow metal.

quantity or quality of aluminum products on Mars and if Settlement B has little or no aluminum in the vicinity, it makes sense to build at site C, where higher quality aluminum is available. Similarly, as depicted in Figure 64, as neither a good deposit of silicon nor deposits of gold have yet to be found on Mars, these issues will continue to figure into the Confederation's development plans.

	Settlement A	Settlement B	Settlement C
Iron	X	X	
Water	X		X
Aluminum	Low-Grade		X
Silicon		Low-Grade	
Fluorides		X	
Nitrates	X		
Gold			
Uranium			X

Figure 64: Resources and Where They Are First Found

New settlements help to drive the hunt for high quality supplies of materials, but the demand to transplant one industry or another from the earth Embassy will be so strong, and the seats to Mars so few, that there will be no lack of good *economic* choices about which businesses and organizations to move next to Mars. As was stated a few pages back: "The plan is not to manufacture the easiest or the most needed product on Mars, but to *encourage the manufacture on Mars of those items it makes the most economic sense to manufacture there.*" If a *cost* can be assigned to each item imported to Mars, then our Martian colonies can begin to see what makes sense to "create" or manufacture on Mars. If there were scores of potential entrepreneurs on Mars, these people by themselves could orchestrate Martian production. As this will not be the case during the early years, the Fleet hierarchy will make the decisions about what manufacturing to begin on Mars and what should continue to be imported from earth.

Likely early candidates for manufacture on Mars are water, rocket fuel, and metals. The availability on Mars of the resources necessary to a given manufacturing process is a prerequisite to establish the industry. For while it makes sense to import a few items at first, if the vital resources needed to make rocket fuel on Mars are available (and our knowledge of Mars ensures that this is a virtual certainty), then it makes sense to create a rocket fuel manufacturing capacity there. As we expect that the raw materials for water, rocket fuel, and metals will be available almost immediately on Mars, only processing equipment needs to be imported before crude manufacturing of such products can begin. Note that each of these products is also bulky and quite *heavy*. Since the largest part of the cost to import these items to Mars would be to lift them into orbit from earth, it makes sense to manufacture these necessities on Mars as early as possible. As should be clear, there will be a great many "expensive" items imported to Mars and limited manufacturing capacity during the early years. While this equation will change as more people arrive, it should also be clear that the availability of resources for any given manufacturing process will help to determine when an operation can begin there[2].

The Nuclear Assumption

We now introduce another important concept for this work. We make the "nuclear assumption" in these three volumes, which means that nuclear energy provides the majority of the energy for Mars and much of the energy to run the transporter vessels. The nuclear assumption enlists three subsidiary ideas.

First, is that the mission to Mars cannot expect to use or develop nuclear energy on earth. There are many who loudly trumpet the utility of nuclear energy, especially for use in space. While the mission to Mars agrees with this as a statement of engineering, it must take the practical political position that no government anywhere would support the mission to Mars to the extent that it uses or experiments with nuclear technologies on earth, at least until it has proven itself over the course of decades. Even then, the cost to comply with international protocols on earth would likely be of such a magnitude that little of practical effect could be undertaken except in space or on Mars where, of course, good engineering practice rather than international protocols would be followed.

Thus, the second aspect of the nuclear assumption: that the mission to Mars will be able to procure radioactive materials and to make use of these materials to supply much of its energy once it has people in space. One early source of uranium might be the inner asteroids (discussed next). Alternatively, Deimos might yield uranium. Finally, Mars will almost surely have uranium deposits. The nuclear assumption therefore posits that energy will be reasonably available on

---

2   A rich source of creativity will be the need to find new (and sometimes better) ways to manufacture a given product on Mars. The positive psychic benefits concomitant with creating something new to benefit an entire community will serve as a powerful recruiting incentive for the mission to Mars and the opportunities to participate in such a process will be legion.

Deimos and Mars without being a critically important parameter for the structure of new settlements[3].

Finally, the nuclear assumption underlies the discussion in these volumes to the extent that the mission to Mars' efforts in space and at its early bases will not need to be skewed to account for lesser efficient technologies, such as the harnessing of wind, geothermal, or solar energy. Without nuclear energy, these technologies will dominate the Martian economy (not to mention the design of transporter vessels) for at least the first few decades.

It must be understood, almost to the point of being a fourth aspect of it, that mission to Mars success does not rest upon the nuclear assumption. Indeed, chemical rockets and six-month mission times for Mars have been used thus far in discussing its basic attributes. And while the growth of colonies on Mars may change from what was presented in the last chapter to one where the first eight years are dominated by the creation of wind, solar, and Martian geothermal energy, the basic idea of a military mission importing civilians and new businesses will not change. Remember, the main reason we are not on Mars today has to do with the lack of political will and economic factors, not with technology. If, as is unlikely, nuclear fuels cannot be readily obtained by the mission to Mars once it is in space, then these other current technologies will be substituted and transplantation plans to Mars adjusted accordingly.

The Inner Asteroids

Before a mission is sent to Mars, it might prove valuable to send one or more Explorer Transporters to an inner asteroid. An inner asteroid has an orbit that carries it between the earth and Mars. There are thought to be many thousand such asteroids, some of them a few kilometers in diameter. There are many reasons to take a decision to send a mission to an inner asteroid. First, any doubters left after the mission to Mars launches humans into orbit will be eliminated after it sends a mission to an inner asteroid and creates a base there. Three or four semi-permanent bases on inner asteroids will not only bring in hard currency (selling telescope time and trips by non-Confederation scientists), but it will offer mission-related advantages as well. Bases on these asteroids will serve as a test bed for all sorts of ideas for Mars and Deimos. If something works an inner asteroid, it will work on Deimos and on Mars. Having three or four inner asteroid bases will also allow for a rudimentary "rescue" capability. In the event a transporter vessel falters on its voyage to Mars, one or two of these asteroids might be close

---

[3]   Uranium will remain a vital resource, even if not an absolutely crucial resource to the existence of Martian settlements.

enough to the transporter to attempt a landing there or the asteroid might be able to launch a rescue craft. In addition, having bases on asteroids whose orbits take them close to Mars may offer all sorts of "administrative" advantages (to ferry settlers or supplies) for planning Mars missions. Finally, these asteroids may offer resources for rocket fuel and water. This inner asteroid example demonstrates again how sending settlers into space will help to lower costs dramatically. If there are no "gravity costs" for water and fuel to move settlers to Mars because these materials can be mined at an inner asteroid, everything becomes much cheaper. The most important resource, of course, would be the discovery of uranium on a metallic asteroid. Such a discovery would allow the mission to Mars to proceed upon the nuclear assumption that was just discussed.

We now turn to incentives-based programs to "to explore" Mars. Not only will the costs for the efforts described be low, but results will be expected for each mission. A NASA mission searching for life can only hope for a result; a NASA mission studying Martian geology can only issue a report of the most general kind. As you shall now see, mission to Mars "exploration" is a vastly different sort of effort.

Chromium

Chromium serves as a feedstock metal for several types of products manufactured on Mars, including the steel industry. It was hoped that Deimos or Phobos would contain rich chromium deposits. Unfortunately, only small amounts of chromium were found. Relatively early, LY 13, when there were about 1600 people on Mars, the Confederation offered a contract for the discovery of chromium deposits on Mars.

A man at the earth Embassy won the contract with a bid of CS $300,000. Two other people at the earth Embassy and a woman on Mars also bid. Although not the low bidder, Ivan's bid proposal was deemed to be the best, owing in no small part to his PhD in geology from the University of Kiev.

After a few months of work developing specific areological data available at the Mars University/Martian Intelligence Agency library, Ivan stepped aboard a Caravan Transporter for an eight-month trip[4] to Mars. Ivan continued his research aboard the transporter ship, so when he finally arrived on Mars he was ready to implement his plans for three expeditions to locales likely to contain chromium-bearing ores.

---

[4]   Mission parameters for this mission required a slightly longer than normal journey.

Ivan led each of the three expeditions. The first found a small deposit, as did the third. It was the second expedition, however, that found large amounts of chromium ore very close to the surface. Small-scale operations to mine this ore began almost immediately, with Ivan becoming a key manager at the mine. Thereafter, the main problem was finding enough workers to operate the mine, which easily sold all the chromium ore that it could produce.

Uranium

The drive for uranium began in the early days of the earth Embassy. Though there was some uranium found on Deimos—more than enough for early Confederation purposes—larger deposits were needed to facilitate the growing number of nuclear-related projects that were to occur once the Ark Transporter missions began. Accordingly, in LY 15, when there are about 2100 people on Mars, the Confederation offered a contract to locate deposits of uranium-bearing ores.

The winning bid for the contract had nothing at all to do with the price. Rather, the winning bid put together the best proposal to search for uranium within a 1,000-mile radius of the first Martian village. The winning bidder proposed to use surveys conducted by the Martian Lander followed up by ground surveys using Fleet marine escorts.

Guy LaPrenz would eventually receive a handsome stipend for leading the survey teams, but most of the "costs" for this mission were assumed by the Fleet. The Fleet hired LaPrenz to be the project manager, which angered members of the Fleet. Politics being what they are, however, the Confederation bowed to pressure from the earth over fears that Martian uranium could find its way into the hands of terrorists or that having too much "military control" over Martian uranium could be dangerous. With a civilian project manager, the Confederation could accommodate its "allies" on earth.

Of course, this tactic was only a partial solution. There were some Leftists on earth who strongly suspected that there were people on Mars living in freedom and having fun; the Leftists aimed to put a stop to that. Uranium was one front in their war on Mars.

LaPrenz worked for several years as the project manager. Three small mines were developed and uranium ores on Mars became the primary fuel source for the planet's development. Using nuclear fission meant that lack of energy was not an issue for any of the mission to Mars colonists; nor was it an issue when post-mission-to-Mars immigration to the Red Planet began. At least as important, the uranium was used to create a new fleet of space ships that could make the Mars-to-earth run in 30 days. Their nuclear-powered engines transformed Mars from a place almost unimaginably far from earth to one that seems just a little further off than Australia is from England.

Nitrates

The presence of large nitrate deposits on Mars has long been known. These nitrate deposits—still mostly unexplained—are a great resource to agriculture and for the introduction of other kinds of terran plant life to Mars. Although there has been no problem enriching Martian regolith and making productive cropland, a large, natural supply of nitrates would allow for much faster and much larger-scale introduction of plants, trees, and new cropland in the shelters and—soon—onto Mars' surface. In addition, plans to cap a large crater and transform it into a nature preserve would be greatly facilitated if large quantities of nitrates and processed fertilizers could be introduced to amend the Martian "soils" in the nature preserve. Therefore, in LY 23, when there were about 36,000 people on Mars, the Confederation offered a contract for the discovery of additional nitrate deposits on Mars.

The winning bidder for the nitrates contract was Ivan, the same man who had won the chromium contract ten years before. Ivan had been mining and "prospecting" on Mars since his arrival. He sold out his interest in the chromium mine in order to get back onto the planet's surface. Although the discovery of chromium had made Ivan a rich man, his true love was to be in some back canyon on Mars. This love was only enhanced by the warming Martian temperatures and the thickening atmosphere. Although protective suits were still required to survive, brief exposures to the Martian atmosphere had become possible. Ivan even made it a joke to take a few breaths of the atmosphere occasionally, threatening everyone who saw his antics never to tell his wife.

Nothing could have been more boring to Ivan than looking for nitrates. "Shit-detail" he would mumble to himself in Ukrainian. As the chromium deposit had assured his fortune, however, Ivan's six-month, CS $150,000 contract to search for nitrate deposits was to him little more than a paid vacation. Ivan's wife was frugal, however, so she made him promise to use marines as much as possible and not to waste the money hiring civilians or renting expensive equipment.

Marine expeditions cost money, however, even if you are able to persuade the commander to schedule an expedition. Ivan made five week-long expeditions searching for his nitrates. On every expedition, six to ten marines accompanied him, costing him CS $100 per marine, per day. Though these expeditions cost him over CS $40,000—not counting the cost of supplies—Ivan loved every minute of each expedition.

Two of the expeditions found significant nitrate deposits. Both were almost 500 miles from the nearest human settlement, however, making the immediate value of the deposits hard to judge. Ivan had negotiated rights to claim ten acres of land on his nitrate finds. Rather than try to develop something so far from a human settlement, Ivan claimed his ten acres of nitrates and then immediately sold it.

Ivan continued his prospecting, walking across most of the known Mars, and making several other lucrative—and to him distracting—finds.

Special Silicon Deposits

In LY 26, when there were 41,000 people and three glass manufacturers on Mars, the Glass Manufacturer's Trade Association decided to sponsor a geological survey to identify specific types of silicon-bearing ore for the glass industry. Rudimentary glass manufacturing machines were still being used on Mars and the glass manufacturers realized that certain types of silicon would greatly increase the efficiency and flexibility of these machines.

The consortium selected a geologist from the University of Mars to lead surveys to some of the more promising rock formations already known. As Martian settlements had already spanned almost the entire 600 by 600 mile section of Mars to be settled by the mission to Mars, none of these surveys was to be more than 75 miles from a Martian village. Although Professor Chen would leave off teaching for one semester to conduct these surveys, much of his work was actually done in his office at the university. He was hired by the consortium for CS $650,000, but would have no rights to any of the silicon found. Approximately half of the money bid was to be used to hire sub-contractors to conduct surveys for him in the vicinities of their individual towns and settlements. Chen also engaged the newly formed 2nd Marine Regiment to participate in three ten-person expeditions to examine the more promising surveys. Chen's final product to the consortium would be a report detailing five "Likely" and ten "Suspected" silicon-bearing formations. The report would be used to develop options for mining.

Unfortunately for the consortium, their contract failed to include a provision for discoveries outside of the main scope of work. As it turned out, during Chen's second expedition with the marines, he came across some platinum-bearing ores. Now, there had already been large deposits of platinum found on Deimos, driving down the value of platinum on Mars and even depressing prices on earth, as several thousand pounds of platinum were being shipped back to earth each year[5]. Still, Chen's discovery—which he hid from the accompanying marines—allowed

---

[5]   Some people just insist on shipping their precious metals back to earth!

him to stake a claim to several platinum mines. Chen eventually made several hundred million dollars from these claims.

Of course, the glass consortium should not have been too sorry. After years in court, they also received some rights to the platinum, worth several tens of millions of dollars. Moreover, the silicon ores Chen detailed for them so well in his final report formed the basis of glass manufacturing on Mars for several decades.

Sour Grapes

From time to time, we play into the hands of the Nay Sayers. "These are simply modified exploration missions!" they whine, trumpeting like rogue elephants that there is no conceptual difference between a BGBS mission to Mars and the ideas here. The ability to make fine distinction being one of the hallmarks of a good education, the mission to Mars responds that there are at least three important differences between the efforts described in this chapter and the traditional idea of "exploration missions" to Mars.

The greatest difference will be that the mission to Mars will have "exploration" serve a practical and defined purpose rather than a generalized and non-utilitarian purpose such as the "search for life". As each of the scenario described, there were specific resources sought and specific objectives to be attained, with each objective tied to a reality-based need. A NASA mission to Mars might have a "Mission List", but little or none of it will tie into any reality other than the bureaucracy game back in Washington.[6]

Second, the cost differential between these two types of missions must also be seen as important. Paying a Mars-based prospector is one thing. Sending a team from earth to look for something on Mars is orders of magnitude more expensive. Moreover, the Mars surveyors are actually looking to cut costs—the cost to make cheaper or better glass—while the NASA mission looks for a hook to fish out more of Uncle Sam's money.

Finally, there is a huge difference between the kinds of society-building activities described in this chapter and exploration to amass a database. Each of the surveys described was conceived to assist in the development of the Martian society and economy. No such claim could be made for NASA-style missions that would primarily work to understand Mars "geological" history or in the search for life. Consider that there is a great deal of information known about Antarctica, but the ability of Antarctica to support a human society—for better or worse—is not substantially improved by any new data gathered there.

---

[6] If we find life, we will get more money. If we find "A", we can ask Congress for X. If we can find "B", we can ask Congress for Y, etc.

A surveying operation on Mars may indeed prepare itself like a group of NASA scientists might. Supplies, transportation, and operations will need to be planned in either case. The fact remains, however, that the differences will be enormous. The NASA mission will be funded by a line item in a NASA budget, perhaps smuggled through a sleepy congressional oversight committee, with a vague or highly flexible mission statement. The other will be funded by the Fleet or private enterprise to obtain answers to specific questions, about a specific area on Mars, or about a specific problem. Focus defines the mission to Mars survey. A romp in the Martian woods describes the NASA program.

Readers with an open mind will understand that the mission to Mars program is not a half-price NASA knock off, but a program with a completely different agenda, focus, and mode of operation. Most readers will understand this; those that don't probably never will.

A Year in the Life of Ares Aresovich

Yes, I am a pirate.
Two hundred years too late
The cannons don't thunder
There's nothing to plunder
I'm an over forty victim of fate.
Arriving too late, arriving too late ...

Mother, Mother Ocean,
after all the years I've found,
my occupational hazard, dear.
My occupation's just not around.
I feel like I've drown ...
Gonna head uptown ...

—Jimmy Buffet, *A Pirate Looks at Forty*

This chapter paints a picture of life on Mars as an off-shoot of western civilization. It will describe the environment and activities of a human society on Mars through the eyes of a Martian settler: Ares Aresovich, a Russian-born member of the mission to Mars.

The key point in all of this is the idea of a *pioneering society*. The vista presented—like so much of the mission to Mars—is likely to differ from most readers' preconceived notions. Mission to Mars' settlements will not resemble a Russian science camp in Antarctica. We do not envision a littered camp of shivering, huddled "astronauts", glad to have been selected for Mars, but happier still when the day finally arrives they are to return home. Even early on, mission to Mars settlements are to be bustling and reasonably spacious, with more than a few idyllic, pastoral settings. The first few years of settlement will be regimented, with people concerned about creating the basic infrastructure of life on Mars; and each *new* settlement will start out much the same way, even after the rest of Mars has become reasonably well-developed. Still, the artist is painting here with soft colors and he hopes to render a human face on what might otherwise be perceived as a sterile environment. Remember, American pioneers often lived in one-room cabins. Abe Lincoln slept on a mattress of leaves. And, of course, nothing here is guaranteed, as the third day of an Italian vacation guarantees you Florence. The mission to Mars anticipates that the feeling of "being on a different planet" will be dwarfed by the thoughts and experiences of an exciting life in a dynamic, new frontier; where the reality of being on Mars is often hidden and usually experienced by choice, as an adventurous trip onto the planet's surface. In this chapter, we begin the real discovery of "life" on Mars.

A Review of Terms

Thus far, the book has used the terms "bases" and "settlement" rather loosely to describe locations of human activity on Mars. It is only in using the term "village" that it has provided much specificity. To add both subtlety and meaning to the terms, we now define these three terms with greater precision.

These three terms fall into two categories: locations with local government and those without it. It should be understood that Martian villages would have a local government. In contrast, a base or a settlement will not have a local government, because they will fall under the control of the Fleet or a private corporation.[1]

Most new locations on Mars will fall under military administration. A local commander will be appointed and given the mission to develop the site. This commander will have broad powers over disputes and mission execution. As the number of residents increases, however, the military administration of the new base will transition to a point where normal, western-style local government can begin. Once the Martian base or settlement institutes this local government, it will be called a village.

The customs, traditions, and methods of Martian local government will be pioneered at the earth Embassy. At the appropriate time, working models of this Martian local government can be transferred to Mars. As political circumstances on Mars develop, these models will also develop and will become better defined,

---

[1]   For example, a Martian university or other commercial venture. Some of these shelters will have dormitories or residential areas.

with duties split between the Confederation, the government of the republic, and the local government.

A gray area exists between the definitions of a base and a settlement. A base will primarily be a military operation, while a settlement will have a much larger civilian side. The base on Deimos, for example, will be run by the military, as it serves as a port of entry and departure for spacecraft docking from and launching for earth. Even with a significant mining or manufacturing presence on Deimos, the lack of any permanent staff there means that it will remain a base, under military control. A settlement, on the other hand, will include places like power plants, universities, etc., where a non-governmental entity has established a shelter and which may run independent of local government authority, as colleges and large manufacturing facilities are mostly independent of local government on earth.

More About The Mission to Mars "Village"

We return to the idea of the Martian village, now seen through the eyes of Mr. Aresovich. As you recall, the village design envisions creating 1000-person human habitats on Mars that will eventually be interconnected to form Martian towns and cities. This design will allow Martians to keep much about what is good about modern urban life while discarding that which is not. The clever reader will already understand how these village constructs might integrate into ideas about sheltering on Mars. Since shelters have to be built on Mars to keep out the Martian environment, designing settings to accommodate a reasonably low number—here 1,000 souls—will quickly be seen to have its advantages.

Ares discovered that the mission to Mars' lifestyle on earth and on Mars offered *better* social conditions than was to be found in either the crowded, uncaring western societies he had lived in as a student or the conditions he experienced in Mother Russia.

Figure 65 identifies several of the most important attributes about Martian village life that was particularly attractive to Ares.

The mission to Mars created its villages without the negative attributes of American society by structuring a society to live in small, closely knit communities. When he first experienced it, Ares could not believe how fast paced and stressful life was in American suburbs. Unlike life in a Martian shelter, where most of life's important activities are centered one-quarter mile from home, American suburban life had to be lived on a daily course run twenty or more miles from home to job and to practice field. Unlike Martian settlements, American zoning laws prevent most residential areas from having retail or commercial enterprises within walking distance. While the American practice may have been a good idea in the last half of the twentieth century, it was not how Americans lived during most of their history and seems no longer to be necessary. The existence of small, rural communities in America, indeed across the globe, reflects the utility of having walking-distance access to most of life's important activities. Said another way, it has become very *convenient* on Mars that a *convenience* store can be located in residential neighborhoods, as general stores used to be within walking distance of those living in small, rural towns. What may have been necessary when industry was dirty—requiring a separation from residential neighborhoods—is not true on Mars, where "high-tech" and service businesses can be down the block without harmful effect.

Attribute of Martian Village	Corresponding Attribute of American Suburbia
Comfort	Long Commutes and Poor Urban Designs Stress Out Life in Suburbia
No Crime	Violence and Anti-Social Behaviors
Convenience	Gridlock On Highways
Conservation	Waste of Terran Resources, Especially Land
Community	Neighbors Disconnected From Each Other
Continuity	Home Towns Grow Strange For Those Who Are Away Even a Short While

Figure 65: Attributes of Mission to Mars' Life Contrasted With Attributes of American Suburbia

Ares knew that the other reason for stress in suburban America was the over-reliance on automobiles. The victory of the automobile industry over local transportation authorities—say about 1950—meant that private developers and urban planners created huge individual lots for development and expected similarly large parking lots because people in America *drive cars everywhere* rather than take a bus. When land was so cheap around major American cities, suburbs could be created that catered to automobiles, which could commit space for vast parking areas and a huge infrastructure of streets and highways[2]. On Mars, where people

---

[2]   Drive around any modern suburb and see how much land is wasted for parking and "landscaping" that would not really be necessary if there were access by bus or tram to facilities and businesses.

walk or use buses, a medical building with five doctors is fifty feet away from the strip mall, instead of five hundred feet as it is in the US. And neither facility requires much parking[3].

On Mars, Ares lives in close proximity to his job, his neighbors, to farms—to everything—in his shelter. He does not, however, feel cramped. Martian villages have been cleverly designed. Without a need for automobiles, there is no reason that people cannot feel very comfortable in a modern, "walk about" world. Automobiles will become part of the Martian economy but, as may appeal to automobile enthusiasts, are primarily meant for Mars' "open road", rather than smelling up the inside of Martian shelters. To be sure, there are "golf carts" and other vehicles to tote heavy loads, but because Martian villages are designed to be reasonably small, self-sufficient entities, a car is as unnecessary for most people as they are in Manhattan. Thus, Martian "cars" help service longer-range transportation systems or "Sunday driving", rather than "commuting" within domed shelters.

Figure 66 shows how a typical Martian village might be planned to integrate residential, industrial, commercial, agricultural, and other activities in a small area, where everything needed for daily life is within walking distance. The center of the village houses commercial and industrial activities. From the center of the village, twelve residential avenues radiate out, each avenue home to 80 inhabitants. The breadwinners walk to work in the village center. Caregivers and shoppers walk to the village center to make their purchases. One or two lots in the village are devoted to some industrial purpose. At least half of the total acreage, the land "between the spokes" is devoted to agriculture or trees for eventual harvest.

---

[3]   Just to ensure there is no misunderstanding, the issue of "buses" versus "autos" is *not* one of freedom. It is a political question about which kind of infrastructure a society decides to create. American society has decided that it will invest tens of billions of dollars annually to maintain a system of roads not found anywhere else on earth. If this money were instead invested in bus and rail systems, America would have fewer cars and more buses. Instead of sprawling suburbs and heavy traffic, America would have much more compact residential and other developments, surrounded by farms and still pristine woodlands.

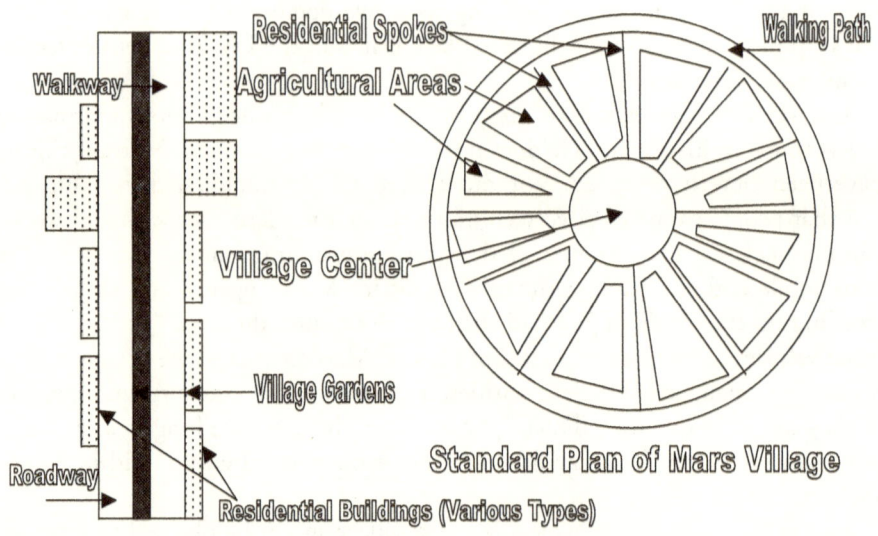

Figure 66: Details of Martian Village

As villages are interconnected, more variety is added to the community's economic and social life. Though most people still walk to get around, more taxis and buses come on line to move people from one place to another, interconnecting one village to the next. There is no reason why the process of interconnection could not continue indefinitely, with each new village adding its own flavor to the whole. As New York City is a place with hundreds of different neighborhoods, so a Martian city will be a collection of individual villages, growing as large (100-, 500-, 1000-interconnected villages?) as necessary.

Ares lives near the end of a spoke in his village. It takes him no longer than ten minutes to walk to any point in the village. As in any pioneering society, there are not many glitzy consumer goods available to him, though the "Wal-Mart" in his village allows for internet purchases with pick ups usually within two or three days[4]. Ares takes a bus to one of the neighboring villages (closest is 50 miles, another 150 miles) every month or two. One neighboring village has a McDonald's restaurant and another has a really fine music store, both of which interest Ares a good deal. The ride is bumpy, as the road between villages is not paved, but Mars' low gravity makes for a quick ride.

---

[4]   The Wal-Mart carries almost everything manufactured on Mars or imported from earth. A central distribution point ships the orders out immediately, so finding transport space is the main determinant in getting a product delivered.

It is easy to see how the other problems of modern suburban life listed in Figure 65 (two pages back) are absent from life on Mars. The problem of traffic gridlock is non-existent on a world that does not rely on automobiles. While crime and violence will never be eliminated from any human society, life in a small village—on earth or on Mars—does not provide the anonymity required by most criminals. Sex offenders, for example, thrive in environments where no one knows who they are. Even where a sex offender preys on his own family, the opportunity to get away with it "at home" is reduced when everyone in the neighborhood knows everyone else. A society of Martian villages has no analogue to the American problem of disappearing open spaces[5]. The cultural habit of settling people in villages will probably carry over to the time when shelters are no longer needed on Mars. Cultivating environmentally- and Mars-friendly habits early will mean that even with a population of several billion humans, Mars can remain a world with a well-conserved, open environment. The problems of suburban alienation (people living as strangers next door to each other) can be much ameliorated if humans walk to work and to shop, greeting each other on the streets, and living in close proximity one to another. Finally, the matter of development—landscapes changing so rapidly that one loses touch even with one's hometown—can be addressed if a community is laid out for the long term. In modern America, farms are lost to mall development and new housing areas. A business district loses its importance or becomes boarded windows when the Wal-Mart arrives in town. Much of America changes nearly every decade and rapid change fosters a deep sense of alienation as towns disappear before ones' eyes. On Mars, this need not be the case. A home may be remodeled, an ice cream parlor can replace the clothing store, a shoe manufacturer may move into an industrial site that previously had manufactured soap, but the essential character of a Martian settlement can remain the same for decades or even generations. When the Wal-Mart arrives and begins to out-compete the local hardware store, the former hardware store is remodeled for offices or a new service. Sad in an economically progressive way, but the building itself retains vitality when someone new moves in. Even after the shelters come down, there is no reason to believe that the face of a Martian village need change dramatically from generation to generation, even assuming a growing economy. People will remodel homes and businesses to fit into existing structures without destroying the harmony of the village atmosphere. For those

---

[5]   The most important environmental problem in the US is the disappearance of habitat for native plants and animals. On Mars, natural "environments" will not needlessly be consumed, if humans live in comfortable, well-planned communities that minimize land requirements.

who simply insist upon a new lot or building, they can locate in the new village going in twenty miles away.

The First Shelters on Mars

The first humans on Mars did not immediately start to create 1000-inhabitant communities. Such engineering projects were out of reach for several years. Rather, the first humans sought to improve caves or other natural shelters against the Martian weather and radiation. Three or four prefab walls and some inventive sealants made for an airtight room to serve the purposes of the first shelters on Mars.

The realities of Martian atmospheric pressure, temperature, and weather meant that the first settlers needed protection from low temperatures, low atmospheric pressure, and also from those famous Martian dust storms. The first shelters on Mars were pressurized and warmed environments to accommodate a human community, secured well enough to make catastrophic failure of the shelter a remote concern. Some of the initial shelters on Mars were simply adapted nursery greenhouses, manufactured at the earth Embassy, packed into containers, and erected by hand or using minimal amounts of automation. As larger and more sophisticated sheltering systems were transported to Mars, the greenhouses were put to other uses, sometimes even serving as individual homes.

As soon as was practicable, the manufacture of shelters and sheltering materials started on Mars. Engineering native Martian resources was not the problem, rather, the problem was the shortage of machines, tools, and supplies. This was anticipated, of course, and the transportation of people and supplies to offer solutions was a priority. Within a few years of first landing on Mars, acre-sized shelters had become a standard issue item for every new community. These one-acre shelters were designed to fit together seamlessly, allowing them to fit the terrain and the wide varieties of uses—including agriculture—to which they were put.

Eventually, manufacturing on Mars reached the point where custom-crafted domes became standard. Forty two of the forty nine villages created by the mission to Mars' "50,000" used custom-crafted domes from the beginning, meaning that most new Martians—transporting aboard an Ark mission—arrived to find that their new village already was capped for habitation. For these people, there was no ugly assortment of shelters, agglomerating gradually over time to form a village, as was the case with the first seven villages on Mars.

Ares' village uses a plastic sheet to cover a Martian "valley" that is part of a fairly large crater. Ares works as an electrical technician in the center of his village. He is not yet rich. Both his home and his workplace are converted Quonset huts, capable of retaining an atmosphere for a short time in case of a catastrophic failure of the village shelter.

Ares' village "dome" was a pre-fabricated, modularized shelter system, designed at the earth Embassy, but manufactured on Mars. It is already one of the oldest domes on Mars and his village intends to upgrade the dome system over the next ten years. Even Ares' dome, however, is a highly sophisticated piece of engineering compared to the first shelters created on Mars.

As mentioned, these shelters were designed to handle most emergencies with minimal risk of catastrophic failure. Standard practice has always been to seal the domes with a thin sheet of water ice, either from the inside to reduce sublimation into the atmosphere or externally. Villages deal with pin hole leaks in the domes mostly by making note of them. These leaks usually seal themselves, as water vapor from inside the dome quickly condenses to ice over the pin hole, sealing it. Additional water sprayed onto the area creates a strong seal. Larger holes are more problematic, usually the result of a vehicle collision or some accident next to the shelter wall, but such incidents are relatively rare and many shelters incorporate a six or eight foot wall of native brick to reduce the problems of wear and tear at the dome's lower levels. Even the case of larger holes in polymer fabric is usually easy to correct[6].

Fire is the greatest threat to a village dome. Fire melts the water ice and the dome's polymer fabric, but here Mars helps its hardy settlers. A fire close to the dome fabric itself is most often extinguished by the cold and the Martian atmosphere. In a village mostly devoid of wood, a fire distant from the dome may not grow hot enough to melt more than a moderately sized hole[7]. The domes would not survive artillery attacks but, in general, will be resistant to most human error or natural phenomenon.

Emergency shelters are also scattered across the landscape. These are usually nothing more than an airtight room or structure with an emergency supply of oxygen. Some are permanently located by roadsides and some are transportable—Martian Porta Potties—that move with construction and other crews. One example of an emergency shelter is found at many microwave towers (used to facilitate cell telephone communications and navigation). Each microwave tower

---

[6]   Fire hoses spray water, which turns to ice in the breach, and sheets of semi-rigid plastic are used as an emergency sealant. It takes a long time for air to escape from a large dome (150–200 acres), so direct threats to humans inside the domes are minimal.

[7]   A fire fed by chemicals may grow large enough to melt a large part of the dome. Again, a properly engineered solution—citing facilities that use flammable chemicals outside of villages—may put a few workers and the plant itself at risk, but not an entire village.

has an emergency/temporary shelter built along side of it. As with Ares' Quonset hut home, the idea is not to be able to survive long periods, but simply to have a place of short-term refuge in case of emergency.

The Shape of Things To Come ...

While the term "village" has been used repeatedly to describe Martian settlements, the reader should not assume that the centers of such villages will be devoid of character. Just the opposite. After just a few years, these village centers will be built like a giant "shopping malls", complete with the plastic familiarity of the American suburban mall. Parts of the "shopping mall" complex might be made of Martian adobe and parts with high-tech materials developed at the earth Embassy and manufactured on Mars.

Ticket contracts with the Travel Corporations provide settlers their own homes on Mars. These homes are likely to resemble mobile homes or possibly smaller adobe-style structures, made from Martian mud. And although anyone may contract for the construction of a home according to his own specifications, during the early years there will likely be insufficient resources to build custom homes. On the other hand, there are not likely to be any apartments for a few years. Not only will the low population not require apartments, but the existence of individual homes will help offset the quasi-military lifestyle that will dominate the first Martian bases.

Each individual or family will be encouraged to grow her own fruits, flowers, trees, and vegetables. Larger fields for government or commercial growing ("the land between the spokes") will be within walking distance. There is another reason, beside oxygen and wood and food, why the settlements and bases should be filled with plants. At least as important will be the psycho-spiritual blessings of a beflowered world. Imagine seeing a flowering dogwood as you step off the rocket after six long and sometimes frightening months in space. Not only is the tree a reminder that this mission to Mars is not a prison sentence, or exile on the remotest of islands, but an adventure awaiting the few who have the wisdom to take it.

Although he owns his home, Ares must nonetheless pay a small monthly "rent" to the shelter condominium. Some of the rent goes to general clean-up around the shelter, but most of it goes to pay for the upkeep of the shelter dome. We could again make a point about how the mission to Mars will achieve its goals at a far lower cost than BGBS methods. BGBS shelters of comparable size to what have been discussed here would likely cost several billion dollars and an equally large sum to transport to Mars. Mission to Mars versions, because of cramped budgets and local manufacturing, will be available at one or two orders of magnitude less cost. To be sure, these are costs, but costs borne in by the individual settlers and part of an economic structure that will pull equipment and people to Mars.

Home, Sweet Crater

As happy as Ares has been in his village, he is thinking of moving to another village about 250 miles away. This village has been built under a capped crater, a convenient way to create a spacious 1000-person village. The village is still relatively new and there is a great demand for workers. Ares would leave for the higher wages, to scratch an itch for adventure, and to experience life in a setting eight times the physical size of his home village.

The crater idea is simple enough: stretch a plastic cover over the top of a crater and fill it with oxygen. A structural support system may not even be needed—though a plastic cover a quarter- or half-mile diameter will weigh many tons[8]—if atmospheric pressure alone inside the dome can support the cover. In a short time, humans on Mars might be living in their own "Swiss valley", complete with terraced agriculture up the sides of the crater. The "valley" could support a human village living little different from a village on earth, except for the gravity and the blurry skies. If the settlement were designed to be primarily agricultural, the crater might be rather large. On the other hand, if a more urban environment is intended, the crater need not be big at all. For a 1000-person village living in an urban settlement (remember, all the early villages on Mars will have some agriculture, both for its soothing emotional effect and to help maintain the oxygen levels), the crater may not even be 1000 feet in diameter.

There will be many advantages to capping a crater. First, would be the ease of construction. Not only might the crater serve as "walls" for the shelter, but they could support most of the weight of the cap. Next, would be the great psychological benefit of humans working "on the Martian surface" without pressure suits. The ability to walk around the settlement without such suits, playing baseball and watching their children play soccer, should be obvious. The more engineering-inclined reader will appreciate that the problem of radiation will be greatly reduced, as the crater walls—depending on their slopes—will shield much radiation. Thus, a "capped crater" settlement will allow humans to feel at home on Mars, while participating in the history-shattering event of creating a new human village on Mars.

---

[8]   Human-designed polymers will have the strength they exhibited on earth, but will only weigh 38% as much, simplifying support engineering.

An Economic Scenario: Year 2075

Ares has two brothers living in his village. Mike is a writer and Mack has a small machine shop where he creates machined steel parts. Mike writes scripts for soap operas, mostly for earth TV. Being a Martian, he and his writing are viewed as eccentric, which helps him keep working. Ideas for scripts come up via the internet to Mike from Los Angeles. Mike has three earth days to flesh the ideas into a script, which he does with a gusto that pleases the producers. After the script is sent back to Los Angeles, earthbound writers polish off the Martian edges and hand the script to the actors. Mike does not write to win awards, but with over 1000 television channels in English available on the internet, there is a significant demand for people who can put a sentence together. Mike also writes for a Martian TV soap opera. Here, too, the work is just potboiler stuff, but the show is popular on Mars and it also has a small, but loyal, viewer base on earth. Mike makes a good living writing, though lately he has been intrigued by the high price of mangos on Mars.

Mars and Deimos now grow virtually all the fruit consumed on Mars, but no one has yet planted mangos. Mike knows very little about fruit trees, but only a few businesses on Mars were started by people with great expertise in the field. Mike is thinking about buying a few mango trees to see if he could make a go of it on the side. It will cost Mike a few weeks salary to get started and it will take several years to see if he can get any fruit, but even a thousand mangos a month[9] would pay him more than the comfortable living he now makes writing nonsense. Mike is intrigued.

Mack was surfing the net a few weeks ago looking for new business. The pump manufacturer at Luna Station was looking for a contractor to supply one thousand pump cases. Mack decided to put in a bid over the internet. Mack was disappointed that his bid was not accepted, not because of the price, but because the quality of Mack's Martian steel was too low. It was a typical problem for Martian manufacturers trying to sell off-world. Although many grades of steel are produced on Mars, Martian steel is still not of high enough quality for some jobs. There is simply no need to invest in the kind of equipment necessary to produce some grades of steel on Mars, as it will be many years before Martian steel manufacturers will be able to keep up with demand for even the simple grades they make. So Mack created a half-hour video (a telephone call was not possible) for the pump manufacturer, sent it over the internet, and hoped the gal took the time to view it. Mack had only a vague understanding of how the pump would be used, though he suspected that they would be used in space or near space

---

[9]   Mike thinks he can sell the mangoes initially for CS $6 or $7 apiece.

conditions. Mike argued that the specifications were too strict and that his low-cost pump case would be both cheaper than a pump with more expensive steel and every bit as good.

Mack was lucky. On Luna, Cindy had a few Martian friends, was intrigued by the video, and decided to take a chance. Besides, the price could not be beat. So Mack put in another order for 505 stainless, which would arrive in about two weeks. Mack knew that the iron, chromium, and all the trace elements came inexpensively from some asteroid and that this steel was far cheaper than anything he could have gotten from earth. Once the steel was in, Mack worked 18-hour days, increased his helper's hours from 20 to 45 hours per week, and got soft-handed brother Mike to help him run a machine or two. In fifteen days, Mack had created 988 pump housings and was ready to ship them off to Luna.

Mack had to finish the job quickly, because the only fast ship to Luna was going to leave three weeks after he got his steel. There was no other ship due in to Luna for months. As it was, the fast ship would take four months to get to the moon. The time delay, however, was common. Even earth manufacturers could not deliver goods to the moon without weeks of delay. There was simply not enough traffic running between the earth and the moon to justify more than one or two ships per month. The end user was building for Mars and the moon, however, and so the long lead times for the pumps had already been taken into account. Mack loaded up his "golf cart" and his trailer to drive four times the 18 miles to the lift pad. Two stevedores were buddies from school, so he knew the boxes would be loaded. Besides, all Martians knew how important off-world shipments were. There would be no question about Mack's pump housings reaching Luna in good shape.

Parks

The drive to create "wilderness" on Mars will be a two-pronged attack. The easiest and most convenient will be the creation of 10–1000-acre sites, enclosing suitable-sized ravines and valleys with a Kevlar cover. The other kind of wilderness area will be specially selected surface sites, where forests might be established early, with a hope that very hardy animal life (worms, insects, etc.) might soon follow.

The smaller wilderness areas should be designed to be self-sustaining biospheres, even though most are likely to have life-supporting "regulators" annexed to them. A one-thousand-acre tract makes a wilderness area two miles long by two-thirds of a mile wide. Parcels this size can house a considerable variety of flora and fauna. A careful design can create forested glens, wetlands, streams, and meadows to provide habitat for all sorts of birds, fish, and amphibians. After a time, larger mammals such as deer and possibly even some predators can be introduced. Not only might these areas be used year-round for hiking, boating, fishing, etc., but a reasonably lush environment would also allow for some limited hunting, probably using shotguns, .22 caliber rifles, or bows and arrows. Thus, a full experience of the outdoors might be created for immigrants to Mars and, perhaps more importantly, for the native-born Martians who would otherwise not have such experiences of earth's bounteous nature. As discussed below, once recreational travel becomes readily available, these wilderness areas could become tourist destinations.

These areas would certainly cost much less to create than human settlements. Suitable terrain will need merely a cap on top and a wall at one end. Heated mostly by the sun, a circulatory system can be created to add external carbon dioxide to the system as trees and other vegetation cause the internal oxygen content to rise. The cost to bring in fish, amphibians, and small mammals need not be high. Though it may take many years to allow the trees to mature, the refuge will likely be able to sustain itself just a few months after opening. Staffing of the preserve need not even be a necessity, since the terraforming biologists living at the larger settlements can use the preserve for research and can provide advice to help maintain an environmental equilibrium. Some maintenance will be required on the life-support annex and irrigation/water-circulation systems, but given the likelihood that modest construction activities (trails, lakes, special habitats, etc.) will continue for many years, this limited maintenance can be contracted to the construction crews.

Even at an early date, the preserves can provide some revenue through the sale of entrance fees, hunting and fishing licenses, timber harvests, stays in a cabin, and possibly through commercial fishing, crabbing, or clamming. In fact, the cost to create such wilderness areas can likely be borne by those who develop it for "outdoor" usage by humans. As much of the conservation and research on animals in the various states in the US is mostly served by hunting license fees, so the cost of the preserves can be met through the appropriate usage fees and by the bequests of magnanimous humans who will see the value of these kinds of shelters. Lakeside cabins could be made available for rent or, especially in locations where a mega-reserve can be created, tourists can mimic the earthbound activities of wilderness camping or trips to a national park. In addition, some of the costs can be underwritten by terraforming activities, since not only will the shelters at some point come down, but these preserves can house creatures for release onto the Martian surface, when the surface environment is conducive to survival.

There are two such reserves within twenty miles of Ares' village. He has spent time and some of his money fishing in two lakes. The bass fishing is good, but though there is some trout fishing, Ares does not particular enjoy standing alongside a hundred other anglers during the two days of trout season, scheduled right after trout from the hatchery are released. For much the same reason, Ares rolls his eyes whenever one of his brothers talks about going bow hunting for deer. A lottery awards only two or three licenses each year and the deer are too easy to find.

The second front of this war for nature will be waged on Mars' surface. Though Mars will fight an intense battle against the hardy lifeforms invading her domain, progress will be inexorable. In as short a time as a few decades, there will be recognizable plant species living on its surface and, as we will discuss two chapters on, specially prepared ground can also give rise to groves of trees.

Ares gets out onto Mars' surface about once a month. The cost is not insignificant, but the pleasure of being the first human to step on a piece of ground is well worth the cost of renting a Mars suit. Usually, Ares goes with his brothers or his friends and they stay within a mile or so of the shelter. It is interesting to see the new plants that have sprouted up since the last time they were out. About once a year, Ares pays to go on a tour. The tour guide caravans two or three vehicles together, taking groups to points of interest within a hundred miles of their village. Usually, there is also a stop at another village as part of the weeklong package. As will be discussed below, people understand the desire to get out onto the planet's surface, so taking weeklong vacations every few months is something that most employers allow and even help plan for, through a combination of easy vacation policy and compensating work on Saturdays and Sundays.

If Mars' atmospheric pressure can be increased in the first few decades, then the question becomes not whether plants can survive on its surface, but what kinds. As soon as plants grow in an area, the next agenda item will be to introduce animal life. This will be more difficult, as most animal life will be repelled by the low-oxygen, high-carbon dioxide atmosphere, but doubtless there will be a few pioneer species. Once even a few animals live on Mars' surface, the question as to animals will resemble that with plants: What will be the next species to be introduced onto the ever more-friendly Martian environment? The advantage of all of this, of course, is that unlike the capped reserves, the creation of new ecosystems on Mars' surface will be determined by "natural" processes, not by how much another cap will cost; so once the process of claiming Mars' surface begins there will be no stopping it. This does not mean that there will be no problems. Indeed, it is almost certain that there will be many failures, as species are introduced that have no natural enemies and thus overwhelm areas, as rabbits have become a nuisance on New Zealand or zebra mussels in the Great Lakes. These problems can be solved, however, over the course of time by those terraformers who learn to love Mars both for its original beauty and for the terran beauty that they will bring to a formerly red planet.

The Martian "Tourist"

It is possible that the Fleet may establish a work routine at new settlements that will facilitate opportunities to explore Mars' surface. Given that each new settlement will be a "military operation", there is no reason that duties might not be scheduled for six or seven days a week, with one week off after three or four weeks of work. Such schedules allow the worker to focus from start to finish on a discrete task at the new settlement, followed by sufficient time off to allow for the individual exploration of and travel on Mars. Not everyone will enjoy such a schedule, of course, but the twin ideas of focused work followed by weeklong periods of relaxation and personal development may offer advantages in the "military" environment of each new settlement. Not only might such a schedule serve to raise morale, as people work hard for their about-to-arrive vacation, but there is great value in having people get onto the planet's surface in that *they may easily find something valuable to the overall success of settling Mars.*

Thus, a nascent tourist industry on Mars that can help people come to know Mars as their own world. While "earth tourists" will only arrive later, the seeds of this industry can be planted early. And they will come, whether it is to explore the planet for themselves or as scientists hoping to solve the mysteries of life on Mars or the depth of its dust. What will likely be of far greater importance to the financial and mental health of the Mars colony, however, will be travel across Martian terrain to places of extraordinary beauty, interest, and places where new cities are rising. This tourist industry will service both settlers and the terran visitor. Once the Confederation begins its Caravan Transporter missions, there is no reason why persons willing to pay $5–10 million should not be allowed to travel as a tourist to the Mars settlements[10] Tours of the settlements can be offered, the tourist can stay in a newly constructed home or in a hotel, or the Confederation may choose to create temporary quarters for its purposes. One or two excursions into the Martian wilderness can be given as well as some trips to outlying settlements used by the settlers for recreation. It should not take much incentive or planning before one or two commercial travel enterprises are established, which can provide a top-notch travel experience, as well as attract additional capital to the Martian colonies.

---

[10]   Seats for tourists should be provided from the quota allocated to the Confederation. As indicated earlier, visitors paying king's ransoms to see Mars can help prevent hard currency shortfalls.

Coda

The chapter began with the suggestion that life on Mars would not resemble scenes portrayed by Hollywood. As one of the great attractions of the *Star Trek* television series was its uniquely upbeat portrayal of life in space, so the reality of the mission to Mars will be that humans will not live in space caves or in cramped domes rattling in the Martian winds. Rather, human beings will live a life where Mars becomes a positive attribute some choose to incorporate into their lives, rather than as an ugly reality of a corporate monster. For those living in a capped crater, yodeling may become the pastime, as they create a rural life in a village seemingly high up on a mountain somewhere. For those inclined to sing *New York, New York!*, home will be a shopping center settlement with bustling stores and bodegas, where pedestrians find vistas of the Mars on promenades at the city limits. Naturally, one of the great recreational activities available to all will be to travel outside the shelters, to see what exists in their strange new world of slowly rising temperatures and monumental scenery, but this Mars, this red playground, will be something sought out, not a taskmaster lording over impoverished souls. Settlers will relish their lifestyle, offering challenges and circumstances not found in scrubbed, well-lit American bedroom communities. But their pioneering exis-tence will have the warmth of a Currier and Ives print, where one will find few ghosts of a terrible Mars. Martian settlers will live a life that offers them both financial security and a once-in-a-millennium chance at adventure, on a world that will be both pleasantly mysterious and maternally secure.

Economics 103

The [trans-Appalachian] West was predominately agricultural. The abundance of cheap land was the lure which drew to it the great majority of its inhabitants. The first-comers had found themselves cut off from the rest of the world by stretches of unpeopled forest and mountain. Their isolation was even greater than that of the back-country folk of the coast states, and the privations and perils of frontier life were intensified. Theirs was the task of building society from its very foundation stones. Thrown upon their own resources, their first productive efforts were devoted to securing the rudest necessities of existence.

But with astonishing rapidity the clearing about the pioneer's cabin widened, and the huts which clustered here and there upon the river banks grew into towns. The very isolation of the West acted like a duty upon imports; the difficulty of intercourse with the remote East was a stimulus and protection to the manufactures which it was compelled to set up for itself ... These [factories] led the people to dream of a self-sufficing economic province, a western world, maintained by exchange between town and country, the farmer supplying the wants of the townsman in the way of food and receiving in return the products of the craftsman's art.

—*Political and Social Growth of the United States, 1492–1852* by Homer Carey Hockett

There are a number of economic prospects on Mars that will not fall neatly into one chapter or another. Still others make a chapter too long. As we near the end of this volume, this chapter presents a few such economic oddities.

Transportation

The development of transportation capabilities on Mars will be of major initial concern to the Confederation and its Fleet operations. Transportation needs on Mars, however, will be different from transportation needs on earth. For the first few, fragile decades, transportation will be divided into four categories: transportation inside of the shelters, long-range transportation, short-range transportation, and hauling.

Transportation within the shelters need not soak up huge amounts of resources. Likely, something as simple as a golf cart can be used to move objects and people the short distances to be covered within the first shelters. These means will become more and more sophisticated, of course, as the shelters become interconnected and the Martian society richer.

The Mars Landers will transport people long distances on Mars. These trips may or may not be via Deimos. Of course, not everyone desiring to travel a few hundred or thousand kilometers will be able to schedule a Mars Lander. Thus, additional means of travel—besides walking expeditions with the marines!—will need to be planned. A few ideas to this end will be presented in this chapter and another idea will be found in the next chapter.

The third means of transportation relates to short trips onto the Martian surface. At first, modified golf carts might be used. This idea will be discussed both here and in later chapters.

Finally, hauling heavy loads will be uncommon for many years, both because the lower Martian gravity will mean that there will be fewer "heavy" loads to haul, but also because the equipment needed for such hauling will be prohibitively expensive. This problem will gradually get solved, but it serves as yet another example of the difference in cost and philosophy between settlement and exploration missions. NASA would find cost no object to ship ten-ton land rover behemoths to Mars for its laser-focused exploration missions. In the strongest contrast, the mission to Mars will "work around" literal and figurative boulders to find solutions other than moving them.

Vehicles

Electric or small engine "golf carts" can be assembled on Mars, even in the early years. Many of the parts, especially the small engines, would need to be imported, but if one of the first businesses brought to Mars on a Caravan Transporter in LY Ten was Golf Cars Assembly Corporation, dozens might be built every year[1].

---

[1]   As with so many early commercial ventures on Mars, a portion of the output would be stored for later floods of settlers.

The carts will not be very powerful, might only be moderately useful for farm-ing, and the first ones will be only used *inside* the shelters. Equipping the golf carts for use outside of the shelters would require a level of infrastructure and expertise beyond just the assembly of parts manufactured at the earth Embassy. The standard "surface" models, with most parts made from moldable polymers and a primitive suspension system, would be road-bound and probably danger-ous to take over most Martian terrain. Truly safe overland vehicles will need to be custom built and imported whole from the earth Embassy. By LY 25, however, Martians may be able to create sufficient industry to build most of the parts for a "surface" golf cart. This will not be because of the demand for golf carts so much as because of the progress of the marsmover industry, the next topic.

Marsmovers will be vehicles designed to move "dirt" or rock. The first one or two will need to be imported from the earth Embassy. Thereafter, the need for this type of construction vehicle will be so great that their assembly and even parts manufacture on Mars seems assured. As will be common for Martian manufac-turing, Mars will begin by assembling marsmovers from parts made at the earth Embassy. Trying to supply all the Martian bases from earth, however, would be prohibitively expensive. Despite having complicated hydraulics, rugged metal bodies, pressurized cabs, powerful suspension systems, etc., it will be necessary to make the massive effort to move the marsmover infrastructure to the Red Planet. Understandably, these vehicles will be expensive and have large operating and maintenance costs. The never-ending construction projects for arriving settlers, businesses, and others will nonetheless induce such demand for marsmovers that it will outstrip supply for many decades.

	When Introduced (LY)	First Assembly Process On Mars (LY)	First Year 50% of Parts Manufactured on Mars (LY)
Golf Carts	2	8	15
Marsmovers	3	15	15
Winnebagoes	4	20	25

Figure 67: Vehicles

The last category of vehicle has been named "Winnebagoes" for reasons that will become clear in the next chapter. In effect, this series of vehicles will com-monly be called "trucks", although most will have pressurized cabs and some capability to serve as living space outside the shelters. The difference between a "pick-up" with a backseat for sleeping and a "Winnebago", with room for cooking and sleeping, will be one of degree. Parts for the first few vehicles will be built at the earth Embassy and assembled on Mars. As the golf carts will serve some of the same functions, Mars can get by without trucks for many years. Once long-haul capacity becomes critical, however, trucks will be needed. With a golf cart manufacturer or two on Mars and an assembly line for marsmovers, the ability to manufacture a substantial number of parts "locally" will be a matter of bringing in the proper businesses from the earth Embassy. Most of the other infrastructure necessary will already support the golf cart and marsmover businesses. As trucks become more sophisticated, larger and larger "crew" and "family" compartments will be added, to allow people to live extended periods on the surface, away from the settlements.

Johnson & Johnson

Five years before the first manned mission to Mars, Johnson & Johnson, a world-wide manufacturer of personal and home care products negotiated a series of agreements with the Confederation and with several individuals. In exchange for the right to advertise that, for ten years, Johnson & Johnson would be the exclusive supplier of several products on Mars, Johnson & Johnson agreed to a number of things, including:

1) an annual cash payment of US $2 million to the Confederation;

2) sponsorship of at least one person or "employee" per product, who would manufacture Johnson & Johnson products at the earth Embassy;

3) sponsorship of an active research program to manufacture Johnson & Johnson products on Mars;

4) technical assistance for manufacturing at the earth Embassy and on Mars.

The cost for American-sized bottles of shampoo at the earth Embassy was CS $2. The cost for these same bottles on Mars was CS $102, as each one weighed about one pound[2]. The requirement that all earth Embassy businesses file an

---

[2] During the first eight years on Mars, the Fleet issued three bottles of shampoo per year to everyone living there. While there remained a demand on Mars for shampoo, many of the first settlers—including the first women on Mars—simply cut their hair very short. Those who insisted upon more than three bottles of shampoo paid CS $102, which covered both the CS $2 retail price charged at the earth Embassy and the CS $100 cost to ship the shampoo to Mars.

annual business plan with the Fleet allowed Johnson & Johnson to make detailed plans about its move to Mars. One of these plans estimated the cost to manufacture shampoo on Mars at CS $10 per bottle, with costs expected to fall about 10% per year until they were in line with earth Embassy prices.

The soaps and shampoos manufacturer going to Mars was licensed to use the Johnson & Johnson label on her products. Most people at the earth Embassy and later on Mars were fully satisfied that these products were of as high a quality (though usually a little more expensive) as most Johnson & Johnson products.

During LY Six, the Confederation started negotiating with Johnson & Johnson and the earth Embassy soaps and shampoos manufacturer about creating an operation on Mars, which the Fleet wanted to begin operating in LY Ten. The manufacturing processes were to be developed and employees to be trained at the earth Embassy. Plans to move the business to Mars would integrate with other Confederation plans to move businesses from the earth Embassy to Mars. As it turned out, this Johnson & Johnson affiliate moved in LY Ten and had sufficient suppliers in place to begin production late in LY 11.

The plant on Mars would turn out 100,000 bottles of shampoo per year. Starting with LY 11, shampoo was no longer issued by the Fleet. The new cost for shampoo on Mars was CS $12 per bottle, a cost slightly higher than planned. Annual consumption was expected to be about 15,000 bottles. This consumption would grow between 20–50% per year for many years. The plant would stockpile rather large quantities of shampoo for the future, eventually growing to almost 200,000. By LY 30, consumption on Mars began to exceed 100,000 bottles per year. Having the extra 200,000 bottles of shampoos in storage would allow for some flexibility to meet this demand, as part of the Martian "rubber band economics" that played out again and again on Mars.

The "Cost" to Move to Mars

This section will continue to develop the theme of the "low cost to Mars", now from the perspective of an owner of a Domino's Pizzeria franchise[3]. As was asserted in an earlier chapter, a BGBS cost to move a real pizzeria to Mars would no doubt be tens or even hundreds of millions of dollars. Even the simplest of ideas, such as our hypothetical of "Mrs. Chen" opening a Chinese take-out at a NASA-run

---

[3]   The idea of the Domino's Pizza will be a recurring theme in these three volumes. It was selected to help illustrate not only the idea of *settlers* and a *society* moving to Mars, but also to provide concrete economic examples of ideas that otherwise would remain "theoretical".

Mars is stipulated to be beyond the ken of an organization like NASA. In the strongest contrast, the cost to move a Domino's Pizza franchise from the earth Embassy to Mars will not only be low, *it will be a cost borne by the business itself.* Low-cost threads like these will continue to be woven together in this these three volumes, finally to reveal a fabric where the cost to "government" to achieve its goal of transplanting a human society to Mars is not only low, it will be almost zero. The issue, as suggested earlier, is not money, but the raising of capital. But let us return to the pizzamaker.

	Weight in Pounds	Percentage Paid By Business	Transport Cost at $100 per Pound (CS $)
Earth Equipment	1000	100%	$100,000
Mars Equipment	500	100%	0 ($50,000 cost on Mars)
Supplies	4000	25%	$100,000
People	1000	0%	0
		Total Cost:	$250,000

Figure 68: Moving a business by weight and cost

The Domino's business will need to move some equipment, supplies, and people if it is to do business on Mars. We assume here that some of the Domino's suppliers will already be on Mars and some will arrive or will be created within a few years of the Domino's arrival. As with most theoretical discussions in this book, we will simplify the ideas to make the larger points.

Figure 68 above lays out the cost to move the Domino's business to Mars.

This Domino's Pizzeria, like all businesses at the earth Embassy, will have been planning its move to Mars for many years. The business will have been required to update this plan annually, which the Fleet will review[4]. Accordingly, the Domino's has planned to conduct operations on Mars by bringing equipment that weighs

---

[4]   As in any human activity, proximity to necessity will drive this process. Thus, during the first few years of operation at the earth Embassy, the "plan" to move the business to Mars may be quite stark and the Fleet review may be cursory. As the business moves closer in time to actual transport to Mars, these plans will be expanded and the Fleet's review will become much more critical.

1000 pounds, with additional equipment to be purchased on Mars. The Domino's has determined that it can operate on Mars with five full-time employees, who weigh a combined total of 1000 pounds. Finally, Domino's believes that it can begin its operations on the Ark Transporter and on Mars if it takes 4000 pounds of supplies with it[5]. Once on Mars, some supplies will be imported from earth, but some of its suppliers will also be in place, allowing many of the pizza supplies to be purchased from sources on Mars.

Of the four "components" of the Domino's business, the owner need concern himself mostly with the movement of equipment to Mars and the purchase of equipment made on Mars. The business will not bear the cost to move the employees to Mars. The employees themselves will pay for their tickets. As for supplies, the cost will be shared with the Travel Corporations and the Confederation. Here is why.

The Travel Corporations will have a continuing obligation to feed people until their arrival on Mars[6]. As the Domino's will operate during the flight to Mars, Domino's will be paid by the Travel Corporations to provide a "ration" to those on the Ark Transporter. On Mars, the Confederation will pay the Domino's to provide a given number of meals or pizzas to members of the Fleet on duty[7]. This contract will also help to pay for the supplies taken to Mars with the business. As the Domino's plan develops over the years, the business owner will have strong incentives to minimize what must be imported from the earth and to maximize what can be manufactured on Mars, as all business owners try to maximize their profits.

---

[5]   The actual weight of supplies for Domino's will be higher, but it will be offset by galley food not taken onto the Ark Transporter. For simplicity's sake, revenues and profits from operations on the Ark Transporter will offset the cost of some of the supplies taken aboard the Ark. See next footnote.

[6]   Settlers on the Ark Transporter will eat Domino's Pizza in place of food from the galley. This food will be part of a ration for those on the Ark, so the Travel Corporations will pay Domino's. In addition, those on duty with the Fleet while on the Ark will have their meals paid for by the Fleet. Some of these meals will be bought from the Domino's pizza (a "pizza ration" to take advantage of the presence of the Domino's during the flight to Mars). Of course, people will also be able to buy food "out of pocket" to satisfy the occasional urgent need for a midnight pizza. They, too, will pay Domino's for food. Thus, Domino's will likely only have to pay about 25% of the cost of supplies. The other 75% will be covered by these consumers.

[7]   Especially during the early years on Mars, the Fleet will provide such "commercial" meals to those on duty both to help keep up morale under austere conditions and as a way to help the new businesses stabilize their finances.

The sum total of the costs to the business owner to move his Domino's business to Mars is CS $250,000, the sum indicated in Figure 68. The Confederation will have a program to make loans to those businesses transplanting themselves to Mars from the earth Embassy, so that all businesses can fund their move to Mars. In fact, part of the annually updated plan must refer to the sums to be borrowed from the Confederation for the move, as well as the business' expectations of contracts with the Confederation, such as contracts for pizzas discussed above. We will assume here that these loans will be "interest-free" although this program will be subject to Confederation policy and fiscal restraints. In any case, it will not be a severe disruption to the Domino's to have to pay the Confederation back. In fact, as Figure 69 depicts, if the pizzeria charges an additional CS $1 per pizza, it can repay the loan in 14 years. If it charges an additional CS $2 per pizza, it can repay the loan in just over seven years. While the business owner will want to stretch out repayments on a no-interest loan, the Confederation will likely have a policy that forces early repayment, after which it charges market interest rates.

Pizzas Sold Per Day	Days in Year	Years Surcharge in Place	Total Revenue Raised
50	365	14	CS $255,000

Figure 69: Revenue Raised From $1 Surcharge on Martian Pizza

The reader should note that while the assumption that the business sells "50 pizzas per day" on Mars may be a reasonable approximation for our analysis, the actual circumstance may be much different. More likely, the Domino's business will have a huge demand and the pizzeria will be constrained not by consumer demand, but by the ability to find the necessary supplies.

The cost for pizzas on Mars will be very high at first, but will drop as more and more people and businesses arrive. The owner may initially charge CS $25 per pizza, as people are willing to pay almost any price for an old, familiar treat. The price will reflect both the cost of importing some of the supplies needed for the pizza as well as the strong demand for the product. Over time, however, demand will drop as other "fast food" and restaurant businesses arrive on Mars. Costs will drop, too, as more supplies start coming from Martian sources. Eventually, the cost for a pizza on Mars will not be entirely different from the cost charged at the earth Embassy or at other places on earth. In addition, as this Domino's franchise owner will likely want to expand his business to new villages being established on Mars, he will have an incentive to charge reasonable prices, to ensure a good reputation in the marketplace.

Anyone with an economics or business background can understand that a hundred different details were omitted from this scenario. The major points about how a community starting on earth might transplant itself to Mars, however, should be better understood. That plan is feasible. All that is required is for a sufficient number of dedicated people to undertake the task to create such a community.

Domino's: Settler versus "Explorer"

BGBS-think will now pose a question straight from its playbook: "How are you going to find a pizza maker that also wants to go to Mars?" The suggestion is that it will take many hundreds of thousands or even millions of dollars to entice such a person to the mission to Mars. As usual, BGBS gets things backwards.

The pizza maker we have been discussing is not a Domino's franchisee enticed to a Martian future. Rather, a person who has a strong desire to get himself into space and to Mars will play pizza maker to satisfy his Martian dreams.

Most people who went west into the American frontier did not go there to become a shopkeeper. They went there because they had an itch to be free or to see the west. Once there, they found employment as best they could. The mission to Mars will simply *organize* the economic efforts of people whose larger goal in life is to go into space and to Mars. Some people will count it an advantage that there will be virtually limitless opportunities to express oneself economically, at the earth Embassy and on Mars, as every sort of business will be needed; adding a second or third "candlestick maker" to the economic mix will simply help to keep costs low.

What is the competition? NASA holds its consensus-building meetings, working on a plan to get a pizza maker to Mars until some President or Congress or new administrator changes direction. Then, things must be reworked and redesigned. Not only is it hard to keep individuals interested when faced with such uncertainty, it also keeps costs very high. Nor is there any hope of lower costs via an economy when one "explores" Mars. Everything needed for an exploration mission will need to be imported from earth, by an expensive government bureaucracy. Lift costs for NASA will continue to be $10,000 per pound, versus zero lift costs for the stuff manufactured on Mars by settlers.

There will be costs at the mission to Mars. Some of these costs will be higher than planned and there will be setbacks that will eat up money, too. The structural differences, however, combined with the ability to consider "creative" ideas like charging a pizza maker to fly his equipment to Mars will keep costs realistic and will allow the mission to Mars to move ever forward. Shakespeare wrote that "there is a tide in the affairs of men which, taken at the flood, leads on to fortune; omitted, all the voyage of their life is bound in the shallows and in miseries." The mission to Mars is the tide that will carry human beings to Mars, while NASA forever wallows in the reeds.

Marbillies

Into every corner of the earth ... the English will penetrate.
—Rose Macaulay

This Part of the book ends with two views of human society "beyond" early Martian settlements. While this chapter is not specifically directed toward "radical" libertarians, they have clearly inspired it[1]. As with so much in these three volumes, a serious consideration of the project's ideas and possibilities leads one inexorably to thoughts similar to those found here.

Mars-e-ba-goes

If you have traveled the US, especially the west, you will have marveled at and sometimes condemned, those 30–50 foot behemoths plying the heat waves of the American summer. Whatever you care to call them: campers or recreational vehicles or your favorite curse, these monstrosities announce a lifestyle that is both free and adventurous.

You know the scrip by now. If there is a self-sustaining Martian economy that makes what it needs from native Martian resources, anything can be built on Mars. Having 100,000 and more people on Mars means that there will be a sufficiently diverse economy to create the kinds of vehicles discussed here, if only for specialized purposes.

---

[1] Libertarians and their ideas are discussed in great detail in this work, especially in Volume II. This chapter was very late in the original drafts, but there remains a need to address this chapter to "libertarians", both to set the table of ideas and to portray them as potential heroes of the mission to Mars' revolution.

We now ask the reader to make a leap of imagination, to take her understanding of "recreational vehicles" to a point where she might envision a few design changes to prep them for Mars. Imagine it. Mobile, self-contained living units used to "explore" the planet or, more realistically, used to free the owners from the chains of Martian communities. The result would be a truly exciting life, lived on one's own terms, as self-reliant as can be imagined for a citizen of the 21st-century.

Beyond the mechanisms necessary to provide mobility, the essential ingredients to remake an American Winnebago into a roving Martian habitat is to make the vehicle airtight, to provide it supplies of oxygen, to design it for off-road travel at slow speeds, to equip it with electronic equipment for communications and navigation, and to provide for towing and storage space for a family and for personal possessions[2]. If it is possible to design such a vehicle, then it is likely that such design features can be manufactured on Mars. Thus, an additional facet of the Martian economy might be envisioned, this one to service the needs of people who choose the ultra-free life of a Marbilly.

Living Off The Land

It is the economic facet of most human beings' lives that best describes their daily experience. So, too, will it be with Marbillies. While some of the Marbillies will be retired or independently wealthy and will not need to work, most families living the semi-nomadic life of a Marbilly will need to support themselves. Five easily visualized ways Marbillies might earn a living include: 1) as truckers to transport goods from one settlement to the next, 2) to serve as a kind of general store for other Marbillies, 3) to engage in terraforming, prospecting, or other work paid for by deep-pocket enterprises, 4) to travel from settlement to settlement to work odd jobs[3] or as migrant farm workers, and 5) to create "cropland" and grow crops for sale at a Martian settlement. We discuss each idea in turn.

The easiest notion to grasp would be Marbillies trucking goods from one settlement to the next. In the US, truckers have "sleeper cabs" for their drivers. On Mars, there would be a mobile "living space" for "ma" and the kids. The key to employment as a trucker will be the ability to run a small business. Traits such

---

[2] There is also an issue of radiation exposure, but if one is predisposed to becoming a Marsbilly, one may not view this as a grave issue. In addition, the ever-thickening atmosphere will go miles to reduce this radiation problem to one similar to that of being a chain-smoker on earth. Finally, it is easy to envision using Martian terrain— caves or hillsides—as a shield to limit radiation exposure. Park your Mars-e-bago next to a steep cliff each night and problems of radiation are cut in half.

[3] These communities will be hiring *all the time* for many, many years.

as the ability to engender sufficient trust to be consigned goods for transport, to maintain a good reputation in the community, and to have a good credit rating will also be important. As each "trucker" Marbilly family comes to earn trust through hard work or the previous efficient transport of goods, it can expect to find future opportunities. Similarly, in any given small, Martian community, if the trucker causes a problem, that Marbilly family will have to make amends or write off doing business there. The internet will conspire, both on Mars and the earth, to make "careers" much less important. In a free, Martian society, where citizens are not crushed by taxes and stacks of law books, Martian truckers should find both a living wage and a good measure of respect.

Other Marbilly families could move from family to family buying and selling goods to earn its keep and to keep more stationery Marbillies supplied with essential products. This kind of special delivery trucker would make the rounds over several tens of thousands of miles, buying wholesale what they can sell retail at another stop.

As there will likely be a constant supply of jobs at remote sites, building communications towers, grading roads, building a processing facility next to a mine, etc., Marbillies could be a source of labor for these remote site endeavors. Marbillies might also be prospectors, either working independently or with other prospectors, as described a few chapters back.

Some Marbillies will be migrant farm workers. The vast undertaking of supplying an ample and diverse food supply on Mars will require a workforce to move from farm to farm to harvest this food. The work will no doubt be difficult—it is backbreaking[4] work even in the US where there is a high-degree of mechanization—but the wages on Mars will be higher for these migrant farm workers than they are in America and will be able to support a higher standard of living. Why? Because the labor shortage on Mars will mean that even unskilled labor will be in great demand. These migrants will also differ from itinerant farm workers on earth in another way: the high-tech production of food will mean that migrant farm workers will not only pick tomatoes, they will also service complex irrigation and other equipment necessary to make a Martian farm work. Thus, the Martian farm economy is likely to have rich farmers, Marbilly technicians and laborers who earn a decent wage moving from settlement to settlement as work becomes available, and new immigrants who are willing to work hard in the fields so that they, too, can own a Martian farm in a few years.

One offshoot of this migrant farm community will be the ability to grow crops on unclaimed or undeveloped land. Suppose it were possible to haul with your

---

4   The work will not be so difficult on Mars, where the gravity is so much lower than on earth.

Mars-e-bag-o two or three one-acre plastic shelters, each one a circular structure 250 feet in diameter, to an undeveloped stretch of Mars. Such shelters would likely only "weigh" a few thousand pounds on Mars and might be designed to inflate with a pump. A family might conceivably support itself simply by growing crops in these structures and selling any surplus off in nearby markets. Sludge from the nearest wastewater treatment facility or from a biomass reactor[5] could be purchased to create a fertile medium for growing crops. Use this sludge to amend the Martian soil sufficiently and you create a good environment for plants. Water might be "marbillied" in. Once water is inside the plastic shelter—a self-contained unit—it will mostly re-circulate itself. There will be no need for the constant additions of water until the crops are harvested. If the soil is amended over the course of two or three crops, it should become fertile enough to be valuable in itself.

Now, the shelter, the sludge, the water, portage, are all costs that need be considered. The amount of "seed money" (capital) necessary to pay for all of this is likely to be significant and beyond the financial reach of most Marbillies. Two things, however, may nonetheless make such farmland creation programs possible. The first involves a Martian "Homestead Act", which is discussed in more detail in Volume III of this work. The Homestead Act is designed to provide free equipment to anyone who can set up and create new farmland. Thus, the Confederation could provide these shelters, some sludge, water, etc. free to anyone who can pledge that this equipment will be used to create new farmland. The second way to earn a grubstake for this equipment is to use the proceeds from the sale of the last homestead the Marbilly had created. There should be no lack of land-hungry farmers, so the now-established homestead can be sold to earn the Marbilly a fine return. A farmstead with several one-acre plastic shelters already in production, with highly enriched growing medium, will be a very valuable commodity. Some of the proceeds can go into the pocket of the Marbilly family and some of it can go to purchase new equipment to establish another farmstead.

In all, the financial returns of Marbilly itinerant farming will probably be lower than a sedentary operation, where the farmer owned the land, but that is not the issue here. A freedom-loving man or family could simply decide it wanted more "elbow—room" a few terrain features away from the nearest community and the nearest neighbor. The Marbilly should not find it hard to find ways to make their dream a reality.

---

[5] A biomass reactor converts native Martian resources: heat, light, carbon dioxide, and other nutrients into algae or other biomass. The reactor provides batches of material suitable for composting and use as a soil amendment.

One more oddball idea for Marbillies will help to illustrate the complex web of economic opportunity that may await the Marbilly. That idea is to become a 21st-century "Johnny Appleseed" and to plant trees at selected spots on Mars.

Trees Keep Sprouting Up Everywhere

The idea of "trees on Mars" is discussed several times in these three volumes. The topic here concerns a program to plant trees on Mars' surface. What will be discussed are the economics of planting enough trees as to be an aspect of ter-raforming the planet. These groves will have two purposes. First, small groves of mature trees will become a source of lumber and other wood products. Just as important, however, is to begin the self-propagation of trees on the planet's surface. While there will likely be some "trash" species that are so capable of self-propagation that planting will not be necessary, we focus now on hardwoods and pines which will be planted by humans to allow for traditional wood products to be produced on Mars.

There are several factors to be considered in creating groves of trees on Mars. The first factor will be the relative success of research programs to create suitable trees for the Martian surface. The creation of genetically engineered and selectively bred trees can begin on earth and continue in shelters on Mars, until extremely hardy trees are available. These trees must be able to tolerate prepared Martian soils, limited supplies of water, extremes of temperature, and six-month long sea-sons, while still producing suitable wood for lumber. Trees from the taiga and the far northern woods on earth should make excellent candidates for Martian trees and may require less genetic manipulation than one might otherwise suspect.

The next factor is where trees are to be introduced to Mars' surface. Terraforming operations on Mars will need to have reached a stage where potential sites for trees experience temperatures close to what is found in northern Canada and Russia. In addition, air pressure will need to be high enough that these trees can sur-vive in thin atmosphere[6]. Sites near the Martian equator, especially those at low elevations, might have suitable temperatures and pressures just a few decades after arrival on Mars.

The third factor will be the transformation of Martian regolith into Martian soil. Regolith is crushed rock, without any biological material. Soil is a combina-tion of rock and biomass. The "contamination" of the Martian "biosphere" with lichens, fungi, and bacteria will occur even without Confederation assistance. The mission to Mars will have programs to assist the introduction of these life forms

---

[6]   Low atmospheric pressure is likely to be a greater obstacle to commencing the pro-gram than extreme temperatures.

onto the Martian surface. While there will hardly be any "terran-style" soil even 20–30 years after humans first land, there is nonetheless likely to be considerable amounts of biological activity on Mars' surface. As potential sites for groves of trees are selected, additional steps can be taken to prepare a "soil", either by using chemicals, composted materials, or by transporting Martian nitrate deposits to the area.

Finally, one must consider water for the trees. The mission to Mars considers the reactivation of the Martian hydrosphere to be a long-term project, requiring one hundred years or more before appreciable amounts of rain and snow will fall on the planet's surface. In lieu of rain, an irrigation system can be engineered for sites selected for groves of trees. As with wheat fields in western Kansas and Colorado, these systems need not necessarily be endlessly expensive. Pumping groundwater[7] and piping it to the trees may not only be viewed as a terraforming activity—with an implication that it should be done anyway—but need not be so expensive if the trees are bred to require deep watering only a few times per year.

The reader should now imagine ten or twenty valleys near Mars' equator. These valleys will have relatively mild climates, the densest atmosphere on Mars, a thin but useful soil, and enough wells to pump generous amounts of water for the trees. Imagine a Confederation goal of producing one million mature trees on Mars' surface. At 200 mature trees per acre, 5,000 acres will be needed. At 640 acres per square mile, the reader need imagine a forest of only eight square miles. Finding a suitable eight square miles on a surface area as large as all the land on earth will not be difficult. This is where the Marbillies come in.

Confederation officials will offer a program to "tree farmers" or Marbillies to plant trees and then care for them sufficiently that some of these trees take root and can survive to maturity. Suppose that the Confederation contracts with a qualified Marbilly to plant 100,000 two-year-old seedlings in places described above, providing a suitable environment for these trees. Thus, the following "terraforming" contract: the Confederation will pay the Marbilly one dollar for each tree that survives one year, another dollar for every tree that survives a second year, and a third dollar for each tree alive at the end of the third year[8]. This is not as expensive as it may appear[9] (see Figure 70), though there will be plenty of terraforming money

---

[7]   Underground ice may need to be melted before it can be pumped.

[8]   References continue to be to earth years.

[9]   Continue this example with the following assumptions: 1) that 30,000 of the originally planted 100,000 reach maturity; 2) that the total planting costs CS $180,000; 3) that it costs CS $300,000 to buy the two-year old seedlings; 4) that there were CS $120,000 for watering and other costs to bring the trees to maturity; and 5) that

available. After the third year, there exists a stand of 40,000 five-year-old trees. In
the process of planting the seedlings, the Marbilly has earned a total of $180,000.
While planting 100,000 trees may be a daunting undertaking, even with mechanical assistance, if only 50 trees can be planted every hour, then all 100,000 trees
can be planted in one 2,000 hour work year. If 200 trees could be planted every
hour, 100,000 trees could be planted in three months of 40-hour weeks.

Year End	Payments For Living Trees (US $)	Cumulative Total (US $)
1	80,000	80,000
2	60,000	140,000
3	40,000	180,000

Figure 70: Living Trees

In thirty or forty years, these pines or hardwoods might each be worth CS
$1000[10] for lumber, pulp, etc. This would not include the value of providing
a terran environment on Mars' surface, sequestering CO_2 from the Martian
atmosphere and adding O_2 to it. Once groves are established, the value of "tree
factories" to spread trees naturally across the Martian landscape is inestimable. As
with many mission to Mars economic models, the cost to create this forest would
be tiny compared to the value created (as indicated by the footnote below). In all
this process, the Marbilly would be essential, for someone would need to ensure
that these trees lived—mostly by running irrigation systems—during the first few
years. The best candidate to help create groves with several million trees will be
those committed to a mobile and individualistic lifestyle.

---

each mature tree is worth CS $1000, wood being in great demand on Mars. After
forty years, the return on the original investment is about 5% per year. The reader
is assigned the task of devising all the permutations and business arrangements that
a free-or Confederation-organized market might create to provide large amounts of
wood for the Martian republics, at little or no cost to "the taxpayer".

[10]   The actual value of the trees would depend upon the laws of supply and demand.
Comparable softwood trees on earth in 2150 might be worth US $100 each, while
hardwoods might be worth closer to US $1000. We assume a solid price because of
the initial scarcity of trees on Mars.

Maybe It's Not Such a Crazy Idea After All

The chapter began with stories about American retirees leapfrogging from campground to many-splendored campground in their Winnebagoes. Once travel to Mars is possible—aboard transporter vessels sold off by the Confederation to the private shipping lines—there will be retired persons who choose not to explore Utah, but to drive from one Martian settlement to the next. These people might choose to become Marbillies.

If you do not believe Marbillies to be something feasible for Mars, then you are not a man who has ever gone on a weeklong fishing trip with the guys. There should be no doubt that, given the kinds of self-sufficient, money-rich, consumer-goods-poor society that Mars will become, 20–60-year-old men will create an industry where they go out into the Martian wilderness for a week or two in one of those Mars-e-ba-go contraptions. Even if the costs were extreme, some people would pay to do it.

The anti-Marbilly reader might protest that being a Marbilly would be dangerous. Well, who are you to tell someone else that they cannot risk their life? In a free society, people are not so quick to tell others how they should live their lives or what risks are acceptable and which are not. In today's world, young people are obsessed with "extreme sports" and related activities. There is no reason to believe that the motivation behind skate—and snowboarding might not also find an outlet on Mars, in a self-contained vehicle.

If you can envision life as a Marbilly, you can create the kinds of mental furrows needed to become persuaded of the mission to Mars' ideas. Nothing, of course, could be less BGBS than this idea of a "Marbilly". If you remain a BGBSer, you will either be highly offended by the Marbilly idea or more convinced than ever of the poverty of ideas in these volumes. Such displays by BGBSers help the mission to Mars to draw a clean contrast between their "British Lord settling America" philosophy[11] and the mission to Mars' settlement by frontiersmen.

As we approach the end of this volume, any reader unconvinced about the ideas here, who can still manage a glimpse of our Marbilly friends, living in some clumsy hunk of tin, sheltering the night beside a cliff to minimize radiation exposure, is asked to reread some of the earlier chapters in this book. If you can see the Marbilly and his chubby, disheveled wife living in their Mars-e-ba-go, return to some of the more difficult chapters and see if a second or third reading does not make NASA, astronauts, and multi-million-dollar, US-flagged spacesuits fade from view.

---

[11]   An analogy that the reader will find in Volume II.

As with most of this work, the main point is not to metastasize a thought into a Mars daydreamer's paradise or to "design the specs" of a working Mars-e-ba-go. The point is to create a picture of what could be, offering a modest technical skeleton on which to hang the picture. It is not important whether Marbillies are simply tour-guides shuffling accountants out to a nearby peak or whether they truly live off the land by creating farmland or putting up communications towers. The larger point is that on a free world, built not with NASA-beltway bucks, but with the ideas and sweat of a determined, self-selected group, there will be some people who choose a Marbilly-like life of ultimate freedom. This lifestyle choice will serve the larger interests of the Martian community. Marbillies, indeed, will "explore" Mars for us, at no cost, making NASA's multi-billion—dollar search for life both ridiculous and unnecessary. They will put up communications towers, build new roads, haul goods, and create new farms filling in the gaps on a world damning the torpedoes and rushing headlong forward.

Did we mention gold? Prospectors in Mars-a-ba-goes could find themselves living a very rewarding lifestyle …

Primakistan

Success is on the far side of failure.

—T.J. Watson, founder of IBM

This chapter was found near the end of the original 1000-page tome. It illustrates a great many of the political ideas that had been discussed at length in the original drafts, including those ideas relating to the Martian Confederation of independent republics. While telling this story earlier rather than later in the series cannot but guarantee the existence of a few knowledge gaps, we pray the reader will accept these as minor flaws. The story's purpose here is to portray the mission to Mars' political structure in the context of a political crisis on Mars. It is a story about a Mars that has begun to grow rich under Confederation control. In fact, greedy eyes from earth are now casting toward an increasingly wealthy Mars. Someone on earth understands that Martians have been successful. The story shows how the Confederation might react when confronted with political and military threats to its efforts on Mars. We apologize in advance to those who are compelled to dismiss the low literary quality of this chapter, as we insist that there are important ideas and lessons here.

The Primaki Crisis

The Raja of Primakistan had been involved in a series of wars, none of which had helped the Primaki people, but all of which seemed to bolster the Raja's political power. The nation was well off because of its oil and because its population was highly intelligent. It controlled an important share of earth's computer and ultra-net economy. Now, the Raja started making strange speeches, vaguely threatening Mars and arguing that the Primaki people had a right to the rich resources being uncovered there.

On Mars, the eighth republic to gain its independence was the republic of Rajastan. It had taken many years for this republic to gain independence because its early draft constitutions contained religious principles that were not easy to reconcile with the Confederation's Constitution. There were two dramatically more sensitive issues, however, that had made Rajastan's independence politically problematic. First, its population was more ethnically monolithic than any other Martian republic. Two-thirds of the republic traced its ancestry back to Primakistan. The second problem related to the first: it was not clear what relationship existed between Rajastan and Primakistan. Was Rajastan essentially a colony of Primakistan? The composition of Rajastan's population could not help but be relevant to its political leanings. Any one of these matters, by itself, would have been relatively insignificant. Put them together with what was going on back on earth, however, and the Confederation President and its Congress became suspicious of the not-yet-independent-republic and its political agenda.

Fleet leaders had been warning for several years that Primakistan could be a threat to Mars. Those on duty at the earth Embassy warned that Primakistan might be capable of launching an invasion of Mars with up to 1000 (albeit small) ships; but ships that could be armed with several dozen nuclear-tipped missiles. More alarming yet was that the Primaki had developed an ultranet bomb thought capable of wiping out 90% of all computers on Mars. Most Martians laughed. The unbroken Martian peace and its frontier lifestyle were such wonders, such tokens of heaven on Mars, that few could imagine a threat developing one hundred million miles away. The Fleet itself acted, however, by increasing from 10 to 25 the number of combat ready ships in service and by arguing for legislation that required the other 275 spacecraft on Mars to become "combat capable"[1] within ten years. Since every pilot with a license for space was by law a Reserve Pilot in the Fleet, Fleet commanders began an expanded program to call these pilots into confederate service, to ensure both their loyalty to Mars and their ability to serve in the Fleet. As expected, few of these pilots wanted to be in *any* Fleet, but almost uniformly they were loyal to Mars and highly capable pilots. At the Fleet Military Academy on Olympus Mons, thinly veiled Primakistan military formations and doctrine were used as "the enemy" to train cadets.

Confederation policy stated that no ship could come within 50 million miles of Mars[2] without a formal registration with Martian authorities. This sleepy

---

[1]   This requirement would allow for military electronics to be "plugged into" the ship's systems and for the easy mounting of weapons systems.

[2]   To account for the motion of the two planets, the policy would actually be quite complex (the planets would sometimes be very far away from each other and, at other times, very close). That complexity is ignored here.

regulation had always been enforced as routine "paperwork". No ship from earth had ever been denied registration and it had been used mostly as a way to enforce the visitation and immigration fees that Martians imposed upon earthers. The day that 1000 Primaki spacecraft left earth's orbit, however, was a day that allowed Martians to understand the importance of this seemingly innocent regulation.

And on that day, the Confederation demanded an explanation of the Raja's government. Their answer was diplomat-nonsense. In response, the Confederation militarized all space and aircraft and called several hundred very grumpy civilian pilots to duty. Quietly, the Confederation also asked the Prime Minister of the Republic of Rajastan of its intentions in the crisis. The Prime Minister assured the Confederation that it had nothing to do with the Primaki government and that his government had no desire to get involved in the problem one way or another. Confederation envoys reminded the Prime Minister that in the event Mars and the Confederation were forced into war, Rajastan could not *be* neutral, but that it must side with the Confederation and with the other Martians[3]. The Prime Minister nodded weakly and later put out a communiqué asking "both sides" to be calm.

Confederation leaders had long discussed the "military problem" of the inner asteroids. There were a few hundred marines permanently stationed on four of the asteroids, but in most respects, these rocks were only defensible if the Fleet deployed most of Mars' military forces to them. If the Confederation sent its Fleet to defend them before it could confirm the final destination of the Primaki armada, Mars could go undefended. On the other hand, if an enemy power occupied any of these inner asteroids, that asteroid's orbit would eventually take the enemy dangerously close to Mars.

Mars "controlled" sixteen inner asteroids. Ten more had "unauthorized" earth stations on them that were a constant source of dispute between Confederation diplomats and earthers. Some of these inner asteroids came very close to Mars. The Martian military knew that if the Primaki military seized one or more of these inner asteroids, they would be in an excellent position to launch nuclear missiles against Mars; missiles that would go undetected virtually until the last moment.

The Fleet recommended to the Confederation President that one infantry regiment from each of the independent republics be mobilized. Seven infantry regiments would be sent to the inner asteroids to "clear" them of any Primaki troops. The eighth, the regiment from Rajastan, was to be sent to Phobos "to assist with the development of a training base there". The President agreed and

---

[3]   Both the Confederation's Constitution and the Constitution of the Republic of Rajastan contained appropriate provisions to this end.

the order was given. Naturally, the Governor-General of Hellenica, the largest republic, insisted that her own general be appointed expedition commander to the inner asteroids. No one at the Fleet, in the Confederation, or in any of the other republics liked the idea, but everyone understood the politics. General Armani was therefore appointed commander of the Martian Expeditionary Force.

On those issues pertaining to combat, the Confederation regulated all Martian military forces. Thus, all infantry regiments on Mars were armed with a standardized assault rifle and these regiments were supposed to have one particle gun per squad. Most of the regiments were short of these particle guns (they were expensive and the republics were slow to buy them), so the Confederation faced the problem of finding more guns for its federated troops. The Confederation President, pursuant to his emergency powers, levied financial allotments on each republic, "to pay for such items and such expenses as the emergency required"[4]. Part of the money was used to buy eight hundred new particle guns. Four regiments had their particle guns before they shipped out. The other three regiments would receive their guns as quickly as they could be manufactured.

Doctrine for military operations, a standard for communications equipment, the format for military orders, and other matters immediately relevant to General Armani, also fell under the Confederation's jurisdiction. In fact, many of the troops had trained together during exercises on Phobos. Much more importantly, many of the leaders in these regiments had trained at the Confederation's Military Academy and knew each other professionally and as friends. The integration of the infantry regiments from seven different "countries" thus proceeded apace.

There were, however, differences between these regiments, as non-combat aspects of military service remained in the hands of the independent republics. The dress uniforms of the regiments were all different (battle dress protective suits were identical). In a parade held the day before the expeditionary force departed for the asteroids, there were seven different uniforms on the parade ground. No uniformity, but the colors were pretty. Worse, the pay system for each regiment differed from that of every other regiment. This caused no little anxiety and problems in morale. Confederation regulations stipulated a base pay for each grade, but the different republics taxed military pay differently and most supplemented the relatively low wage stipulated by Confederation regulations. New Provence, the poorest of the Martian republics, made no supplementary payments and its troops felt resentment toward the "rich" troops in the other regiments. Another difference was the training regimen for rifles and particle guns in each regiment. This made no practical difference. An infantryman could either hit a target with

---

4   Another provision of the Confederation's Constitution and of the constitution of each independent republic.

his rifle or particle gun or he could not, but the always pungent General Armani fumed over and over about how his boys and girls from Hellenica were ready, but that he was not so certain about the training in some of the other regiments. The two weeks of ground training before they left for the asteroids showed that some of the units had training deficiencies, but such deficiencies exist in any army, where one infantry battalion is better trained than another. The political die being cast, however, the armada set off.

The seven regiments boarded 38 very crowded Fleet vessels and set off for the inner asteroids. Everything went wrong, but everything always goes wrong when an army starts to operate.

The remainder of the 260 Fleet ships remained near Mars, to be fitted with weapons and electronics. None was ready when the armada left, but most would be ready for combat within six months. Their pilots grumbled.

Nothing yet has been said about the Rajastan regiment, because the Prime Minister of Rajastan refused to allow his unit to mobilize. "I have only the Imperial Guards" (an infantry regiment) and the Holy Warriors of the Raja (a combined arms regiment of infantry and engineers) to defend this republic. I cannot allow two-thirds of my fighting strength to be sent away and leave the republic unguarded". The President of the Confederation was afraid that ties—whatever they were—back to Primakistan were beginning to be felt on Mars. The next day, the President mobilized ten other regiments from the seven other republics and told the Prime Minister that he would use force, if necessary, to enforce the Confederation Constitution. "You must allow your Imperial Guard Regiment to be mobilized and transported for duty on Phobos, as required under your and the Confederation's constitutions." The Prime Minister knew all about the provisions that required the republic to render military assistance in a time of declared Martian emergency. Still, he argued, "You do not even know what the Primaki are up to". "That is correct, sir, which is why we need to get ready now. I need your Imperial Guard regiment to be sworn into confederate service. Immediately."

All of Mars waited. The huge majority was willing to fight the Rajastani Martians if necessary, though most also thought that the Rajastanis would submit in the end. Even those who had come from Primakistan, after all, were Martians now. Moreover, there was a large minority in Rajastan that had no connection at all with this earthly "Raja". The path for the Prime Minister seemed clear.

In the end, it *was* clear. The Imperial Guards were taken into confederate service and sent—some said in disgrace—to Phobos.

Accordingly, the President cut back the mobilization order from ten regiments to five. He also sent the Fleet Marine Regiment[5] to garrison Deimos. Rajastan's Imperial Guard regiment was given the duty of creating and running a training program that ensured the five mobilizing republican regiments were fully trained to fight. Once each regiment had completed training on Phobos, it returned to Mars and was placed at a strategic location—all in or very near their home republics—to be ready if there was to be any fighting on Mars itself. The President coordinated with the republics to accelerate—on a voluntary basis—the military training for the 31 other regiments not yet called into service. Most of these were not infantry units, but were engineer, military police, and a few old "field artillery" units. The political consensus was that the 14 regiments on duty (seven going to the inner asteroids, the Marines on Deimos, the Guards on Phobos, the five republican regiments called into confederate service) and three more to be described below were sufficient for the circumstances (note Figure 71). Moreover, because of their now-accelerated training cycles, if any or all of the other 31 regiments on Mars were needed, they could be brought into confederate service very quickly.

Name	Number	Station	Notes
Martian Expeditionary Force (infantry regiments from the independent Republics)	7	Inner Asteroids	
1st Marine Regiment	1	Deimos	
Rajastan's Imperial Guards	1	Phobos	
Infantry Regiments called into Confederate Service	5	Confederate Service	
Heavy Attack Regiments	3		Newly Created Marine Regiments
Other Republican Regiments	31	Home Republics	In Reserve (abiding by accelerated training programs)

Figure 71: The Martian Order of Battle

---

[5]   At this stage in Martian development, there was only one marine regiment. It was composed of both full-and part-time marines, much as the original marine regiment had been constituted back at the earth Embassy.

Finally, and most controversially for Mars, the president ordered a planet-wide conscription to create a new Confederation "heavy attack division". This unit, too, would be used to defend Mars if necessary. It was designed to fight with highly advanced weapons, resistant to the ultranet bomb and most other conventional weapons the Primaki were thought to have. The conscripts of the heavy attack division—organized into three new Fleet marine regiments—were also to be trained on Phobos, under the auspices of the Rajastani Imperial Guards. "Conscription" was not really necessary, but the President had insisted that it be called "conscription", to emphasize his right under the Constitution to order a Mars-wide draft. As it turned out, there were more than enough volunteers to fill the new units. The heavy attack division was organized within a month. Its commander, a rival of General Armani, expected the division to be combat ready within six months and at a peak of military efficiency in nine.

The War With Primakistan

Even with the fastest Martian ships[6], it would take General Armani a month to occupy all of the larger inner asteroids. Given the general's pomposity, it took him two months before the first one was occupied to his satisfaction. He was moving to his fourth asteroid when all communications with the expeditionary force were lost.

The only indication of what had happened was that large numbers of shape-shifting worms and viruses were received on almost every electronic channel on Mars. The attack caused significant damage on the Red Planet, but the basic Martian military defense system managed to stay intact because of its military defense hardware and software. The verdict about the expeditionary force, however, was clear. It would have been attacked by ten times the electronic force that had reached Mars. The expeditionary force's computer, command, communications, and other capabilities would have been destroyed.

There was panic on Mars. Politicians and military officers went on the ultranet-TV to explain that "loss of communications" was not the same as "everyone being killed". Likely, the regiments were now holding up on the asteroids. Not able to help Mars now, but not helpless. Still, the panic continued.

Two things now concerned the Fleet. First, its ships might be vulnerable to this or a variant of this "ultranet" bomb. Second, was that in three months, Mars itself would become vulnerable to attack from an inner asteroid that today was tens of millions of miles away. Either case meant that the Fleet had to win a victory

---

6   Two month ships, between Mars and earth.

or Primakistan—what ever it was that they wanted—would be in a position to dictate terms.

> *General Armani thought the fighting rather odd. Old-fashioned gunfights by one-tenth of his troops; everyone else awaiting the outcome, to see if they were to die. There were enough spacesuits for only some of his troops and most of those were being used by soldiers with old-style rifles. The mission had not envisioned so many thousands of Primaki soldiers or that so many asteroids would already have been occupied. Armani knew for certain that there was fighting on six other asteroids besides the one he was orbiting. He suspected that there was fighting on twenty more. Worst of all, he sensed that his troops were losing. It had not been the Primaki whose computers went down. They still had communications and particle guns. After the bomb, the Confederation troops had only rifles firing old-fashioned projectiles that jammed easily in the cold of space. The Martians gave as good as they got, but it was one-sided from the start. Armani felt the particle beam hit his ship, but there was nothing he could do. No more suits. No communications. His flagship starting to drift away helplessly from its asteroid. As he heard the atmosphere whistling out of his ship, Armani understood his fate …*

It made no *military* sense to bring the other 31 regiments into confederate service, but the President wanted to make a political gesture, so the orders went out. The President understood that everything now rested on the epaulets of the Fleet's leaders and crewmen. The President had been briefed about a few primitive radio signals from the expeditionary force. The reports were grim. Fighting everywhere. Most fights lost. What the hell did the Primakis want? Why was there no help from the US or the UN on earth?

Space is not like an earth nation's airspace. There is no "radar" that can see everything. No *Star Trek* sensors to detect a sentient life form ten light-years away. There were many Primaki space vessels that the Fleet *could* see, but the Fleet could not see all of them.

Division	Number of Ships[7]	Station	Mission
I	16	Phobos	Defend the moon
II	7	Deimos	Defend the moon
III	122	A hemisphere 20 million miles from Mars	Screen Forward to Detect the Enemy
IV	32	Martian Orbit	Defend Mars/Defeat Enemy Fleet
V	78	Low Martian Orbit	General Reserve/Defeat the Enemy Fleet

Figure 72: Defenders and Divisions

To deal with the fog of this war, the Fleet divided its ships into five divisions (see Figure 72). Division I was to "defend" Phobos. Division II was to "defend" Deimos. These were tiny divisions, almost useless if there was going to be a fight, put there to give some resistance if the enemy came at a moon suddenly. Most ships were in Division III, a screen of spacecraft on the edge of a globe of vacuum, one hundred million miles separating some of the ships. These were good, fast ships, but they were not the best in the Fleet. The last two divisions constituted the main fighting forces. Division IV was to defend Mars and Division V was to provide a general reserve, to meet the main enemy attack when it developed. The reserve pilots complained about the strategy, but they complained when they got paid, too.

Division III would alert the Fleet commanders to where the attack was coming. Divisions I, II, and IV would serve as the main defenders, to slow any attack, while Division V would move to defeat the enemy Fleet at the critical point. As a last resort, 50 nuclear tipped missiles were shot into a Martian orbit. They would be available to take out clouds of enemy ships, if they ever showed themselves.

The biggest surprise came from Prime Minister of Rajastan himself. He made a personal appeal to the Raja of Primakistan, to call off his attack on Mars, in the language he had spoken as a boy. The answer from Primakistan was cryptic: *"Help is coming"*.

Martians were getting suspicious of Rajastan's Prime Minister. *"More likely you are here to grab up half of Mars for yourself, you & #! @)$...."* The sentiments were widely shared on Mars.

---

[7]   Some ships not yet fit for military action.

Then, word came of Primaki ships fighting their way through that outer screen of Confederation ships. Two Confederation ships were destroyed and three Primaki ships were damaged. It was also reported that several hundred ships were now visible on Fleet sensors.

That is when Martians got another shock. The Prime Minister of Rajastan knew all about the Raja; it was because of the Raja's father that the Prime Minister had left Primakistan. The Raja's answer was not good enough for the Prime Minister and it was not good enough for the people of the newly independent Rajastan. There was nothing in the Confederation Constitution that *prevented* a Martian republic from making war on an earth nation and that is exactly what happened. The Rajastani government bought up a fast ship being built in one of its factories, armed it with three nuclear missiles and launched it toward earth. The Confederation President insisted that the ship be brought into Confederation service, as every other space ship on Mars had been brought into service. The Prime Minister ignored him. The Fleet got orders to intercept the ship, but there was no order to shoot it out of space. The Fleet watched it speed off toward earth. A Fleet ship—a very slow one—was ordered to pursue, but it quickly got so far behind the Rajastani ship that it would take weeks to catch up.

Division V was put into action. It moved toward the Primaki armada in coordination with Division IV, the defenders of Mars. About three million miles out, an intense ultranet bomb struck both divisions. This time, the Fleet was ready. Damage was substantial, but the Fleet was still fit to fight. Martian space technology then showed itself. The bigger, better-armed Martian ships shot up one Primaki ship after another. The Primaki ships were hardly interplanetary vessels at all; anything with rockets that could haul troops or cargo had been ordered into the Primaki fleet. Likely, a good percentage of that fleet had been lost on the way to Mars. After a two-hour battle, the Primaki ships scattered. Confederation commanders ordered a general advance to destroy the remainder of the enemy ships, but the mop up was called off the next day. Over three hundred Primaki ships were reported destroyed as compared with just six Martian ships. There is no way to know how many Primaki had been killed, but likely thousands or even tens of thousands had died, as compared with 916 Martians. Throughout Mars, the people stopped to rejoice. Immediately, a fifty-ship mission was sent off for the asteroids, to see what was left of the expeditionary force.

Just as quickly as the war had been lost, now it was won. Grumpy civilian pilots, in uniform against their will, knew it all along, of course.

Then, the war was lost all over again.

Six hundred Primaki ships swooped down on Phobos. It did not matter that half were destroyed before they got there. Tens of thousands of Primaki soldiers landed. The Imperial Guards, assisted by two regiments in training, fought for a week; a few hundred more Martian troops were ferried in during the fighting, but the result was predictable. Also predicable was the Primaki announcement that twenty-two nuclear missiles were aimed at targets all across Mars.

So what did the Martians think about that?

The Fleet moved in with every ship. The Fleet commander announced that he would shoot down any missile that was launched. The Prime Minister of Rajastan announced that his ship was nearing the earth and that he would destroy Primakistan's capital city if Phobos was not returned to Martian control in 24 hours. The Confederation president ordered an evacuation of all towns on Mars. Though the atmosphere only contained 4% oxygen and the CO_2 content still made the atmosphere poisonous to humans, the President was not going to be bullied into anything. People got out of the towns and into one of the thousands of scattered shelters and outlying facilities found everywhere on the planet. The President told the Primakis that Phobos was going to be broken apart with nuclear missiles if they resisted the Fleet marine regiment that was going to start landing in a few hours.

The US President announced that he was sending troops to assist the Martians and that his forces were on the highest alert, "prepared, if necessary, to strike at Primakistan itself". The UN Secretary General announced that she was prepared to seek Security Council resolutions, if Primakistan did not cease all aggressive acts against Mars. No one on Mars could fail to note the irony of these johnnies-come-lately, as neither act was likely to help at all in the crisis on Mars.

Ensign James McDonough, Platoon Commander, 2nd Platoon, Company B, 2nd Fleet Marine Regiment, volunteered to take the first six marines down in a shuttle pod. He was the only survivor, for as they were about to land, twenty nuclear missiles hit Phobos. It took the next wave of marines, coming in twelve hours after the nuclear attack, four days to find him and to pull him out of the wreckage. There were very few Primaki soldiers left to surrender to the marines.

Primakistan itself was not attacked. The ship's commander did not fire and the Prime Minister later denied giving any order to shoot the nuclear missiles. A court martial proceeding against the commander was begun, and then quickly dropped. The Fleet ship that had finally caught up with it escorted the "Rajastani" ship back to Mars.

Over two-thirds of the Martian Expeditionary Force had died in confederate service; half of Phobos' defenders had been killed. There were investigations for years, within the Confederation and by the independent republics, but few officers were ever brought up on charges. In the public's mind, the troops had simply been bushwhacked. Wrong? Yep. Bad feelings toward Primakis? You bet. Cashier leaders who went through that hell? No way; not that nice lieutenant from my republic, anyway.

Part V

Of Humans and Histories

... faith, coz, wish not a[nother] man from England.
God's peace, I would not lose so great an honor ...
Rather, proclaim it, Westmoreland, through my host
O, do not wish one [man] more!
That he which hath no stomach to this fight,
Let him depart, his passport shall be made,
And crowns for convoy put into his purse.
We would not die in that man's company
That fears his fellowship to die with us.
This day is call'd the feast of Crispian;
He that outlives this day, and comes safe home,
Will stand a' tiptoe when this day is named,
And rouse him at the name of Crispian.
He that shall live this day, and see old age,
Will yearly on the vigil feast his neighbors,
And say, "To-morrow is Saint Crispian."
Then will he strip his sleeve and show his scars,
And say, "These wounds I had on Crispian's Day."
Old men forget; yet all shall be forgot,
But he'll remember, with advantages,
What feats he did that day. Then shall our names,
Familiar in his mouth as household words,
Harry the King, Beford and Exeter,
Warwick and Talbot, Salisbury and Gloucester,
Be in their flowing cups freshly rememb'red.

This story shall the good man teach his son;
And Crispin Crispian shall ne'er go by,
From this day to the ending of the world,
But we in it shall be remembered—
We few, we happy few, we band of brothers;
For he to-day that sheds his blood with me
Shall be my brother; be he ne'er so vile,
This day shall gentle his condition;
And gentlemen in England, now a-bed,
Shall think themselves accurs'd they were not here;
And hold their manhoods cheap whiles any speaks
That fought with us upon Saint Crispin's day.

—William Shakespeare in *Henry V*;
Henry's speech to his troops on the morning
of the battle of Agincourt, where a bedraggled and
outnumbered English legion slaughtered
a haughty French army.

Frontiers and Free Men[1]

Everything that Yahweh had said, Samuel then repeated to the people who were asking him for a king. He said, "This is what the king who is to reign over you will do. He will take your sons and direct them to his chariotry and cavalry, and they will run in front of his chariot. He will use them as leaders of a thousand and leaders of fifty; he will make them plough his fields and gather in his harvest and make his weapons of war and the gear for his chariots. He will take your daughters as perfumers, cooks, and bakers. He will take the best of your fields, your vineyards and your olive groves and give them to his officials. He will tithe your crops and your servants, men and women, of your oxen and your donkeys, and make them work for him. He will tithe your flocks, and you yourselves will become his slaves. When that day comes, you will cry aloud because of the king you have chosen for yourselves, but on that day Yahweh will not hear you."

The people, however, refused to listen to Samuel. They said, "No! We are determined to have a king, so that we can be like other nations, with our king to rule us and lead us and fight our battles."

—*I Samuel*, Chapter 8

History teaches us that humans enslave each other. It has only been since the rise of the modern, western democracy that government has played even a small role in the prevention of political slavery. Indeed, through the ages it has been government itself that has been the primary supporter of and active agent for slavery. Unfortunately, the trend continues, if today in much more subtle form. Americans

---

[1]   Pardon here and in the title of the next section the use of sexist language for literary purposes.

may think themselves free, yet pay government's 50% tax on all they earn. And these are the taxes you see. As for the corporate taxes and endless volumes of government regulations, it bespeaks the master-servant relationship Samuel fore-told, not freedom.

It has not always been this way. Prior to 1776, and for a century afterward, Americans mostly escaped taxation. The American Revolution was about not paying small fees on imports and for government services. It has only been in the past 100 years that most people have begun to pay income taxes in the US. Yet today, there is no place on earth where one might escape the objectionable creep of government. No wonder some yearn for the open skies of Mars ...

We have not much treated a central idea of the mission to Mars. The idea of *frontier* and its impact on human societies is something that the 21st-century—too busy with 24-hour sports television and video games—has forgotten. Frontiers and freedom are so central to this business of Mars that an introduction in this volume could hardly be avoided. Indeed, ideas of frontiers and freedom capture the whole in a way that talk of Ark Transporters and Fleets and Confederations cannot. There are reasons to go to Mars. The real question is whether human beings in the modern West are free enough to be allowed to try to get there. The power brokers on earth will certainly work against this effort, for it not only means an expansion of the very idea of freedom, but—like the love you set free—it asks that you trust the unchained to serve a greater purpose.

A rather simple truth is that if you distrust freedom and a bold, new frontier, you will oppose the mission to Mars; and that if you love freedom and truly desire that people have a place to express themselves and to fulfill their dreams, then you will support the mission to Mars. In either case, your basic political values will show, for all the world to see.

Free Soil For Free Men

Frontier creates the conditions for freedom, but the condition merely capitalizes upon a spark buried deep in the genome, the spark that kindles freedom's fire. Poetical as it sounds, of course, frontier is not paradise. New territories imply primitive conditions, requiring a grubstake of forbearance and poverty that many today would shun. Indeed, in our own cable-television age, one might better imagine early Americans "going east, young man" from the mountains of New York or Virginia to the developed and more comfortable seacoast towns where street lights and refinement might make for a merry backdrop to while away a life. Of course, it was not gas lamps and fancy cloth that people of that era sought. What they sought was freedom.

In one of those old, tattered volumes, with heavy, heavy paper, now assuredly "out-of-date" for the Elitists, having been published in 1967, one American history text tells us:

"Old America seems to be breaking up and moving westward", remarked an Englishman who joined the throng. Some were Kentucky and Tennessee frontiersmen, restless spirits who had begun to feel crowded as their states became increasingly populous. Others were small farmers from the back-country of Virginia and the Carolinas who fled the encroachment of slavery and the plantation system. Still others came from the middle states, New England, and foreign countries … they took the turnpike to Pittsburgh or the national road to Wheeling and thus reached the river, or they took one of its tributaries such as the Kanawha, the Cumberland, or the Tennessee. Downstream they floated on flatboats bearing all their worldly goods. Then, leaving the Ohio at Cincinnati or at some place farther down, they pressed on overland with wagons, handcarts, packhorses, cattle, hogs.

Once having arrived at his destination, preferably in the spring or early summer, the settler built a lean-to or cabin for his family, then hewed a clearing out of the forest and put in a crop of corn to supplement the wild game he caught and the domestic animals he had brought with him.

These frontier folk knew loneliness and poverty and dirt, suffered much from the forest fevers and malnutrition, commonly had a lean and sallow look. Yet they were on the whole remarkably proud, bold, and independent. They were "half wild and wholly free."

Though a tiny percentage of the whole, there will be people aplenty eager to join the mission to Mars. They will ink a signature even as they are told that some of them will die along the way. Mars will not lack for occasions to mark the grave of an unfortunate who has met an untimely death. Still, people will come. For the real conquest of Mars will not come from new, shiny structures fastened into the red dirt, but from the grimy determination of people who decide to rebuild a city after a catastrophic fire or with an earth Embassy that launches another ship on time in the shadow of the wreck that had burned out only a few miles after launch.

Just as the Southern planter or the Northern industrialist did not settle the American west, so today's ESPN anchorperson will not sign up for the mission to Mars. Common sense informs us that many will consider the economic incentives, but that the overriding factor will be that which has always driven mankind: the urge to be free. People living in middle-class luxury with finely-coiffed hair, those whose generous salaries have dulled their senses, unable to recognize the siren song of encroaching taxes, government regulations, and political correctness, will not hear the mission to Mars' call. And the frontier will only *call*. It will require those with an ear for freedom to respond. Free land on Mars will be a double entendre. The land will be acquired for a marginal *financial* sum, but there will be *free land* on Mars because the Red Planet will be free.

Frontier's Effect On Individuals

We begin the discussion about the effect of a frontier on individuals by submitting the hypothesis that if there is no frontier, a society tends to stagnate and its people tend to become domesticated. The Einsteins, the Wyatt Earps, and the mountain men tend to drown in a life of boredom or, if they are unlucky, to be killed by agents of a stagnating society, unwilling to tolerate people who live outside the cultural norm.

The biography of Benjamin Franklin by H.W. Brands suggests a powerful illustration of the effect of frontiers on one individual. In most respects, Benjamin Franklin's place in history is diminished by his appearance on the stage of the American Revolution, where he is lumped together with other "Founding Fathers", as if that was where he staked history's claim. Had there been no American Revolution, had the US been set free by Britain in 1876 rather than having to fight for its freedom in 1776, Benjamin Franklin might have been remembered as the American Leonardo. Not only were his business exploits magical in themselves: a highly successful printer and editor of *Poor Richard's Almanac*, but he was a scientist and inventor of the first order. The stove, the glasses, the lightning rod, the work in electricity, afforded him entry into Britain's Royal Academy, as it did to Isaac Newton, Joseph Priestly, and other names of science's pantheon. Capping this marvelous resume are Franklin's less well-known political exploits as a loyal Briton and citizen of Pennsylvania. He was the author of a proposed union of the British colonies, a movement that led to a "continental congress" in Albany and a written plan of union that was submitted to the British Parliament. He was also the founder of the Pennsylvania militia and served as a "commander" in the field when defenseless backcountry settlements were being attacked during the French and Indian War. Less important, but still noteworthy is how Franklin created America's first public library by setting out a subscription for members and how he founded the first fire company in much the same way. In virtually every endeavor, Franklin was a man of amazing resourcefulness and success, very much imitating da Vinci of a few centuries before.

One of the most moving sections of Brands' biography is the story of Franklin's visit to the village of his father. Josiah Franklin was a man mostly estranged from

Benjamin, not for reasons of the spirit, but for reasons of the mind. The one was a staunch Puritan; Ben was a freethinker and wedded to the Age of Reason. In some respects, Ben can be imagined wondering whether Josiah was his real father at all. Then comes the visit to Ecton in Northhamptonshire, where Ben's father grew up, and Franklin's introduction to his uncle, Thomas:

> *Thomas Franklin … was a conveyancer, something of a lawyer, clerk of the county courts, and clerk to the archdeacon in his visitations; a very leading man in all county affairs, and much employed in public business. He set on foot a subscription for erecting chimes in their steeple, and completed it … He found out an easy method of saving their village meadows from being drowned, as they used to be sometimes by the river, which method is still in being; but when first proposed, nobody could conceive how it could be; but however they said if Franklin says he knows how to do it, it will be done. His advice and opinion was sought for on all occasions, by all sorts of people, and he was looked upon … as something of a conjurer.*

Our last detail is the most thought provoking. Thomas Franklin died exactly four years before Benjamin was born. William Franklin, Benjamin's son, "already struck by the similarity between his father's career and his great-uncle's, commented that had Thomas died four years later, those who knew the two might have supposed a transmigration of souls."

It is probably not too much an exaggeration to suppose that if instead of a transmigration of souls, there had been a transmigration of bodies, the name *Thomas Franklin* would be renown and that of Benjamin Franklin lost to us. Frontier had allowed one man to play on the larger stage of human history, while the cramped girdle of British society had forced Thomas to be the great man of a tiny village. In the freedom of frontier America, Benjamin was able to transform the course of an entire nation; in the glens of England, Thomas was only able to transform the course of a tiny stream.

The Old Santa Fe Trail

An informative interaction between individuals and government can be seen through an examination of the years before the American Civil War, when American traders from Missouri began to trek west to Santa Fe, a largely forgotten province of Mexico. A lucrative trade was growing up—partly because New Mexico was so isolated from the rest of the Spanish Americas—but there were many obstacles to using the trail. Among them were the depravations of renegade bands of American Indians, the lack of trading and other facilities along the Santa Fe Trail, and the lack of drinking water. This excerpt from *The Santa Fe Trail* by David Dary explains the actions and the reactions of the people wanting to use the Trail:

> *By the spring of 1830, Missouri traders knew the federal government would not provide military escorts for the Santa Fe-bound caravans. With the Indian attacks of 1829 still fresh in their minds, the traders preparing for the spring journey to Santa Fe realized they would have to protect themselves. From experience they knew that if they traveled in one compact, well-armed body and corralled their wagons at night, they could protect themselves from Indians. Though their organization was almost military, it was very democratic. Each man had a vote, and the officers of the caravan were chosen by open balloting. First, the men elected a captain of the caravan, nearly always someone experienced in trail travel, often a wealthy trader who knew the ways of the Indians. Next, the men elected a lieutenant for each division of the caravan, the number of divisions depending on the caravan's size. Each lieutenant's duties were to ride in advance of this group of wagons and inspect the road and the crossings and warn the teamsters driving the wagons of rough spots on the trail. Each lieutenant also supervised the corralling of wagons when the caravan camped each evening. One authority observed that the first wagon to reach a campsite would park at an angle. The second wagon would then pull up at the same angle, next to the first wagon, stopping with its near hind wheel against the front wheel of the first wagon. This process was continued until the enclosure was completed. "It was sometimes in the form of a square—one division to each side if the caravan was composed of four divisions. But it was as often in a circle or an oval. The wheels were frequently chained and locked solidly together." Thus was constructed a sort of temporary fort or stockade. In case of attack it afforded a defense, and the animals were sometimes driven into it. The encampment was made where the grass was sufficient for the animals of the caravan. Guards were always set at night, and every man was expected to take his turn at guard-duty.*
>
> *In addition to the election of a lieutenant for each division, a clerk, three judges to try offenses, an officer of the guard, and even a chaplain were elected along with someone to guide the caravan ...*

The Missouri traders did not rely upon the far away federal government and they could not rely on their own elected officials in Missouri, who had no jurisdiction in the unorganized parts of the Louisiana Purchase. Had there been bureaucrats and Leftists in those times, no doubt arguments would have been made that the traversing of this unorganized part of America and into "Mexico" was illegal and that it should be halted[2].

People will always flock to places where there is work, freedom, and opportunity. Thus, the Missouri traders developed their own system to traverse the harsh lands of present-day Kansas and New Mexico. Their quasi-military organization allowed for easy defense against renegade American Indians, created a primitive infrastructure for finding and distributing water, and enforced discipline to facilitate 900-mile journeys by wooden wagons carrying two tons of cargo across roadless lands. The development of this "popular shadow government" enabled traders to make repeated, lucrative, and mostly uneventful trips back and forth between Santa Fe and their homes in Missouri.

The lessons of the Santa Fe Trail also speak to us about how difficult and perhaps futile individual corporate plans—"private ventures"—into space might be. The caravans that crossed Kansas were highly organized and quasi-military expeditions, allowing people to achieve goals that otherwise would have been unattainable. People of the era understood that individual traders alone on the trail would have been killed by renegade Indians, died of thirst along the way, been utterly incapable of repairing broken wagons, fording treacherous rivers, or otherwise have been at the mercy of chance events that a larger party could overcome.

More importantly, the lessons of the trail also teach that disciplined efforts by large groups of people, organized to overcome the obstacles before them, can make otherwise impossible tasks ordinary. Traders did not await developments in Washington to organize themselves for journeys to Santa Fe. They organized themselves and created a 19th-century infrastructure capable of achieving their personal and financial goals. Their solutions and modes of travel were both inexpensive and demanding on those willing to make the journey. After an infrastructure was put into place along the Santa Fe Trail—later enhanced by military activities, government-sponsored portages, and government-subsidized railroads—smaller groups and individuals could make the journey in greater comfort and even in style.

Are humans today so much less brave, less motivated, and so wedded to comfort and luxury that they could not band themselves together to build a bridge to Mars? This book's premise is the opposite. History teaches us that the much less free and more socially dependent people living in New Mexico[3] did not develop

---

2   The borders of the Louisiana Purchase and the political status of the land outside of it were unclear until after the Mexican War of 1846–1848.

3   It is beyond the scope of this book to enumerate the contributions of the Spanish-speaking New Mexicans to the development of the Santa Fe Trail. Indeed, their contributions were immense and not well publicized. The matter, however, is one of

the Santa Fe Trail. Rather, newly freed, fiercely independent Americans, driven by reasonable visions of adventure and economic gain, developed these links. The mission to Mars rests its hopes upon the idea that brave and independent people can still be found today to develop a 21st-century version of the Santa Fe Trail, get themselves to Mars, and to build there a rich and fulfilling life.

The Effects of Frontier on Economic Life

One of the clearest ideas the mission to Mars proffers is that the creation of a frontier on Mars will create opportunities beyond the imaginings of most people; opportunities that will become realistic possibilities for people who otherwise might live out their lives in Third World misery or First World ennui. The idea is simple: new lands will present resources previously untouched by mankind. Whether mineral, entertainment-oriented, or the soon to be created ecosystems, these resources will create unparalleled opportunities for those arriving to populate this next frontier.

This section excerpts several ideas from the book *The Ordeal of the Union, 1852–1857*, by Allan Nevins, where we use the example of life in *ante bellum* America to suggest what will occur on Mars.

> *Huge vacant areas were to be settled, [but] nobody doubted that business would successfully exploit the wealth of the continent. The imperial domain of farmland, cattle ranges, timber, and minerals roused the appetites of a people already materialistic, and quickened the hopes of a society already sanguine ... Year by year, the advance of technology and invention strengthened the country ... [T]he underlying forces of the industrial revolution were simply irresistible. The wealth of natural resources; the energy of the people; the flow of invention; the provision of cheap labor by heavy immigration; the supply of capital by savings; gold-and-silver discoveries, and European investment—all this combined like a chain of bellows to make the forge roar.*

---

degree. The political rights of the New Mexicans were not protected by an effective constitution and the hierarchical nature of the powerful Catholic Church supported a society that was highly stratified into rich and poor. New Mexican rich had little incentive to risk their lives to develop trade with the Americans and the wretched poor had no means, until the trade with Missouri began and they could find work as muleskinners or cooks. Clearly, Americans developed the trade, even if New Mexicans later contributed mightily to its success.

Only the cynic would argue that Mars' currently inhospitable climate would block the repetition of similar processes there. Of course people will need to wear pressure suits. Of course plants and animals will need to be introduced over the run of several centuries. Once the temperature and atmospheric pressure has been adjusted, neither of these propositions need be overwhelming and, indeed, the reality will be that people will not be kept off Mars' surface.

Now consider the impact of "novel conditions" to be found on Mars on the very process of invention itself.

> *Invention seemed an American pastime. By 1850 American sewing machines, telegraphs, reapers, Colt revolvers, vulcanized rubber, pressed glass, and circular saws were famous. On each a whole industry had been founded. In 1854 the number of patents was one-fifth greater than in 1853, and in 1856 it was one-third greater than in 1855 ... [T]he United States led the world in the invention of farm machinery ... American plows, cornplanters, wheat-drill, revolving rakes, cultivators, and other horse implements were excelled in no part of the world ... All of this American ingenuity was not difficult to explain. For one reason, while most workers in old countries merely perfected the patterns used by their fathers, the novel conditions of the American scene encouraged workmen to strike out unconventional patterns. The variety of circumstances met in the raw continent tested men's resourcefulness, and the free atmosphere of social and economic life placed a premium on boldness.*

It would be trite to point out that necessity is the mother of invention, but trite does not imply anything other than a writing style. The truth cannot be avoided that where there is a need to improvise, there will be an explosion of inventions. It was true before in history and it will be true again on Mars.

Social Stability In A Pioneering Setting

This mission to Mars argues that its social fabric, too, will have advantages over that of the modern west. One of these advantages will result from the kind of pioneering society that life on a frontier will require. In small communities—in a society with a relatively low population—relations between people will differ from those in the more anonymous societies of the modern West. A close analogy might be to say that life at the mission to Mars will resemble life in a small town, though even this is not accurate, since small town life today rarely is on the cutting edge of historical and cultural progress.

At the earth Embassy, and especially on Mars, relationships between adults will not allow for the kind of immature behavior that is often associated with American "freedom". Whereas freedoms pair with responsibilities, irresponsible behavior denotes license, not freedom. We examine here two examples of immature behavior. Of greatest interest, perhaps, will be that marriages will not fall apart for the trivial reasons they sometimes do in the US. The social pressures to "get on with things" on Mars will rarely allow a couple to split up because they have decided they "have fallen out of love" or succumbed to some other psychobabble excuse for divorce. Stated more positively, decisions about interpersonal relationships will be shaded by the important subplot of settling Mars. And then there is the old standby: in a society that may at first have many more males than females, if I lose this wife I might not be able to find another.

Social responsibility toward one's family is also likely to be reinforced in Mars' pioneering society. Opportunities for young adults to live on their own may not be quite so prevalent as they are in the modern West. While there will be high wage jobs, "housing shortages" are likely to exist for many decades. This does not mean there will not be enough homes, but simply that the construction of new homes and apartments so that young males can have "bachelor pads" will be a lower priority than other kinds of infrastructure. The same will be true for unmarried women. Thus, the likelihood on Mars that two and even three generations will live together in the family "homestead". Some accommodations for singles will occur at Martian universities and on job sites in new villages, but these will be, respectively, dormitory situations and makeshift living arrangements in transient camps. Again, the nature of life on Mars will create social incentives for families to live together to an extent that is uncommon in the US. When America had a frontier, families tended to live together, sometimes even after a child was married. The same will be true on Mars. And given the weight of evidence that people in family settings tend to be happier, the result is likely to be a healthy Martian psychology.

Life in a pioneering society is clearly not for everyone. The ability to be irresponsible—so common in the US—will be much less common on Mars. As those who understand that being free requires that people take responsibility for that freedom, so the free society on Mars will exhibit both unprecedented opportunities and a requirement that people live their lives responsibly. Still, the creativity and the unleashing of human potential means that people on Mars will solve problems in ways that will amaze most people on earth. And, if history is any teacher, the degree to which better solutions are found and how larger problems are solved will be beyond our poor ability to imagine.

The Mission to Mars Answers Yali's Question

"Why is it that you white people developed so much cargo and brought it to New Guinea, but we black people had little cargo of our own?"

—Yali's question to Jared Diamond in *Guns, Germs, and Steel*

We follow hard on the heels of a discussion about the effects of frontier on individuals and small groups to move to discuss the effects of frontier on entire cultures. We turn now to a book called *Guns, Germs, and Steel,* Jared Diamond's Pulitzer Prize winning effort that offers hypotheses about the development of human societies. In it, the author makes a number of interesting and insightful observations, among them that an answer to Yali's question can best be understood in terms of four factors: 1) the availability in any given area of native plant and animal species for domestication, 2) the geography that either favored or hindered the spread of new ideas and technologies, 3) the relative positions of a continent that also hindered or helped spread technology, ideas, and domesticated plants and animals, and 4) the size of a continent and its population. These assertions are not new. What may be new, however, is the relentless logic that Diamond brings to bear on concepts already well accepted.

While acknowledging the value of the book's ideas, the mission to Mars would offer this additional insight into Diamond's work: that the existence—or absence—of a frontier also contributes mightily to a proper answer to Yali's question. This chapter explains why this idea of frontier is so important and how it relates to the mission to Mars' major themes.

Mr. Diamond's Factors

Diamond's answer to Yali—and a major thesis of his book—is that *all* successful societies travel essentially the same path[1]. Europe has "cargo" because it got a head start. New Guinea had little "cargo" because its culture "stagnated" for many thousands of years. Given time, all successful societies end up at about the same place, even if some have steeper courses to run.

The first part of *Guns, Germs, and Steels'* argument is that "white people's" (Eurasians') ancestors had access to a much broader suite of wild plants and animals receptive to domestication than did the people of New Guinea. The more types and varieties of plants and animals that could be domesticated and turned into crops and herds, the more a given culture could survive droughts, depletion of hunting grounds, and other calamities.

Diamond's second point cites the enormous role of geography either to assist or to hinder cultural diffusion. In Yali's case, the relatively small size of the island of New Guinea and its isolation from the rest of humanity afforded those living there little opportunity to advance beyond the Neolithic, until they were suddenly forced into the 19th- and 20th-centuries with the arrival of westerners and others. Similarly, China's political and economic life was stunted by the vast deserts, steppes, mountains, and oceans that separated it from other cultures. In contrast, Eurasian culture benefited from the geographical connectors between India, Babylon, Egypt, and Europe, etc.

The third point is a little more difficult, but is in many ways a restatement of the second. Diamond concludes that Africa and the Americas were slow to develop due to their north-south orientations. He reasons that since climates tend to be similar along east-west orientations, people are generally able to expand their culture[2] along an east-west track. This is not the case for north-south orientations, where climates are much more likely to change and where—because of climatic differences—domesticated plants and animals may not thrive. Early societies are unable to adapt to such changes because it requires the abandonment of familiar patterns of agricultural and social organization. Diamond cites the great empires of equatorial Africa and of the Incas as two societies that could not expand outside of their original homelands. Even these relatively advanced societies could not overcome the north-south orientation of their respective continents because insurmountable climatic barriers existed to the north and south of their empires. In contrast, climate changes from Spain to India, though at times dramatic, were

---

[1]   The definition of successful is "a society that continues to survive".

[2]   Because of climatic stability, infant cultures are better able to transfer their suites of domesticated plants and animals from one climatically similar locale to another.

nonetheless small enough to allow the transmission of cultures and ideas from one area to the next.

The final factor is self-evident. The size of the continent and a continent's population would dictate how much social experimentation and progress could occur. In New Guinea, with its relatively small size and population, little social experimentation was possible. In contrast, Eurasia, being the earth's largest land mass and having the most people, was able to experiment continually and with great success.

The advantages provided Eurasian societies allowed them to travel further down the paths of civilization and of "cargo" than those of New Guinea.

Many points made in *Guns, Germs, and Steel* can be debated. Its biggest problem is that its argument appears many times to be circular: these cultures were "successful", so Diamond argues that certain attributes of that culture were important. Other cultures were slower down the path of development or, indeed, succumbed to other cultures, so he seeks common points in these cultures. Correlation, however, is not always causation[3]. The cause and effect linkages suggested by Diamond sometimes seem tenuous. His logic might suggest, for example, that England (only recently "in touch" with civilization) should be primitive and that Egypt (long at the center of civilization) should be extremely advanced—a point that this chapter's discussion about *frontier* hopes to address.

The inclusion of such a long analysis of ideas about human history would have little relevance in a book about Mars, were it not a good introduction to the interaction of frontiers and human progress, a point which Diamond addresses only implicitly in his work. The England/Egypt dichotomy mentioned above can be explained as a question of *frontier* and which culture had access to it. The mission to Mars takes the position that—all else being equal—it is access to a frontier which allows one culture to advance and that the lack of such access implies stagnation.

Frontier

The importance of frontier for the happiness of a population and the progress of a culture is a thread running throughout the mission to Mars. This major idea is very much like most of this work: some will grasp an idea intuitively, some will require a modicum of persuasion, and some will dismiss the idea outright. The purpose of this admittedly non-authoritative section is to present the idea for those who will immediately intuit it and to attempt to persuade a few others.

---

[3] The classic correlation between shoe size and knowledge. It does not follow, however, that small infants are less intelligent than adolescents or adults.

The social and cultural mechanics of a frontier are simple to understand. A frontier offers a challenge not to be found in the home society. People of ambition, those without social standing or skills, and those who see the economic or political opportunities not available at home all tend to gravitate to a frontier. Thomas Paine[4], for example, a failure in everything he tried in England, became a great success in America—England's frontier. Like Thomas Paine, people in a frontier manifest possibilities that do not exist in more settled areas. Arnold Toynbee, a great historian of the 20th-century once wrote that "Civilizations ... come to birth and proceed to grow by successfully responding to successive challenges. They break down and go to pieces if and when a challenge confronts them which they fail to meet." When there is a frontier, there is a buffer zone that both invigorates a culture and protects it from too sudden a contact with new challenges. When a frontier is lacking, or is an insufficient buffer, the results can be as severe as the overthrown of Egypt by the chariot-driving Hyksos.

A frontier offers possibilities and a dynamic not only for individuals, but for entire societies. Thus, access to a frontier serves both as a place of challenge to mold stronger individuals and also as an agent to mold stronger cultures. Established societies benefit from contact with a frontier both because the harsher aspects of frontiers challenges a culture to adapt to these harsh circumstances and from the distinctive opportunities they present. Figure 73 lists four major cultures from history and suggests that nations with a frontier tend to out-compete those without one.

Name of Civilization	Date	Competitors Who Lacked Same Degree of Contact With Frontiers		
Babylon	1500 BCE	Egypt	India	China
Rome	400 BCE	Egypt	Babylon	Greece
England	1750 CE	Italy	Germany	Austria
America	1880 CE	England	France	China

Figure 73: Cultural Success Spurred By Frontier

Building upon this idea, we see in Figure 74 that once these cultures began to lose contact with a frontier, a period of cultural decline began.

---

[4]   American Revolutionary War era author of *Common Sense* and *The American Crisis*.

Name of Civilization	Date Frontiers Disappeared	Date Cultural Decline Began
Babylon	1000 BCE	500 BCE
Rome	200 CE	400 CE
England	1900 CE	1950 CE
America	1900 CE	?

Figure 74: Connection Between Cultures and Frontiers

We have historical information about the gradual decline of Babylon, Rome, and the British Empire. Observers of contemporary America can form their own conclusions.

Diamond himself offers the subplot about frontier through his discussion of cultural diffusion. When cultural diffusion is easy—by virtue of geography—a culture can breathe and grow. When it is not, the result is invariably cultural stagnation. Thus, a frontier like ancient Palestine might receive the blessings of civilizations arising in Babylon and Egypt, but also generate opportunities for traders and new ideas. For example, when Palestine was a "frontier" (before Alexander arrived) it connected Babylon, Egypt, Greece, and the growing colonies in the western Mediterranean. And though a frontier, it was Palestine that provided—providentially or not—an idea of fundamental importance to western civilization: monotheism, an idea that was not fostered by the more "civilized" areas of the Fertile Crescent.

Going back to Figure 74, Babylon remained a vibrant society until the Greeks, Persians, and other surrounding societies in contact with frontiers overtook Babylon. Rome, once a society on the periphery of the Eurasian ecumene, became its master. And when Gaul, Spain, and North Africa were made into cultural imitations of Rome and cut it off from frontier, Rome declined. Similarly, the British were a powerful, even supreme, force in world affairs until their access to frontiers dried up during World War I.

Frontiers have also been in the forefront of human progress and especially human political freedom. As with the ideas of economic and cultural vibrancy, the gradual accretion of ideas about political freedom seem to have been stimulated by contract with a frontier. Britain was still mostly a frontier, far from the centers of civilization in Rome, Constantinople, and Jerusalem, when the *Magna Carta* was penned. Certainly, few readers will need to hear the story of American colonies, thousands of miles away from "civilization" in France and England, fighting to create the world's first government truly "of the people".

Today, frontiers no longer exist on earth. There may be not a few uncivilized locales and "economic frontiers" may remain, but the traditional frontier that has fed the engine of human progress vanished at the beginning of the twentieth century. It was then that the American west lost most aspects of being wild, when much of today's Second and Third World began to gain standing as nation-states, and when the ideas of European imperialism lost most of their savor.

The mission to Mars proposes the idea that human society as a whole will decline unless it can regain the safety valve of frontier. Humanity *has never before* lacked some kind of frontier. While there has been great economic and other progress made during the last hundred years, the question remains how long such progress can continue. At least as important is the question of how much freedom can survive in a world burgeoning with people and increasingly populated by groups willing to use cutting edge technologies to annihilate their enemies. Modern American offers a wonderful example of a society whose freedom has been alarmingly eroded by the complexities arising from population growth and from a war against extremists who will not hesitate to use nuclear, chemical, or biological tools against it.

In the 21st-century and beyond, there may be a valid need to regulate nearly every aspect of the American economy. Such regulation takes on a different perspective, however, if it must occur against a canvas of a Martian economy free of unnecessary regulation. Again, the dynamics of frontier. If light bulbs on Mars are made without the need to regulate every aspect of their manufacture, light bulb tycoons in the US and elsewhere on earth have a powerful argument against those who insist upon regulating them.

Human colonies in space and especially on Mars will serve as a counterweight against "all the totalizing systems of power" as Robinson put it. While it is certainly not clear that democratic institutions and ideas about individual freedom will be replaced on a no-frontier earth, today's trends are mostly unfavorable. At the very least, the "totalizers" tend to drown minority views, blur cultural and regional distinctions, and otherwise homogenize human culture in favor of broad dictates from the top. A mission to Mars society on the Red Planet will give humans a real choice: continue to live in societies bent on ever more regulation *or* in a rough and tumble Martian society of almost limitless freedom. As was the case with the light bulb manufacturers, so legislators, bureaucrats, and other "totalizers" will be forced to take this choice into account every time they contemplate the need for additional restrictions, regulations, and taxes. Indeed, the mission to Mars argues that the greatest reason to support its program is *not* to help manifest the desire some small numbers of humans may have to get to the Red Planet; but rather to create political and economic vistas on a Martian frontier so that freedom can be preserved against those who—however well-intentioned—foster the growth of "totalizing systems of power" on earth.

Back to Yali

We now find ourselves equipped to answer Yali's question in a way that both comports with Jared Diamond's basic ideas and enlightens it with ideas about frontier. Diamond's response is, in most respects, a simple one. As *every* successful human culture follows the same basic path, those that have more "cargo" simply represent themselves to be further along the path that human societies trek. Agree or disagree as you like, accepting Diamond's idea *arguendo* allows one to consider how often societies move further down this path of progress in relation to that society's proximity to a frontier. In Yali's specific instance, the British, Australian, and American "white people" he noted laden with "cargo" were all the products of societies freshly seasoned by a frontier. Yali's "black people" of New Guinea, in contrast, were the product of very traditional societies, cut off by geography from the beneficial effects of cultural diffusion. For not only is New Guinea a rather remote island, but the unique topography of the island makes overland travel almost impossible. The people of New Guinea thus lived countless genera- tions in societies where "the world" was the 1,000 square miles surrounding one's own village and where towering mountains, the ocean, or nearly impassable rivers separated one New Guinean society from another. In essence, the geographical obstacles on New Guinea prevented the existence of a "frontier" on the island to challenge and to stimulate societal progress. Thus, travel along the path of civilization was slow for all of New Guinea's societies.

Naturally, knowledge of the cultural evolution of New Guinea is not a pre- requisite to moving humans to Mars. What the mission to Mars is proposing, however, should be understood not only in technological, but also in sociological terms. Thus, the efficacy of Yali's question. Thus, the power to be derived from the creation of a new human frontier in space and on Mars, open to all humans. Mars daydreamer books can concentrate on the technical, since they all but admit that consideration of human and cultural questions are irrelevant to their never- to-be-realized proposals. As the mission to Mars expects to get to the Red Planet; as it expects to establish a human society there; and as it expects to sponsor a new, egalitarian frontier available to all humans, massive sociological implications linger on its horizon. How this may play out on the grand stage of human history is a subject worthy of consideration. For if these ideas about frontier have merit, then history's new stage will be extraordinary, as Martians compete against no- frontier societies on earth.

Iron-Willed People in Wooden Ships are Going to Mars

I offer neither pay, nor quarters, nor provisions; I offer hunger, thirst, forced marches, battles and death. Let him who loves his country in his heart, and not with his lips only, follow me.

—Giuseppe Garibaldi, Italy's "George Washington"

Someone once said that the Americas were settled by iron-willed people in wooden ships. Contrast this with modern "challenges" where we have iron-hulled ships, but people of wooden character. The American space program presents less risk to its tens of thousands of employees than what they face getting onto a freeway during the morning commute. The West has the technology to do almost anything imaginable, but the will to use that technology can be broken if there is any interference with Sunday football games or having two SUVs in the drive.

One of the more remarkable human dramas, albeit one decidedly off today's Greatest Hits List, is the story of the Spanish conquistadors. In twenty years, these men conquered one of history's largest empires and set the political and cultural foundations for South America and much of North America, foundations which have lasted for five hundred years. Only the Pope, who expropriated Brazil, and the US, who acquired a small part of this land through war and purchase hundreds of years after the conquistadors, reclaimed any part of this land taken for the Spanish crown. The conquistadors are not to be envied for their human values, but that is not the point. Rather, the lessons of the conquistadors are found written on the hearts of anyone who dares dream of wondrous deeds. The rapidity of their expansion, the solidity of the foundations they laid, and the raw determination of these men pay tribute to an inner strength that human beings possess, even if rarely visible today. This is not an accolade to Rambo or

Schwarzenegger, but to the idea that otherwise normal human beings can drive themselves to great achievement. Our television-measured generations can only *imagine* these conquistadors: an entire class of people achieving *in the real world* what a few modern Olympic athletes achieve in an artificial world of weights and running and training. The reader can write the conquistadors off as men chasing whispers of gold or heavenly glory, but that also misses the point and, frankly, reveals a hollow spot in the reader. Indeed, most of us *norteamericanos* trivialize Pizarro's legacy by repeating non-historical platitudes like: "Pizarro conquered the Incan empire with 300 men"[1]. But here historical inaccuracy paints a deeper truth. The Spanish and their will to succeed revealed a zeal that most people living in the early twenty-first century only *suspect* exists. As the mission to Mars works to settle a new world, it intends to rediscover the reservoir of human spirit that fueled Pizarro and his men, as it writes a new history for mankind.

The mission to Mars will reprise the old idea thus: iron-willed people in wooden ships are going to Mars. The implements the mission to Mars will use to settle the Red Planet will not be the stuff of an antiseptic NASA. The mission to Mars will seem "wooden" as it creaks along with unsophisticated vessels and simple plans. There will be no triple-contingency back-up systems. And certainly some of the dozens and scores and hundreds of people who arrive on a cold and storm-swept planet will not die in a warm bed. There will be accidents on earth, Mars, and in space. These accidents will be unforgiving, but the iron-willed people of Mars will not be deterred. They will close ranks and move forward, sweeping away all obstacles before them. They will remember that one-third of the passengers on the Mayflower died before reaching the New World and that another third would not survive their first winter. They will remember that those who lived founded Plymouth Colony, which became the Massachusetts Bay colony and later the US. Those going to Mars will remember the slave ships where hundreds were chained below decks with little water and almost no food, where men and women would lie chained next to corpses for days or even weeks, and where, despite the horrors, people lived to spawn new generations in the New World. Those going to Mars will remember the Chinese immigrant to San Francisco, who braved Saturday nights where it was a pastime, not a crime, to kill a Chinaman, and will remember that those brave men sought simply to earn money to help poor family members in China. They will remember all these things, as they build a new and better world than the one they left.

If they could do it, I *can do it.*

---

[1]   As valid as the idea that Christopher Columbus discovered America.

The psychology of all this can be distilled into just a few ideas. The first idea is that tough-minded people—or people who discover that they can be tough-minded—will decide to go to Mars to live. These people will turn their back on societies seemingly imported from 18th-century France, with its media-aristocracy, measuring people by their heredity, wealth, good looks, or fame. People will go to Mars to escape the MTV-Hollywood-New York-snob world the West has become. They will go for the same reason young Americans join the US Marines: to see if they are tough enough. They will go for the adventure or for God or to make money the old-fashioned way: to start a business in a rough-and-tumble world growing at 10% per year. They will go, however, with a silent understanding that it will be rough. And they will want it to be rough, at least for a while. They will not be disappointed. Ultimately, however, when they find themselves questioning their own sanity, perhaps during month four of a six-month-no-turning-back trip through space, these people of iron wills will resolve to move forward. They will go forward and they will conquer a new world. Though some will die, Mars will be conquered.

A recent movie portrays an American Special Forces mission in Somalia gone wrong. *Blackhawk Down* depicts young American Rangers and grizzled "D-Boys" in desperate firefights against an enraged, well-armed populace in Mogadishu, Somalia. After a dozen and more Americans are killed and seemingly everyone wounded, the men are pulled back and pulled out of Somalia. These men—these iron-willed men—were tested and found worthy of the challenge. Indeed, the Ranger and Delta Force soldiers wanted to go back to "finish the job" they had started. Unfortunately, in today's "How will this look during tonight's news cycle?" world, men of wooden wills decided the dice roll that killed 19 Americans was not worth the game. Wooden-willed men pulled the plug on the men of iron, allowing the dead and wounded to have bled in vain. These men of iron are the type that will go to Mars, not armed with grenades and mini-guns and SAWs, but armed with the implements of a determined peace that will conquer Mars. They will leave the wooden souls for NASA, congressional budget hearings, and weather satellites.

Martian pioneers will brave the horrors of generic shampoo, razorblades that are not sharp, and tampons that bulge and bend. They will not have sporty new automobiles, indeed may have only horses or supercharged golf carts. They will eat food that is sometimes bland and will sometimes have "worse-than-HMO" healthcare. Still, these iron-willed people will accept their wooden ship, knowing the reward to be worth the effort.

They will know that the argument in favor of going to Mars is much stronger than any argument in favor of going from Europe to the New World in 1500. In the case of the US, it took a little over two hundred years before there were hundreds of thriving, self-sustaining communities. There was little hope that a settler would live a life anywhere near as grand as that found in one of Europe's large cities. Even the vaunted hope of a new, freer life in America would have been something promised, not something assured, unlike the cobblestone streets, doctors (however primitive), pubs, and churches found everywhere back home. In fact, the settlers would have likely been guaranteed only that the voyage across the sea would kill many of their number and that, once in America, the natives might try to kill the rest. Mark this section and return to it as you read later chapters and volumes. See if the chances for individual happiness and success will not be far greater under the mission to Mars than it was for those long ago wishing on an American star.

Settling a new world will not be for most people. *Everything* about the mission to Mars will be difficult and, unlike the rich playboy investing a summer to ascend Mt. McKinley, there will be neither quick reward nor a speedy return to lives of luxury or suburban tranquility. Indeed, some will die prematurely as they attempt to settle Mars. Note, however, that the specter of death did not stop Europeans from going to the Americas, nor did it stop European women from following kooky husbands into the unknown. The mission to Mars will not be easy; people will be uncomfortable, some will hate their bosses, and most will be forced to work long, tiring, mindless hours for a goal that awaits them only in the far future. As every couch-mellow will attest, difficulties and physical risk will stop most people; but it will not stop the motivated Martian pioneer. The mission to Mars will not be a fully funded, career-enhancing trip as are our current efforts in space. Rather, the decision to settle on Mars will involve a great deal of work, huge amounts of sweat equity, and the possibility that something will go wrong, to earn the potential settler a quick death. Such a life is not for everyone. The mission to Mars is not, however, about sending wooden people in iron ships, but vice versa.

The mission to Mars has an un-Hollywood-like vision of its person going into space. There will be found few skinny scientists seeking an all-expense-paid sabbatical on Mars. Rather, crusty people will enlist who cannot find any better way to scratch their itch for adventure. You know the kind. They're the ones who stand up in small town meetings, speaking forthrightly for a few minutes until a chubby mayor gavels them out of order. This does not, certainly, exclude scientists and engineers. Indeed, the mission to Mars may usher in the first truly "scientific" society, where everything will need to be understood, computer-analyzed, and tested. These men and women of science, however, will also be of iron constitutions and mostly hard to be around. Some of them will be people who dropped out of Harvard because things were not moving fast enough for them. Others will be greasy mechanics who happen to have the knack to fix *anything*. Once this crew of terribly intelligent, determined, and tough persons is found, it will set out to begin routine manned space flight and Martian settlement. What will really begin, however, will be the recasting of human history away from guns and generals and earthly events, declaring that from Landing Day forward, human beings shall write their greatest stories in space and on other worlds.

A few, like Pizarro, will fight against enormous odds. Fortunately, there will be no physical war; no need to offend twenty-first century sensibilities. In fact, future elementary school texts in watered-down American schools are likely to report the mission to Mars' struggles as a banal reality. *In the early 2000s a small group of pioneers banded together to settle Mars.* The mission's real story, however, may not be unlike Pizarro's:

Finding himself besieged in the Incan capital for weeks by 100,000 enemy warriors, Pizarro gambles his entire 190-man army in an attack on a fortress held by 2,000 men. Somehow, Pizarro's cavalry hacks through the besieging force and his infantry scales 60-foot walls to overcome the citadel's defenders. Pizarro himself is killed in this unbelievably desperate fight, yet the Spanish succeed in capturing the fort. The spirit of the Incan army is broken and, when planting season returns, the Incan army returns to its fields to feed a hungry nation.

The mission to Mars will not replicate Pizarro's gold-lust, but it will find men and women with Pizarro's qualities of fortitude and daring, willing to scale different kinds of heights and able to lay the firm foundations of a new empire, peaceably growing on a new planet.

Short Histories of Iron-Willed People

And Brigham and his staff were learning to manage an emigration while doing their other jobs ... treaties and arrangements with local officials had to be made. Nearly half the camp were sick; they must be ministered to somehow, medicine and care must be got for them, they must be buried when they died. Supplies dwindled; they must be replaced somehow, bought, bartered for, worked for, begged, freighted endless miles going and coming. Weak, shoddy, and ill-built equipment was giving out; it must be restored or replace somehow, more wagons brought up, more stock, more tools, bedding, ammunition ...

—1846: The Year of Decision.

This chapter considers three historical migrations. It offers these examples both as inspiration and as model. Some may choose to be offended by the religious undertone of these examples, but as cracking eggshells is the lot of revolution, the mission to Mars revolution cannot hope to achieve its goal without breaking something. Religion will play no central role in the mission to Mars revolution, so those readers with thin skins are asked to cowboy-up and accept history as history. Others can integrate "God" as they will. These are each stories of "chosen people" being led onto treks they had not previously envisioned, much as the mission to Mars proposes a trek to Mars that most people today cannot envision. Beyond physical similarities to historical events, in this chapter we look to four themes as relevant to the mission to Mars: 1) ideals, 2) leadership, 3) perseverance, and 4) unexpected migration. We will discuss how each of these themes was important to the specific history and how it will be important to the mission to Mars.

They were a full two months ahead of the time when, as the mountain men and the Santa Fe traders knew, it was safe for caravans to cross the prairies. Apart from sudden whirlwinds of sleet out of the north, the snows were over now, but the rains had come. Rain nearly every day for about eight weeks ... it saturated the prairies; after saturation, it turned them into a universal shallow lake ... the season was significantly known on the prairies as "between hay and grass". Prairie craft forbade you to travel before the grass came, but Israel had to travel and so the stock grew weak ... babies howled under the drenched blankets. Everyone who could walk slithered through the mud, "shoe-mouth deep", boot-top deep sometimes, clinging in five-pound masses to each foot.

Six miles was a big day, one mile a not uncommon one. Prairie creeks that would be five feet wide in July were now five rods wide, bottomless, swift, impassible ... [Reaching one, a whole caravan] would have to camp beside it till it should subside or a ford be found, which might be two weeks. If there were no timber, then there might be no fires for two weeks, no cooked food, no dry clothes or bedding except as the sun might come out for an hour or two ...

Hunters ranged the prairies for deer, turkey, grouse, but the season was too early ... [T]he Saints sickened. Frostbitten feet could become gangrenous, knees and shoulders stiffened with rheumatism, last autumn's agues were renewed ...

Sister Ann Richards' husband, who had already served five missions in the United States, was called to a mission in England ... Sister Ann had her two-year-old daughter, Wealthy Lovisa, with her in the wagon—and Sister Ann was big with another child and her hour was near. There was no suitable food for her or Wealthy Lovisa. Many days they could not have a fire, either because night overtook them in the open prairie or because, if they got one started, the rain put it out. But sometimes they managed to keep one going and then Sister Ann could brew a pinch of the tea from the pound which a neighbor had given her before she left Nauvoo. The Word of Wisdom forbade it but she could warm her body and cheer her mind with it, and "through sickness and great suffering [it] was about all the sustenance I had for some time."

Twenty days out from Sugar Creek, her term was full. The wagons stopped and a midwife was summoned, a Gentile whom the Saints had heard about. The hag demanded a fee in advance; Sister Ann had no money; a woolen bedspread would do and "I might as well take it, for you'll never live to need it". Little Isaac was born, and he died at once. The priesthood anointed the small body and buried it; the wagons got started again. Little

Wealthy Lovisa had been sick when they left Sugar Creek, and week by week her strength failed. Presently she was altogether listless on a roll of blankets in the wagon, and could not be induced to eat. Once, however, they passed a prairie farm and Wealthy revived enough to ask for some potato soup. Her grandmother went to the house, but the farm wife had heard the stories. "I wouldn't sell or give one of you Mormons a potato to save your life," she said, and set the dog on the grandmother. Wealthy lived till they got to the Missouri River, and then died. Brigham [Young] told Sister Ann, "It shall be said of you that you have come up through much tribulation."

Moses and the Hebrews

The historical record of the Hebrews in Egypt is clouded by the weather of thirty-five centuries. Nonetheless, a document of unparalleled historical importance, the Bible, offers not a little history and even more insight into the story of the Hebrew exodus from Egypt. Archeology has confirmed many of the main events, even if timelines and details remain sketchy. No matter. It is the motion of these events, not their details, which concern us.

The Hebrews were part of a clan of peoples found throughout the Middle East. Their language and cultural habits were likely so similar to those living around them that it was only their belief in "the God of Abraham" that distinguished them from the other semi-nomadic tribes living in the area.

After Joseph had become a vizier for the Egyptian Pharaoh, he arranged for his family and clansmen to come to the land of plenty, to avoid the famine then lurking behind every Middle Eastern hill. Though the Hebrews were less closely related to the Egyptians in language and culture than they were to the tribes in the semi-arid lands to the northeast, over time, the Israelite minority might have integrated into Egyptian society, were it not for their stubborn monotheism. This belief in "the creator God" kept bright the line between Israelite and Egyptian. After all, the one-god idea had already been tried in Egypt and had already proven a disaster. Akhenaton[1], the experimental monotheist, had presided over a period of military and cultural decline in Egypt. Monotheists—in Egypt anyway—were a class not to be trusted.

Ramses the Great had begun to restore much of Egypt's glory after Akhenaton's disastrous reign. His military expeditions into Palestine allowed for the re-establishment of military buffers between Egypt and would-be invaders from the northeast, but these military expeditions were costly. To pay for them, the people would have to be taxed. Political scapegoats are usually sought where there

---

[1]   A Pharaoh who had experimented with monotheism fifty years before Ramses.

is domestic suffering. Ramses' Egypt would have been no different. The Hebrews, having built their own social and religious fences, would have been particularly easy prey for taxation and the courvee[2]. Thereupon, as a suspect class, the Bible tells us that the Egyptians "did set over them taskmasters to afflict them with their burdens", for the Egyptians were fearful that "when there falleth out any war, they join also unto our enemies, and fight against us".

The story in the Bible's book of Exodus continues with Moses expressing his doubts about his mission; which probably reflected the spiritual doubts of the troubled Hebrew community. The people were suffering for a God they really did not know. Indeed, the lot of some of the Hebrews had improved in Egypt, from one of migrant shepherd to that of city dweller in some of the finest cities in the world. Those Hebrews leading a "comfortable" life in Egypt must have been confused about Moses and probably accepted his message only with mixed feelings. For Moses was a man the Hebrews knew to have been raised in Pharaoh's court, a man with an Egyptian name[3] yet declaring himself a Hebrew[4], with revelations about the Hebrew God, and with the unexpected and probably unwelcome proposal that they leave their homes to return to the desert. Rumors were that this man was a murderer and that he was telling everyone he could do magic. Although the Bible does not give many details about Moses' effort to convince the Hebrews of his authority, the continuing doubt concerning Moses' commission is expressed by the tribulations of a people constantly falling away from God during their sojourn in the Sinai desert.

We all know the rest of the story: Moses interceded with Pharaoh to release the Hebrews, but only after many such efforts did Pharaoh's hard heart finally relent. Moses thus led "out of bondage" a group of people who had only the vaguest notion where they were going and who had only the dimmest conception of their one God. Yet, somehow, the idea of God and the collective desire to build their nation in His honor was something each individual Israelite decided merited emigration out of Egypt, even unto forty years in the desert, and despite opportunities to fall away from their goals and their God.

---

[2]   A draft of individuals for service on government projects.

[3]   "Moses" was a name shared with several other Egyptian Pharaohs: Tut-mose, etc.

[4]   It is not clear when the patrimony of Moses becomes known. Indeed, even after his flight into the desert, the Bible tells us that when asked about the incident at the well, it was told that "an Egyptian delivered us from the hand of the shepherds" (Exodus 2:19).

The Mormons

The Mormons'[5] search for spiritual and political perfection began in 1819, at Cumorah Hill in New York State. A young man, Joseph Smith, claimed to have dug up five golden tablets, which had been written in a strange language and which told an amazing story of Jesus and the lost tribes of Israel, come to the New World. Smith was either insane or a prophet, for these golden tablets disappeared soon after he translated them, never to be found to this day.

Young America shared some, but not all, of the visions of the Mormon Church. Indeed, many Americans were suspicious of the new "Zion", its aspiration to become the perfect godly state, and recoiled at the polygamy practiced by the Mormon faithful. In no small part, Mormon aspirations to include much of the American west in their grand schemes failed in Congress because of these suspicions. Petitioning to become a state in 1849, Utah's petitions failed until 1898, a period unmatched in American history and an experience completely different from a state like Kansas, where statehood was granted but seven years after being organized as a territory.

The Church of Latter-Day Saints has a complex theology that need not concern us. Their division of the world into Saints (individual Mormons), Israel (the collective Mormon Church), and Gentiles (corrupted Christians lacking an understanding of God's purpose), may help the reader understand why the Mormons had repeatedly been driven from their homes. More important for our purposes was that the Mormons kept their numbers and their faith. Indeed, their city in Illinois, Nauvoo, on the banks of the Mississippi, was for a few years—when Abraham Lincoln was still practicing law—the second largest in the state. Only Chicago was larger. Wandering to Ohio, later to Missouri, back to Illinois, the Mormons looked for their Zion, but by 1846 had still not found it. In each place, they were driven out by mobs and local governments.

After Joseph Smith was murdered in jail by a lynch mob, the Mormons faced their greatest challenge. Their leader dead, Mormon heresies arising every month, a stolid leader was needed and it was then that Brigham Young appeared. Rather unexpectedly, Young decided that the Mormons needed to leave the US altogether, to build their Zion in Mexican territory in Utah, and to escape the official pogroms that had helped the mob murder Joseph Smith. The trek was already in progress when war broke out between Mexico and the US.

Though impractical men can plant an idea, it takes a more practical kind to set into motion a vast migration across a thousand miles of prairie and to make a des-

---

[5]   A common term for members of the Church of Jesus Christ of Latter-Day Saints. No disrespect is intended in this use.

ert bloom. In fact, if we can see exactly how important leadership is if we compare Joseph Smith with Brigham Young.

> *Joseph Smith was not an effective administrator. He planned and prophesied and pointed vaguely to far goals, but he was not skilful at delegating responsibility for immediate tasks. Brigham Young was a master of men. If he planned a job, he could pick the man to do it. He was also a visionary, but his visions did not skyrocket into space, for he staked them to the earth.*
>
> *Young was a practical and puritanical Vermonter. He had been a mechanic, a carpenter, and builder. Smith had had no occupation. He was more of a reader than his successor, and he often displayed his knowledge of books. Brigham Young was too busy to be a reader, but he entertained no feelings of his inferiority because of this lack. Toward the end of his days Joseph Smith developed a fastidiousness in dress and a liking for his general's uniform. Not so with Brigham Young; his tastes remained simple. The fact that he did not don the spangles and otherwise parade his person did not evidence any lack of ego on his part. Brother Brigham knew too well that wherever he sat, there was the head of the table. That he was the head was something he did not have to prove to himself or others with each rising of the sun.*
>
> *The primary objective of Brigham Young's headship over the church can be stated in simple terms. He would gather the remnants of Israel together; he would move Zion to the Rocky Mountains; he would drill into the Saints the principles of the gospel laid down by Brother Joseph; he would continue to send missionaries to gather in the chosen from all corners of the earth, and he would build the church so strong that all the legions of hell could not uproot it. Such were the objectives of Joseph Smith. It took a Brigham Young to attain them[6].*

Imagine the scene, then, when Brigham Young himself orders 500 Mormon men to be sworn in as American soldiers at the start of the Mexican War, to serve the state that had tormented them. They left their camp along the Missouri river in Iowa to march to Santa Fe, and across the desert to San Diego and Los Angeles. This march may have been the longest foot march in American military history. This "Mormon Battalion" marched with women-attendees in the ranks and feared that on the Santa Fe road lay in wait members of the Missouri Militia, commanded by an earthly Satan, the very same Sterling Price who had driven the Mormons out of Missouri.

---

[6]   From book by Nels Anderson, *Desert Saints*.

The Mormon soldiers and those traveling to Utah persevered in their immediate tasks. They understood that the suffering of a moment would eventually be replaced by both earthly and celestial rewards. Clearly, they had little in common with those whose course in life demands immediate gratification. The mission to Mars may have little occasion to erect rows of grave markers, as did the Mormons traveling the early season plains, but those who envision a sanitized mission to Mars are kidding themselves. In a pioneering society, there are risks and people die. The lesson of the Mormons is that those who persevere can build great temples to their ultimate success.

Without Brigham Young's leadership, the Mormon Church would likely have withered and died in western Illinois. Even with his inspiration, it is astonishing to contemplate that he was able to move so many thousands across six score hundred miles of undeveloped prairie, mountains, and high plains, retaining control over not only the people, but the institutions that we today know as the Mormon Church.

The Hebrews persisted for forty years; the Saints for forty months, only to find themselves in the Utah desert. Indeed, it took the Mormons forty more years to transform the Utah desert basin into a tidy part of the American landscape. Ultimately, both achieved their Zions. The relevant question for the early 21st-century is whether such people still exist. If so, there should be no doubt of their ability to create human colonies on Mars.

> [T]his was the migration not of certain individuals coming together in a temporary organization while they crossed the plains but of an entire people. The camp of Israel was the Church of Jesus Christ of Latter-day Saints, past, present, and to come. The Mormons carried with them not only their goods but also their church and social institutions—the hierarchy, the various priesthoods, the rituals and sacraments, the co-operative associations, the United Order, the mission system ... Israel had to maintain its nervous system and could support its venture in the west only by constant accessions. It had to be a continuing migration.
>
> So for the sake of many who could go no further, of those still in Nauvoo, and of the as yet unconverted all over the world, facilities of some permanence had to be provided. The problem had to solved at once; Brigham solved it. His little eyes lacked the gift Joseph's had, of piercing the heavens and beholding the glories there, but it is exceedingly unlikely that Joseph could ever have got his people beyond Sugar Creek.
>
> At Richardson's point, fifty-five miles from Nauvoo, they built a permanent camp, which would always have a garrison. Companies coming in from the east would find wood, supplies, blacksmithing tools, experienced

help—and the priesthood making sure that they "accepted counsel", obeyed, kept discipline, and lived their religion. Another one was established farther on, at a crossing of the Chariton River, and here the first crops were sowed. The first companies planted crops, permanent personnel cultivated them, later arrivals would harvest them. There were other farms on the way and other permanent camps on Locust Creek, at Garden Grove on Grand River, and lastly "Mount Pisgah", a hundred and forty miles east of Council Bluffs. At Winter Quarters on the west bank of the Missouri and near Council Bluffs on the east bank much more ambitious camps were built, permanent settlements really, with a vigorous trade, large herds of horses and cattle, and farms of several thousand acres worked by hundreds of the Saints. All these plantations except those on the Missouri made crops ... they were making the land in part support them as they traveled ...

Universal human cussedness, pricked by hunger and doubt, had Brigham and his lieutenants thundering at them a good part of the time ... but, spread out over Iowa, they were laboring strenuously if not concertedly for the Lord. They prepared the permanent farms and wagon shops, dug wells, got the crops planted. They found time to make nails, burn charcoal, shape oxbows, and manufacture harnesses and even wagons as they traveled ...

And Brigham and his staff were learning to manage an emigration while doing their other jobs ... treaties and arrangements with local officials had to be made. Nearly half the camp were sick; they must be ministered to somehow, medicine and care must be got for them, they must be buried when they died. Supplies dwindled; they must be replaced somehow, bought, bartered for, worked for, begged, freighted endless miles going and coming. Weak, shoddy, and ill-built equipment was giving out; it must be restored or replace somehow, more wagons brought up, more stock, more tools, bedding, ammunition ...

All this made a sufficient test of leadership, organization, and public control, not to mention prophecy ...

Well, he [got] them to the Missouri ... the last refugees from Nauvoo on November 27. Through eight months, continuously across more than four hundred miles, the Iowa prairies witnessed such a pageant as no one had seen since the Goths moved on Rome—and moved on it inward from the frontier, not outward toward it. Between fifteen and twenty thousand people uprooted from their land and seeking a new land. Thousands of wagons, tens of thousands of oxen, horses, mules, milch cattle, beef cattle, neat cattle, sheep, goats; chickens, geese, turkeys, guinea fowl, ducks, pigeons, parrots, lovebirds, canaries. Seedlings with their roots bound in sacking, slips from the shrubbery back home, seeds for the harvest to come, disassembled machinery of flour mills and saw mills, a college, the mysteries of heaven, the keys to eternity, the Dispensation of the Fullness of Time. Through sleet and rain, through drouth [sic] and prairie summer, half-starved and half-sick, dispossessed, believing, and faithful unto the last, Israel traveled the unknown, toward the land of Canaan, in God's faith and for His glory and under the shadow of His outstretched hand, to build Zion and inherit the earth.

Zionism and Israel

A more modern migration is the story of those Jews who decided that their people deserved a nation of their own. In the late 19th-century the idea was not new, but the possibility for success suddenly became real for a people who had mostly been driven out of Palestine 2,000 years before. The First Aliyah (the first "going up" to Palestine) occurred during the last decades of the 19th-century, when the earliest Jewish settlers arrived there.

The Zionist movement began in earnest in 1897, however, when the first Zionist Congress adopted practical programs for its ends. Indeed, one of the great leaders of early Zionism, Theodor Herzl, wrote in his diary, "At [the Zionist Congress], I founded the Jewish State. If I said this out loud today, I would be answered by universal laughter. If not in five years, certainly in fifty, everyone will know it." Herzl's diplomatic efforts in Turkey and Britain prior to World War I (WWI) underscored the reality that a practicable plan can truly mark the achievement of a great goal.

Anti-Jewish pogroms in Russia provided the impetus for the Second Aliyah, which added another 50,000 settlers to Palestine just prior to WWI. Kibbutzim—communal farms—began to be created in order to make for a self-reliant Jewish community in Palestine; and the coming Israel was envisioned to be a socialist state, in conformity to the then fashionable ideas of social organization. Howard Sachar reports in *A History of Israel* on difficulties that will surely recur on Mars: "David Ben-Gurion, a nineteen-year-old former student [and Israel's 'George Washington'], succumbed to malaria and nearly perished. A doctor urged him to return quickly to Europe. 'My well-meaning friends all pointed out that this was hardly a disgrace ... Half the immigrants who came to Palestine in those early days took one look and caught the same ship home again.' Indeed, more. Possibly 80% of the Second Aliyah returned to Europe or continued on to America within weeks or months of their arrival."

Palestine was seized by the British during WWI, when their forces rolled the Ottoman Turks back from the Suez Canal. The British government was sympathetic to the Zionists and, as suited wartime political maneuverings, the British Foreign Secretary, Lord Balfour, promised to support the creation "of a Jewish national home". Unfortunately, the British had also traded the Arabs a promise of independence for their support in WWI against the Ottoman Turks. The British were given a League of Nations Mandate after WWI to these ends. The Balfour Declaration and the Mandate provided solid international status to the Zionist movement, barely 20 years after the first Zionist conference.

The British drafted a White Paper on Palestine in 1922, which in part stated that "[d]uring the last two or three generations the Jews have re-created in Palestine a community, now numbering 80,000, of whom about one-fourth are farmers or workers upon the land. This community has its own political organs, an elected assembly for the direction of its domestic concerns; elected councils in the towns; and an organization for the control of its schools. It has its elected Chief Rabbinate and Rabbinical Council for the direction of its religious affairs. Its business is conducted in Hebrew as a vernacular language and a Hebrew press serves its needs. It has its distinctive intellectual life and displays considerable economic activity. This community, then, with its towns and country population, its political, religious and social organizations, its own language, its own customs, its own life, has, in fact, 'national' characteristics."

After WWII, the British gave up most of their colonial empire. The financial and other imperatives of empire were finally recognized for what they were: vast and expensive possessions that were peripheral to the central idea of being British. In fact, the British pulled out of Palestine so suddenly that there was no time to implement a UN program to restructure the British mandate in Palestine. Jewish bombings and the expense of maintaining a 100,000-man army in Palestine had forced the British hand. In the vacuum created by the British withdrawal, in the aftermath of the Holocaust, Zionists proclaimed the state of Israel and 200,000 Arab soldiers attacked. The "'48 War" lasted for several months, but the well-trained, well-organized, and mostly—some from WWII—experienced Jewish brigades first halted and then defeated the more numerous Arab armies.

The end of the "'48 War" marked the end of the work of the Zionists. A communal social structure had been formed, but a religious rather than socialist state was created. What was more significant, however, was that this newborn state had weathered a major storm by turning back an armed expression of Arab resentment. A Jewish state was no longer a dream; instead, there was a Jewish nation recognized by much of the world.

In the years since 1948, there have been other wars which have resulted in Israeli successes, mostly because of the outstanding leadership displayed by its political and military leaders. Included is that same David Ben-Gurion, Israel's first Prime Minister and national leader during the 1956 War. Moshe Dayan and

Ariel Sharon proved to be outstanding leaders during, respectively, the 1967 and 1973 war. In 1973, Prime Minister Golda Meier led Israel back from the brink when several Arab neighbors orchestrated an attack against Israeli that enjoyed initial successes. Without the outstanding leadership of Israelis up and down political and military organizations, the dream of a Jewish state may have been short lived.

Historical Lessons For The Mission to Mars

There is much to be threshed in the preceding histories and the mission to Mars should heed at least four great lessons they teach, as given in Figure 75:

Attribute	Moses	Mormons	Israel	Mission To Mars
Ideals	One God, promised land flowing with milk and honey	A new Zion far from American society	Reclaim a homeland; Holocaust never, ever again	Desire to get into space and to colonize Mars
Perseverance	40 Years in the Wilderness	Crossing 1,000 miles of wilderness to arrive in a desert	100 years of struggle to reclaim land and to forge a peace	Working for years outside of western economy
Leadership	Moses and Aaron	Brigham Young	Hertzl, Ben-Gurion, Dayan, Sharon, Meir	Small corps of dedicated individuals
Unexpected Migration	Exodus to Promised Land	Escape from pogroms in US	Actual creation of the state Jews had dreamed of for centuries	Addresses people's desire to live in space and colonize new worlds

Figure 75: Lessons For The Mission To Mars

We begin with the importance of ideals. The Hebrews of Moses' day yearned to be free and to occupy the land that God had promised them. Most of the

Hebrews apparently trusted in this larger vision, even if at times there was falling away from God or doubts about Moses, because there are no stories of desertions or of return to Egypt. The Mormons struggle to reach their own Zion bespoke an unshakeable belief in their ideals, even to the extent of coming up "through much tribulation". The modern Jewish dream supporting the state of Israel has sustained a steady immigration for a hundred years.

In most respects, what is being discussed in this book is a 21st-century version of the Exodus, the Mormon migration, and the founding of the state of Israel. While the mission to Mars does not compare its goals with the profoundly religious importance of these various migrations, other aspects of their experience will surely be similar. Indeed, the mission to Mars can only succeed if its programs become the means to help people live out their dreams for space and for Mars. In other words, money alone will not allow the mission to Mars to succeed; it will take people who believe in a dream of Mars and space to make these ideas work.

Related to the need of ideas for success will be the need to persevere in the face of difficulties. Moses and the Hebrews offer us forty years in the desert as an example of real perseverance. In like fashion, the Mormons had to persevere to make the Utah desert bloom and the Israelis have had to persevere under circumstances of constant terrorist attacks against them.

Hopefully, there is no need to dwell upon the difficulties the mission to Mars will face. There will be many failures as a small organization attempts what has eluded the grasp of the greatest nation on earth. There will be bad decisions. There will be bad people. There will be great disappointments and the job will require that people show up *year after year*, despite the negatives. It may be trite, but Edison's quotation comes to mind, "Genius is 1% inspiration and 99% perspiration". For forty years, Moses kept his people in the desert rather than return to a "suburban life" in Egypt. Perhaps the ancients had powers of character that modern humans lack. *If they can do it, however, should not we also be able to persevere?*

Next, consider the importance of the leadership. It was only through the leadership of Moses and Aaron that Pharaoh let their people go. It was only through the leadership of Moses and Aaron that the Hebrew people were able to hold fast to their dream in the Sinai. After all, in that Golden Calf moment, the people themselves seemed willing to turn away from their God. Only after Moses demonstrated his anger and the power of God did the people return from their prodigal path. Several of the epigraphs in this chapter have extolled the leadership of Brigham Young, even to the point of contrasting his vast leadership skills with those of Joseph Smith, the Mormon's founder. And there should be no mistaking the importance of Jewish political and military leaders who have led the state of

Israel through a childhood that has included four major wars and a myriad of crises.

The reader should also contrast these histories with the leadership exhibited by the poll-taking leaders of modern, western societies. None of these cited migrations would have succeeded without their leaders. Conversely, "glaring lack of leadership" deserves a prominent place in the pantheon of reasons why the US will not take humans to new worlds. Although the accolade of "leader" is today automatically granted to holders of elected office or heads of other organizations, the leadership essential to the mission to Mars demands that intangible ability to achieve results congruent with one's ideas, even when faced with daunting circumstances. Today's sound bite "leaders", adept at dealing with modern news cycles, need not apply.

Ivory-tower thinkers may pooh-pooh the advantages of leadership, but it will be leaders—like Moses and Aaron—who will ensure the ultimate success of the mission to Mars. As it was not committees that led the people through the Sinai, so it will not be committees that take a human society to Mars. This suggests the importance of leadership during the early years at the earth Embassy. During those early years, it will be easy to be co-opted by NASA or enticed to accept some goal less than Mars. If there are no leaders at the mission to Mars to ensure that the narrow path up the mountain is followed, its entire program may be in peril.

The leadership lesson does not imply a single or a dictatorial leader for the mission to Mars. The most important reason is that its circumstances differ from the histories presented here and that no single person or "dictator" could hope to attract the kinds and numbers of people needed to achieve the colonization of Mars. Moses certainly held great sway, but he had Aaron and God. Brigham Young's leadership over the Mormons was unquestioned and seems to indicate the value of a single leader, but he was a successor to a prophet whose abilities, while not the skill sets needed to cross a desert, had founded the religion. Leadership in the Zionist movement was not centered on any one person, but shifted from time to time as the Zionist community harvested a new crop of leaders every time such leadership was required.

Finally, the reader should consider the remarkably "unexpected" nature of the migrations presented here. Before Moses, no Hebrew conceived of a trek through the Sinai to a promised land. Not until Moses returned to Egypt did the Hebrews begin to see that their entire people could depart a great nation for the desert and thereby begin a journey building God's kingdom. In 1844, the year Joseph Smith was murdered by an Illinois mob, most wise bets would not have been on Utah, but upon the final destruction of the "peculiar people" (Mormons). As for the Mormons themselves, many surely believed that they had finally come to rest at their tidy city on the Mississippi. The resettlement of Palestine by the Jews during the twentieth century was no easier to predict than the other two migrations. Certainly, Hertzl surprised a great many people who had entertained only ethereal hopes of a return to Palestine when he proposed specific diplomatic objectives to achieve those ends.

NASA and the international efforts into space have failed to deliver anything to the common man. This book has begun to lay out arguments suggesting why this has been so and has suggested in stronger terms that another option is available. The mission to Mars may be an unexpected turn from business as usual and the humdrum of another meaningless shuttle launch. As unexpectedly as the Hebrews or the Mormons or the Jews, modern humans can create the means to migrate to Mars and to create a self-sustaining community there. You may not have believed such an option was possible when you first picked up this volume. By now, however, you should begin to see that a migration to Mars of determined people can occur just as quickly as the necessary institutions and infrastructure can be assembled.

Final Thoughts For Volume I:
All You Need Is Love

This is not the end. It is not the beginning of the end. But it is the end of the beginning.

—Winston Churchill, after the battle of El Alamein.

The positive and self-confident tone of the mission to Mars does not imply arrogance about the terrific obstacles to be overcome. Like the early Europeans settlers in America, Martian pioneers will face a never-ending list of problems. Many of these obstacles will need to be addressed during the early stages of planning. Some obstacles will be used by Nay Sayers to conduct a guerilla campaign against the plans presented here. Many obstacles, however, will never reach a conference room of engineers or politicians, but will be faced down by the pioneers themselves, in quiet fashion, teaching us again that "necessity is the mother of all invention". What the mission to Mars proposes is admittedly grand, but humans can achieve great goals, if they are so determined.

Abraham Lincoln's Mission Impossible

When President Lincoln took office in March, 1861, he was faced with the daunting problem of how to react to the secession of several slaveholding states. Lincoln faced his crisis under political and economic circumstances which most people believed would prevent him from solving the crisis of the union. Lincoln's critics believed that either the political realities in the North, an insufficient grant of constitutional power[1], insufficient military forces, or the lack of financial resources would allow the southern succession to stand.

Most of the Presidents since Andrew Jackson had been little more than party hacks, people whose turn had come to hold the nation's top office. These men had little vision and even less interest in achieving any grand purpose. It was thought that Lincoln was from the same mold. No wonder the Southern states seceded. Had Franklin Pierce been the American President in 1861, we would all need passports to cross the Potomac River. One of the most important political figures of the day, William Seward, considered Lincoln to be such a political weakling that he attempted a virtual *coup d'etat* during Lincoln's early days.

If the political problems were daunting, the economic obstacles before Lincoln were likewise vast. After the first few months of the war, it became apparent that the North would require an economic engine that could keep large armies in the field for several years. Napoleon had kept a million Frenchmen in uniform during the early nineteenth century only to bankrupt France in the process. Napoleon's military machine then disintegrated and invading armies could not be met even on France's own sacred soil. Indeed, as the American Civil War dragged on, the South destroyed its own economic engine trying to field half has many men as the North. Lincoln, however, devised a system that deciphered the financial puzzle of solvency during a long war. He spent sixty times the national debt of the time and did not achieve battlefield results overnight. It took the North four years to build up their armed forces sufficiently to defeat the South and 350,000 dead soldiers became part of the steep price the North paid to preserve the union. Still, a sufficient economic structure was created by combining extraordinary determination with wise decision making. In the process, Lincoln's political objective—the preservation of the union—was achieved.

Defying expectations, Lincoln and a small number of people in the North held firm to their political goal of preserving the union. Over time, more and more people came to believe in this political goal. They backed the President, enlisted in northern regiments, bought war bonds, and finally cast their ballots to return Lincoln to office in the election of 1864.

If, against all odds, Lincoln and a small group of people could assert unexpected depths of political will, to achieve a goal most Northerners would have abandoned; and if these people were able, creatively, to assemble the economic power needed to achieve these goals, might not a group of intrepid people, such as still exist in the 21st-century, achieve a much more modest goal; one that requires

---

[1]   Lincoln's predecessor, James Buchanan, did not believe that the US Constitution granted the federal government the power to compel seceded states to return to the union.

only a careful blending of mostly old, some off-the-shelf, and a tiny fraction of just-over-the-horizon technologies; to fulfill a dream to which tens of millions aspire, and which offers financial as well as psychological rewards?

The last part of this volume has informed us that large stories dance across the human stage from time to time. Lincoln's story is heroic, but equally as relevant to the mission to Mars are the stories of the migrations reviewed in the last chapter. As each generation forgets that there are such things as war, so human societies forget that humans have, from time to time, banded together to leave "civilization" for a new "Zion", enduring hardship and deprivation in order to achieve by migration their political goals. So it will be with the mission to Mars. The mission to Mars proposes that the desire to settle Mars, to live in a free, libertarian society, and to be in the vanguard of mankind's New Era in space will motivate sufficient numbers toward its ends. The result will be a human society on Mars. The timeframe will be short; most readers will see these things come to pass in their lifetime.

The Torch Passeth

Tom Brokaw's book, *The Greatest Generation*, tells not a tale about "the greatest" generation, but rather about a generation that faced great challenges. Heroic stories of men landing and dying at Omaha Beach in the face of shattering small arms and artillery fire may seem to small minds to be the genius of a unique generation. But go to any Marine Corps base today and hang around for a few hours. Improbable as it may be to some readers, today's men and women, if asked, would charge those same beaches and face those same cliffs.

Now, the fates have chosen a new "greatest generation". Our generation will replace the WWII generation as the "greatest", because we will conquer a new planet and will usher in a New Era for mankind. After we have come and gone, the human species will never be the same.

As the discovery of the New World changed humanity forever, so the settling of Mars will change it. As the New World took ordinary men and turned them into Columbuses and Pizarros and John Smiths and thousands of extraordinary men and women unknown to history, so will the mission to Mars take ordinary people and transform them into extraordinary human beings. In another era, Pizarro would have been a playboy or a Spanish Mafioso or a faceless mercenary. Sent to the New World, Pizarro forged a new, Spanish Empire in less than twenty years. And though the first English speaking settlement in the Americans, Roanoke, disappeared completely, leaving no trace of the hundreds who had come to that Virginia shore, Englishmen did not repent of their desire to trust their fate to the strange, new land. Nor shall the loss of some few settlers stop the mission to Mars from creating a new civilization, a Martian civilization, and from recasting the human condition in its new image.

Human beings love a challenge. Human beings need a challenge. Human beings will flock to the opportunity to be part of that most magnificent challenge, the one awaiting them in space and on Mars. If these ideas at all appeal to you, consider donating in proper proportion your treasure, talent, and time to an organization dedicated to making manifest the bold ideas in this book; to help the very best of our humanity glow in the sunset of a new human home on Mars.

All You Need Is Love

The purpose of 1000 pages is not to scare off any starchy Martian Socialites. Just the opposite. I know that anyone who *truly* loves Mars will wade not just through this volume, but the other volumes as well. This volume has attempted to present the basic outline of the mission to Mars in an entertaining fashion, grasping for some measure of the whole. But basic questions remain unanswered—the purpose of Volume II—and many details of the program have been put off to Volume III.

Volume II will set the readers through a vigorous program of mental basic training, to help NASA ride off slowly into the sunset. Some of this training was begun in this volume, but questions as basic as "Why should we even consider going to Mars?" and "Why not just wait for NASA to do the work?" remain. It is thoroughly recommended that the reader continue the journey begun with this volume by moving on to Volume II.

Volume III is much like certain sections of Volume I, in that the details of many of Volume I's first-coat ideas are set out for exploration. It is fine to say you create a Mars Bank, but how it will operate and how it will handle the precious hard currency that will bring in raw materials from the world at large cannot be articulated in a short work. Volume III devotes the necessary space to this task. Likewise, the structure of a "Martian Confederation" is laid out in some detail, to show that participation will entail life as a citizen of a western-style society and not that of a cloned automaton for some cult leader.

Volume I was the easy part. There is more intellectual work and fewer droll anecdotes in Volumes II and III. The question for now, however, is whether the reader really loves space and Mars, or whether they are merely your hobby. The mission to Mars proposes to enter the real world, to force people to make a choice about their "love" of space. It asks people to become dedicated to its purpose or to acknowledge themselves as simply couch potatoes awaiting the next rerun of *Star Trek*.

How can you love a planet, truly?

Stephen Covey says in *The Seven Habits of Highly Effective People* that "[p]roactive people make love a verb. Love is something you do, the sacrifices you make, the giving of self, like a mother bringing a newborn into the world. If you want to study love, study those who sacrifice for others, even for people who offend or do not love in return. If you are a parent, look at the love you have for the children you sacrificed for. Love is a value that is actualized through loving actions." While it is common in today's America to turn love into that MTV feeling you have after sex with some attractive person you have known for two weeks, that is not real love. Real love means making sacrifices and holding fast to ideas that may, at times, make demands of you.

People are free to say that they "love space" or "love Mars" without manifesting that statement in the real world. This is the MTV version of love. Real love means dedication, not fanciful daydreams disconnected from daily life. A real love of Mars or space will soon mean that one must make a choice: love as a feeling you have after watching a good movie about Mars or love manifested as actions in the real world, supporting the mission to Mars.

In a very real sense, all the mission to Mars really needs is love. There are tens of millions of potential candidates for participation, infatuated with the idea of Mars and space. All it truly will take to reach Mars is for a few thousand of these to people to manifest their love with the effort and money necessary to plant the first human colonies there. The author believes that such feelings of love exist on earth. He may be wrong. It may be that the "right stuff" went out decades ago and that we must await Uncle Sam's anointing. If we must await an earth government for something people want, then not only have we truly lost our souls as a virtuous society, but it will surely condemn almost everyone reading this book to permanent exile from Mars.

Lincoln was able to solve seemingly insuperable economic problems; so can the mission to Mars. A "great generation" was able to kill off the shadow forces of fascism that were spreading across the world in the 1930s and 1940s; but that generation was great because it met the challenges it faced, not because of any inherent genetic or moral superiority. A mix of good plans and a few motivated people can manifest one of the greatest dreams of all humankind: the dream to move off our home planet, to move into space, and to build a new home in our solar system.

Mars is there, teeming with resources. Its promise is every bit as great as would be an earth half unpopulated, where only a tiny desert separated humanity from new lands of milk and honey. All we need to fulfill the desires of tens of millions is a little love. Our hope is that such love exists in our busy, dissatisfied, modern West and that it will sustain the grand, yet modest, goals of the mission to Mars.

Acknowledgements

Linda Gunning, Vicki Hester, and Susan Davis helped in the preparation of the manuscript. Linda was especially helpful in helping to format the document for the publisher.

Contact Information

For more information about the mission to Mars, please see our website at
www.mtm2025.com. Or, leave an email at marsdan2005@yahoo.com

Index

978-0-595-41587-8
0-595-41587-3